Instructor's Manual
for

Mathematical Structures for Computer Science

Fifth Edition

Judith L. Gersting
University of Hawaii at Hilo

W. H. Freeman and Company
New York

ISBN: 0-7167-4440-6

Printed in the United States of America

First printing 2002

Contents

A Web site for Mathematical Structures for Computer Science, Fifth Edition, may be found at www.whfreeman.com/gersting, accessed using password "logic". The web page contains representative new Example problems (not contained in the text) for most of the end-of-section Techniques. Each Technique that has a corresponding web page example is marked in the text with an icon

Each example on the Web first states the problem. Then succeeding pages develop the solution, much as the student would be expected to write it. As the student navigates the pages, the solution unfolds step-by-step. A compressed audio file is also part of each Web page after the initial problem statement. The audio file contains a first-person stream-of-consciousness thought process about that step of the solution - why it occurred to the narrator to try this, why it looked promising, what knowledge was being called upon to suggest that this should come next, etc. The point is, students see perfect and complete worked-out proofs in the textbook, and often see them performed by the instructor. Yet when a student goes home and tries to produce such a solution by himself or herself, he or she is unsure where to start or how to think about the problem or how to see any pattern to enable a guess as to what to do next. Consequently the student gives up in frustration. The purpose of the audio narration is to share the "secret picture" that mathematicians use to solve problems.

CHAPTER 1: Formal Logic

Chapter 1 is an introduction to formal logic and some of its implications for computer science. Formal logic is nothing more than a systematization of much of what we do daily when we communicate in natural language and when we draw conclusions using informal reasoning. It is certainly the basis for scientific thinking and reasoning.

Because of its formalization, many students seem to find logic a difficult topic. You can point out to the students, however, that this very formality actually relieves them of the burden of undirected creative thinking! There are certain patterns and rules that must be followed in constructing formal proofs, which serve to channel and limit the possible steps to be taken at any time. Students face similar constraints in following the rules of syntax of a programming language; in fact, they are so much more constrained in formal logic that their task is even simpler than in programming. Pointing this out to the students (who have usually experienced some success in programming tasks by this time) can serve as a morale-builder. Or you can also note the similarities to standard high school geometry, but you are on less firm ground here because some students hated their high school geometry class! At any rate, they need some reassurance that symbolism itself is not all that formidable and, by its constraining nature, is actually helpful. I usually point out that one of the purposes of symbolism is to strip away meaning in order to concentrate on form alone, and thus help ensure that we are "pure in heart" in our thinking processes. Students find this a bemusing idea, but they generally see the point.

Of course, natural language statements must first be translated into symbolic form, and it is this part of the chapter that is most fun to teach. It points up how sloppy we tend to be in our natural language statements. (See "John loves only Mary," "John only loves Mary," and "Only John loves Mary" in Section 1.3.) One would think that students for whom English is a second language would have more difficulties than native English speakers with this type of translation exercise, but that doesn't seem to be the case - they all have to work at this, but they also enjoy it. Students often come back later and tell me that after this course, they never take words and sentences at face value again!

Propositional logic is covered is Sections 1.1 and 1.2, while predicate logic is discussed in Sections 1.3 and 1.4. Sections 1.5 and 1.6 give some direct applications of formal systems to computer science; Section 1.5 is a brief introduction to logic programming and Section 1.6 introduces proof of correctness.

The discrete structures course is too often seen as a collection of disjointed topics, so it behooves us as instructors to emphasize the connecting threads whenever possible. The notations of formal logic will be used throughout the book to clarify definitions. Logical reasoning, although not always formal, is the basis for much of the rest of the work in this course and will also stand the students in good stead in many other computer science courses.

Answers that are starred also appear at the back of the textbook.

EXERCISES 1.1

*1. (a), (c), (e), (f)

2. a. T b. T c. T d. F

3. a. T b. F c. F d. F e. T f. F g. T h. T

*4. a. antecedent: sufficient water
   consequent: healthy plant growth
   b. antecedent: further technological advances
   consequent: increased availability of information
   c. antecedent: errors will be introduced
   consequent: there is a modification of the program
   d. antecedent: fuel savings
   consequent: good insulation or storm windows throughout

5. a. 1 and 3 b. 2 c. 4

6. *a. The food is good but the service is poor.
   *b. The food is poor and so is the service.
   c. Either the food is poor or the service is poor, but the price is low.
   d. Either the food is good or the service is excellent.
   e. The price is high but either the food is poor or the service is poor.

7. a. Either the processor is slow or the printer is fast.
   b. The processor is slow and the printer is fast.
   c. The processor is fast but so is the printer.
   d. Either the processor is slow or the printer is fast, but the file is not damaged.
   e. The file is not damaged, the processor is fast, and the printer is not slow.
   f. The printer is slow and the file is not damaged.

8. a. $A \wedge B$
   b. $A \wedge (B \vee C)$
   c. $B \rightarrow (A \wedge C)$
   d. $A \rightarrow (B' \vee C')$
   e. $A \wedge [C' \rightarrow (B' \vee C)]$

9. a. $B \wedge D$
   b. $A \wedge D$
   c. $D \rightarrow (B \vee C)$
   d. $B' \wedge A$
   e. $D \rightarrow C$

10. a. Violets are blue or sugar is sour.
 b. Violets are not blue or, if roses are red, then sugar is sweet.
 c. Sugar is sweet and roses are not red, if and only if violets are blue.
 d. Sugar is sweet, and roses are not red if and only if violets are blue.
 e. If it is false that both violets are blue and sugar is sour, then roses are red.
 f. Roses are red, or violets are blue and sugar is sour.
 g. Roses are red or violets are blue, and sugar is sour.

*11. a. A: prices go up; B: housing will be plentiful; C: housing will be expensive
 $[A \rightarrow B \wedge C] \wedge (C' \rightarrow B)$
 b. A: going to bed; B: going swimming; C: changing clothes
 $[(A \vee B) \rightarrow C] \wedge (C \rightarrow B)'$
 c. A: it will rain; B: it will snow
 $(A \vee B) \wedge (A \wedge B)'$
 d. A: Janet wins; B: Janet loses; C: Janet will be tired
 $(A \vee B) \rightarrow C$
 e. A: Janet wins; B: Janet loses; C: Janet will be tired
 $A \vee (B \rightarrow C)$

12. H: horse is fresh; K: knight will win; A: armor is strong
 a. $H \rightarrow K$
 b. $K \rightarrow (H \wedge A)$
 c. $K \rightarrow H$
 d. $K \leftrightarrow A$
 e. $(A \vee H) \rightarrow K$

13. a. $A \rightarrow T$
 b. $T \rightarrow (A \wedge E)$
 c. $E \rightarrow T$
 d. $A \rightarrow E$
 e. $E \leftrightarrow (A \vee T)$

14.*a.

A	B	$A \rightarrow B$	A'	$A' \vee B$	$(A \rightarrow B) \leftrightarrow A' \vee B$
T	T	T	F	T	T
T	F	F	F	F	T
F	T	T	T	T	T
F	F	T	T	T	T

Tautology

*b.

A	B	C	$A \wedge B$	$(A \wedge B) \vee C$	$B \vee C$	$A \wedge (B \vee C)$	$(A \wedge B) \vee C \to A \wedge (B \vee C)$
T	T	T	T	T	T	T	T
T	T	F	T	T	T	T	T
T	F	T	F	T	T	T	T
T	F	F	F	F	F	F	T
F	T	T	F	T	T	F	F
F	T	F	F	F	T	F	T
F	F	T	F	T	T	F	F
F	F	F	F	F	F	F	T

c.

A	B	A'	B'	$A' \vee B'$	$(A' \vee B')'$	$A \wedge (A' \vee B')'$
T	T	F	F	F	T	T
T	F	F	T	T	F	F
F	T	T	F	T	F	F
F	F	T	T	T	F	F

d.

A	B	A'	$A \wedge B$	$A \wedge B \to A'$
T	T	F	T	F
T	F	F	F	T
F	T	T	F	T
F	F	T	F	T

e.

A	B	C	$A \to B$	$A \vee C$	$B \vee C$	$(A \vee C) \to (B \vee C)$	$(A \to B) \to [(A \vee C) \to (B \vee C)]$
T	T	T	T	T	T	T	T
T	T	F	T	T	T	T	T
T	F	T	F	T	T	T	T
T	F	F	F	T	F	F	T
F	T	T	T	T	T	T	T
F	T	F	T	F	T	T	T
F	F	T	T	T	T	T	T
F	F	F	T	F	F	T	T

Tautology

f.

A	B	$B \to A$	$A \to (B \to A)$
T	T	T	T
T	F	T	T
F	T	F	T
F	F	T	T

Tautology

g.

A	B	$A \wedge B$	A'	B'	$B' \vee A'$	$A \wedge B \leftrightarrow B' \vee A'$
T	T	T	F	F	F	F
T	F	F	F	T	T	F
F	T	F	T	F	T	F
F	F	F	T	T	T	F

Contradiction

h.

A	B	B'	A ∨ B'	A ∧ B	(A ∧ B)'	(A ∨ B') ∧ (A ∧ B)'
T	T	F	T	T	F	F
T	F	T	T	F	T	T
F	T	F	F	F	T	F
F	F	T	T	F	T	T

i.

A	B	C	A ∨ B	C'	(A ∨ B) ∧ C'	A'	A' ∨ C	[(A ∨ B) ∧ C'] → A' ∨ C
T	T	T	T	F	F	F	T	T
T	T	F	T	T	T	F	F	F
T	F	T	T	F	F	F	T	T
T	F	F	T	T	T	F	F	F
F	T	T	T	F	F	T	T	T
F	T	F	T	T	T	T	T	T
F	F	T	F	F	F	T	T	T
F	F	F	F	T	F	T	T	T

*15. $2^{2^4} = 2^{16}$

16. 1b.

A	B	A ∧ B	B ∧ A	A ∧ B ↔ B ∧ A
T	T	T	T	T
T	F	F	F	T
F	T	F	F	T
F	F	F	F	T

2a.

A	B	C	A ∨ B	(A ∨ B) ∨ C	B ∨ C	A ∨ (B ∨ C)	(A ∨ B) ∨ C ↔ A ∨ (B ∨ C)
T	T	T	T	T	T	T	T
T	T	F	T	T	T	T	T
T	F	T	T	T	T	T	T
T	F	F	T	T	F	T	T
F	T	T	T	T	T	T	T
F	T	F	T	T	T	T	T
F	F	T	F	T	T	T	T
F	F	F	F	F	F	F	T

2b.

A	B	C	A ∧ B	(A ∧ B) ∧ C	B ∧ C	A ∧ (B ∧ C)	(A ∧ B) ∧ C ↔ A ∧ (B ∧ C)
T	T	T	T	T	T	T	T
T	T	F	T	F	F	F	T
T	F	T	F	F	F	F	T
T	F	F	F	F	F	F	T
F	T	T	F	F	T	F	T
F	T	F	F	F	F	F	T
F	F	T	F	F	F	F	T
F	F	F	F	F	F	F	T

3a.

A	B	C	$B \wedge C$	$A \vee (B \wedge C)$	$A \vee B$	$A \vee C$	$(A \vee B) \wedge (A \vee C)$	$A \vee (B \wedge C) \leftrightarrow (A \vee B) \wedge (A \vee C)$
T	T	T	T	T	T	T	T	T
T	T	F	F	T	T	T	T	T
T	F	T	F	T	T	T	T	T
T	F	F	F	T	T	T	T	T
F	T	T	T	T	T	T	T	T
F	T	F	F	F	T	F	F	T
F	F	T	F	F	F	T	F	T
F	F	F	F	F	F	F	F	T

3b.

A	B	C	$B \vee C$	$A \wedge (B \vee C)$	$A \wedge B$	$A \wedge C$	$(A \wedge B) \vee (A \wedge C)$	$A \wedge (B \vee C) \leftrightarrow (A \wedge B) \vee (A \wedge C)$
T	T	T	T	T	T	T	T	T
T	T	F	T	T	T	F	T	T
T	F	T	T	T	F	T	T	T
T	F	F	F	F	F	F	F	T
F	T	T	T	F	F	F	F	T
F	T	F	T	F	F	F	F	T
F	F	T	T	F	F	F	F	T
F	F	F	F	F	F	F	F	T

4a.

A	0	$A \vee 0$	$A \vee 0 \leftrightarrow A$
T	F	T	T
F	F	F	T

5b.

A	A'	$A \wedge A'$	0	$A \wedge A' \leftrightarrow 0$
T	F	F	F	T
F	T	F	F	T

17. *a.

A	A'	$A \vee A'$
T	F	T
F	T	T

b.

A	A'	(A')'	$(A')' \leftrightarrow A$
T	F	T	T
F	T	F	T

*c.

A	B	$A \wedge B$	$A \wedge B \rightarrow B$
T	T	T	T
T	F	F	T
F	T	F	T
F	F	F	T

d.

A	B	$A \vee B$	$A \rightarrow A \vee B$
T	T	T	T
T	F	T	T
F	T	T	T
F	F	F	T

e.

A	B	$A \vee B$	$(A \vee B)'$	A'	B'	$A' \wedge B'$	$(A \vee B)' \leftrightarrow A' \wedge B'$
T	T	T	F	F	F	F	T
T	F	T	F	F	T	F	T
F	T	T	F	T	F	F	T
F	F	F	T	T	T	T	T

f.

A	B	$A \wedge B$	$(A \wedge B)'$	A'	B'	$A' \vee B'$	$(A \wedge B)' \leftrightarrow A' \vee B'$
T	T	T	F	F	F	F	T
T	F	F	T	F	T	T	T
F	T	F	T	T	F	T	T
F	F	F	T	T	T	T	T

g.

A	A	$A \vee A$	$A \vee A \leftrightarrow A$
T	T	T	T
F	F	F	T

18. a. $(A \wedge B') \wedge C \leftrightarrow A \wedge (B' \wedge C) \leftrightarrow A \wedge (C \wedge B') \leftrightarrow (A \wedge C) \wedge B'$
b. $(A \vee B) \wedge (A \vee B') \leftrightarrow A \vee (B \wedge B') \leftrightarrow A \vee 0 \leftrightarrow A$
c. $A \vee (B \wedge A') \leftrightarrow (A \vee B) \wedge (A \vee A') \leftrightarrow (A \vee B) \wedge 1 \leftrightarrow A \vee B$
d. $(A \wedge B')' \vee B \leftrightarrow (A' \vee (B')') \vee B \leftrightarrow (A' \vee B) \vee B) \leftrightarrow A' \vee (B \vee B) \leftrightarrow A' \vee B$
e. $A \wedge (A \wedge B')' \leftrightarrow A \wedge (A' \vee (B')') \leftrightarrow A \wedge (A' \vee B) \leftrightarrow (A \wedge A') \vee (A \wedge B)$
$\leftrightarrow 0 \vee (A \wedge B) \leftrightarrow (A \wedge B) \vee 0 \leftrightarrow A \wedge B$

*19. dogs AND NOT retrievers

20. "oil paintings" AND (VanGogh OR REMBRANDT) AND NOT Vermeer

21. (novels OR plays) AND AIDS

22. 1.0, 2.4, 7.2, 5.3

23. For example: (A OR B) AND NOT (A AND B) AND NOT C

24. The conditional expression has the form $(A \vee B)' \vee (A' \wedge B)$.

$(A \vee B)' \vee (A' \wedge B) \Leftrightarrow (A' \wedge B') \vee (A' \wedge B)$ (De Morgan's Laws)
$\Leftrightarrow A' \wedge (B' \vee B)$ (tautology 3b)
$\Leftrightarrow A' \wedge (B \vee B')$ (tautology 1a)
$\Leftrightarrow A' \wedge 1$ (tautology 5a)
$\Leftrightarrow A'$ (tautology 4b)

Therefore the statement form can be written as

if not(*Value1* < *Value2*) **then**
statement1
else
statement2
end if

25. a.

A	B	$A \to B$	A'	$A' \vee B$	$A \to B \leftrightarrow A' \vee B$
T	T	T	F	T	T
T	F	F	F	F	T
F	T	T	T	T	T
F	F	T	T	T	T

b. The statement has the form $A \to B$, therefore

$$\begin{aligned}(A \to B)' &\Leftrightarrow (A' \vee B)' \\ &\Leftrightarrow (A')' \wedge B' \\ &\Leftrightarrow A \wedge B'\end{aligned}$$

The negated statement is "Sam passed his bar exam but he will not get the job."

26.*a. Assign

$B' \wedge (A \to B)$ true
A' false

From the second assignment, A is true. From the first assignment, B' is true (so B is false), and $A \to B$ is true. If $A \to B$ is true and A is true, then B is true. B is thus both true and false, and $[B' \wedge (A \to B)] \to A'$ is a tautology.

b. Assign

$(A \to B) \wedge A$ true
B false

From the first assignment, A is true and $A \to B$ is true. If $A \to B$ is true and A is true, then B is true. B is thus both true and false, and $[(A \to B) \wedge A] \to B$ is a tautology.

c. Assign

$(A \vee B) \wedge A'$ true
B false

From the first assignment, A' is true (and A is false), and $A \vee B$ is true. If $A \vee B$ is true and A is false, then B is true. B is thus both true and false, and $(A \vee B) \wedge A' \to B$ is a tautology.

d. Assign

$(A \wedge B) \wedge B'$ true
A false

From the first assignment, $A \wedge B$ is true. If $A \wedge B$ is true, then A is true. A is thus both true and false, and $(A \wedge B) \wedge B' \rightarrow A$ is a tautology.

27. a. P is $A \vee A'$ Q is $B \vee B'$
    b. P is $A \wedge A'$
    c. P is $A \vee A'$ Q is $A \wedge A'$

28. a.

A	B	$A \oplus B$
T	T	F
T	F	T
F	T	T
F	F	F

b.

A	B	$A \oplus B$	$A \leftrightarrow B$	$(A \leftrightarrow B)'$	$A \oplus B \leftrightarrow (A \leftrightarrow B)'$
T	T	F	T	F	T
T	F	T	F	T	T
F	T	T	F	T	T
F	F	F	T	F	T

29. a.

A	B	$A \vee B$	A'	B'	$A' \wedge B'$	$(A' \wedge B')'$	$A \vee B \leftrightarrow (A' \wedge B')'$
T	T	T	F	F	F	T	T
T	F	T	F	T	F	T	T
F	T	T	T	F	F	T	T
F	F	F	T	T	T	F	T

b.

A	B	B'	$A \wedge B'$	$(A \wedge B')'$	$A \rightarrow B$	$A \rightarrow B \leftrightarrow (A \wedge B')'$
T	T	F	F	T	T	T
T	F	T	T	F	F	T
F	T	F	F	T	T	T
F	F	T	F	T	T	T

30. a. $A \wedge B$ is equivalent to $(A' \vee B')'$

A	B	$A \wedge B$	A'	B'	$A' \vee B'$	$(A' \vee B')'$	$A \wedge B \leftrightarrow (A' \vee B')'$
T	T	T	F	F	F	T	T
T	F	F	F	T	T	F	T
F	T	F	T	F	T	F	T
F	F	F	T	T	T	F	T

$A \rightarrow B$ is equivalent to $A' \vee B$

A	B	$A \rightarrow B$	A'	$A' \vee B$	$A \rightarrow B \leftrightarrow A' \vee B$
T	T	T	F	T	T
T	F	F	F	F	T
F	T	T	T	T	T
F	F	T	T	T	T

b. $A \wedge B$ is equivalent to $(A \to B')'$

A	B	$A \wedge B$	B'	$A \to B'$	$(A \to B')'$	$A \wedge B \leftrightarrow (A \to B')'$
T	T	T	F	F	T	T
T	F	F	T	T	F	T
F	T	F	T	T	F	T
F	F	F	T	T	F	T

$A \vee B$ is equivalent to $A' \to B$

A	B	$A \vee B$	A'	$A' \to B$	$A \vee B \leftrightarrow A' \to B$
T	T	T	F	T	T
T	F	T	F	T	T
F	T	T	T	T	T
F	F	F	T	F	T

31. $(A \wedge B)'$ has the value F when A and B have the values T. However, any statement using only $\to$ and $\vee$ will have the value T when A and B are both T.

*32. $A \wedge B$ is equivalent to $(A|B)|(A|B)$

A	B	$A \wedge B$	$A\|B$	$(A\|B)\|(A\|B)$	$A \wedge B \leftrightarrow (A\|B)\|(A\|B)$
T	T	T	F	T	T
T	F	F	T	F	T
F	T	F	T	F	T
F	F	F	T	F	T

A' is equivalent to $A|A$

A	A'	$A\|A$	$A' \leftrightarrow A\|A$
T	F	F	T
F	T	T	T

33. $A \wedge B$ is equivalent to $(A \downarrow A) \downarrow (B \downarrow B)$

A	B	$A \wedge B$	$A \downarrow A$	$B \downarrow B$	$(A \downarrow A) \downarrow (B \downarrow B)$	$A \wedge B \leftrightarrow (A \downarrow A) \downarrow (B \downarrow B)$
T	T	T	F	F	T	T
T	F	F	F	T	F	T
F	T	F	T	F	F	T
F	F	F	T	T	F	T

A' is equivalent to $A \downarrow A$

A	A'	$A \downarrow A$	$A' \leftrightarrow A \downarrow A$
T	F	F	T
F	T	T	T

34. a. In order for $A \wedge B$ to be true, we would want to know that both parts are true; if one part has an unknown truth value then it is unknown whether this is the case. In order for $A \vee B$ to be true, we would want at least one part to be true; if one part is false and the other part has an unknown truth value, then it is unknown whether this is the case. Finally, if the truth value of A is unknown, then the truth value of A' is also unknown.

b. $(F)' \wedge N = T \wedge N = N$
c. $N \wedge F = F$
d. $(N)' \vee (F)' = N \vee T = T$

35. a. If A has a truth value of x, $0 \le x \le 1$, then A' is the "opposite condition" and will have the truth value that represents everything A is not, namely $1 - x$. For $A \wedge B$, both conditions must hold, which means the lower of the two truth values is all that can be achieved. For $A \vee B$, either condition can hold, so the higher of the two truth values can be achieved.
b. $1 - 0.12 = 0.88$
*c. $\min(0.12, 0.93) = 0.12$
d. $\max(0.12, 1 - 0.84) = \max(0.12, 0.16) = 0.16$

36. Machine D is either clean or infected. In either case, by statements 3 and 1, respectively, C is infected. Because C is infected, then A is infected by statement 2. By statement 4, B is infected (since C is not clean). By statement 3, because B is infected, D is not clean. The conclusion is that all four machines are infected.

*37. If Percival is a liar, then his statement is false. Therefore it is false that there is at least one liar, and both Percival and Llewellyn must be truth-tellers. But this is impossible since we assumed Percival is a liar. Therefore Percival is a truth-teller, and his statement is true. Because he said, "At least one of us is a liar," Llewellyn must be a liar. Therefore Percival is a truth-teller and Llewellyn is a liar.

38. Merlyn's statement is of the form $A \rightarrow B$, where A stands for "I am a truth-teller" and B stands for "Meredith is a truth-teller." If Merlyn is a liar then statement A is false; therefore statement $A \rightarrow B$ is true, but Merlyn, as a liar, would not have said a true statement. Therefore Merlyn must be a truth-teller. Then the statement he makes, $A \rightarrow B$, must be true, and statement A is true as well. Therefore statement B must be true, and Meredith is a truth-teller. So both Merlyn and Meredith are truth-tellers.

39. Rothwold's statement is of the form $A \vee B$, where A stands for "I am a liar" and B stands for "Grymlin is a truth-teller." If Rothwold is a liar, then his statement $A \vee B$ is false, and the statement $(A \vee B)'$ must be true. By De Morgan's laws, A' and B' must both be true. But A' is the statement that Rothwold is a truth-teller, which is not true. Therefore Rothwold must be a truth-teller, and his statement $A \vee B$ is true. Statement A, however, is false because it says that Rothwold is a liar. So statement B must be true, and Grymlin is a truth-teller. Both are truth-tellers.

EXERCISES 1.2

*1. $(M \rightarrow F) \wedge F' \rightarrow M'$ - mt
2. $(B \rightarrow A) \wedge B \rightarrow A$ - mp
3. $S \wedge L \rightarrow L$ - sim
4. $(S \rightarrow R) \wedge (R \rightarrow B) \rightarrow (S \rightarrow B)$ – hs

5. The hypotheses have the form $(C \to P) \wedge P'$. By mt, the conclusion is C'; the car was not involved in the hit-and-run.
6. The hypotheses have the form $(W \vee L) \wedge (W \to F)$. No conclusion can be made (note that $W \vee L$ does not mean that you have W).

*7. The hypotheses have the form $(B \to P) \wedge P$. Only P, you will be paid tomorrow, can be concluded, using simplification. (B cannot be concluded.)

8. The hypotheses have the form $G \wedge T \wedge (G \to R)$. By mp, the conclusion is R, we need to rake the leaves.

9.

1.	A	hyp
2.	$B \to C$	hyp
3.	B	hyp (deduction method)
4.	C	2, 3, mp
5.	$A \wedge C$	1, 4, con

*10.

1.	$A \to (B \vee C)$	hyp
2.	B'	hyp
3.	C'	hyp
4.	$B' \wedge C'$	2, 3, con
5.	$(B \vee C)'$	4, De Morgan
6.	A'	1, 5, mt

11.

1.	A'	hyp
2.	B	hyp
3.	$B \to (A \vee C)$	hyp
4.	$A \vee C$	2, 3, mp
5.	$(A')' \vee C$	4, dn
6.	$A' \to C$	5, imp
7.	C	1, 6, mp

12. $A' \wedge (B \to A) \to B'$

1.	A'	hyp
2.	$B \to A$	hyp
3.	B'	1, 2, mt

*13. $(A \to B) \wedge [A \to (B \to C)] \to (A \to C)$

1.	$A \to B$	hyp
2.	$A \to (B \to C)$	hyp
3.	A	hyp
4.	B	1, 3, mp
5.	$B \to C$	2, 3, mp
6.	C	4, 5, mp

14. $[(C \to D) \to C] \to [(C \to D) \to D]$

1.	$(C \to D) \to C$	hyp
2.	$C \to D$	hyp
3.	C	1, 2, mp
4.	D	2, 3, mp

*15. $A' \wedge (A \vee B) \to B$

1.	A'	hyp
2.	$A \vee B$	hyp
3.	$(A')' \vee B$	2, dn
4.	$A' \to B$	3, imp
5.	B	1, 4, mp

16. $[A \to (B \to C)] \wedge (A \vee D') \wedge B \to (D \to C)$

1.	$A \to (B \to C)$	hyp
2.	$A \vee D'$	hyp
3.	B	hyp
4.	D	hyp
5.	$D' \vee A$	2, comm
6.	$D \to A$	5, imp
7.	A	4, 6, mp
8.	$B \to C$	1, 7, mp
9.	C	3, 8, mp

17. $(A' \to B') \wedge B \wedge (A \to C) \to C$

1.	$A' \to B'$	hyp
2.	B	hyp
3.	$A \to C$	hyp
4.	$(B')'$	2, dn
5.	$(A')'$	1, 4, mt
6.	A	5, dn
7.	C	3, 6, mp

18. $(A \to B) \wedge [B \to (C \to D)] \wedge [A \to (B \to C)] \to (A \to D)$

1.	$A \to B$	hyp
2.	$B \to (C \to D)$	hyp
3.	$A \to (B \to C)$	hyp
4.	A	hyp
5.	B	1, 4, mp
6.	$B \to C$	3, 4, mp
7.	C	5, 6, mp
8.	$C \to D$	2, 5, mp
9.	D	8, 9, mp

19. $[A \to (B \to C)] \to [B \to (A \to C)]$

1. $A \to (B \to C)$ hyp
2. B hyp
3. A hyp (using deduction method again)
4. $B \to C$ 1, 3, mp
5. C 2, 4, mp

20. $(A \wedge B) \to (A \to B')'$

1. $A \wedge B$ hyp
2. $(A')' \wedge (B')'$ 1, dn
3. $(A' \vee B')'$ 2, De Morgan
4. $(A \to B')'$ 3, imp

*21. $(P \vee Q) \wedge P' \to Q$

1. $P \vee Q$ hyp
2. P' hyp
3. $(P')' \vee Q$ 1, dn
4. $P' \to Q$ 3, imp
5. Q 2, 4, mp

22. $(P \to Q) \to (Q' \to P')$

1. $P \to Q$ hyp
2. Q' hyp
3. P' 1, 2, mt

23. $(Q' \to P') \to (P \to Q)$

1. $Q' \to P'$ hyp
2. P hyp
3. $(P')'$ 2, dn
4. $(Q')'$ 1, 3, mt
5. Q 4, dn

24. $P \to P \wedge P$

1. P hyp
2. P hyp (just writing it again)
3. $P \wedge P$ 1, 2, con

25. $P \vee P \to P$

1. $P' \to P' \wedge P'$ Exercise 24
2. $P' \to (P \vee P)'$ 1, De Morgan
3. $[P' \to (P \vee P)'] \to [(P \vee P) \to P]$ Exercise 23
4. $P \vee P \to P$ 2, 3, mp

26. $[(P \wedge Q) \to R] \to [P \to (Q \to R)]$

1. $(P \wedge Q) \to R$ hyp
2. P hyp
3. Q hyp (using deduction method again)
4. $P \wedge Q$ con
5. R 1, 4, mp

*27. $P \wedge P' \to Q$

1.	P	hyp
2.	P'	hyp
3.	$P \vee Q$	1, add
4.	$Q \vee P$	3, comm
5.	$(Q')' \vee P$	4, dn
6.	$Q' \to P$	5, imp
7.	$(Q')'$	2, 6, mt
8.	Q	7, dn

28. $P \wedge (Q \vee R) \to (P \wedge Q) \vee (P \wedge R)$

Rewriting the conclusion, the argument is

$P \wedge (Q \vee R) \to ((P \wedge Q)')' \vee (P \wedge R)$ by dn

or

$P \wedge (Q \vee R) \to [(P \wedge Q)' \to (P \wedge R)]$ by imp

1.	P	hyp
2.	$Q \vee R$	hyp
3.	$(P \wedge Q)'$	hyp
4.	$P' \vee Q'$	3, De Morgan
5.	$Q' \vee P'$	4, comm
6.	$Q \to P'$	5, imp
7.	$(P')'$	1, dn
8.	Q'	6, 7, mt
9.	$R \vee Q$	2, comm
10.	$(R')' \vee Q$	9, dn
11.	$R' \to Q$	10, imp
12.	$(R')'$	8, 11, mt
13.	R	12, dn
14.	$P \wedge R$	1, 13, con

29. $P \vee (Q \wedge R) \to (P \vee Q) \wedge (P \vee R)$

Prove

$P \vee (Q \wedge R) \to (P \vee Q)$

Rewriting the conclusion, the argument is

$P \vee (Q \wedge R) \to ((P')' \vee Q)$ by dn

or

$P \vee (Q \wedge R) \to (P' \to Q)$ by imp

1.	$P \vee (Q \wedge R)$	hyp
2.	P'	hyp
3.	$(P')' \vee (Q \wedge R)$	1, dn
4.	$P' \to (Q \wedge R)$	3, imp
5.	$Q \wedge R$	2, 4, mp
6.	Q	5, sim

The proof for $P \vee (Q \wedge R) \to (P \vee R)$ is similar.

30. $A' \to (A \to B)$

1.	A'	hyp
2.	A	hyp
3.	B	1, 2, inc

*31. $(P \to Q) \wedge (P' \to Q) \to Q$

1.	$P \to Q$	hyp
2.	$P' \to Q$	hyp
3.	$Q' \to P'$	1, cont
4.	$Q' \to Q$	2, 3, hs
5.	$(Q')' \vee Q$	imp
6.	$Q \vee Q$	5, dn
7.	Q	6, self

32. $(A' \to B') \wedge (A \to C) \to (B \to C)$

1.	$A' \to B'$	hyp
2.	$A \to C$	hyp
3.	B	hyp
4.	$B \to A$	1, cont
5.	C	2, 4, hs

33. $(A' \to B) \wedge (B \to C) \wedge (C \to D) \to (A' \to D)$

1.	$A' \to B$	hyp
2.	$B \to C$	hyp
3.	$C \to D$	hyp
4.	$A' \to C$	1, 2, hs
5.	$A' \to D$	3, 4, hs

34. $(A \vee B) \wedge (A \to C) \wedge (B \to C) \to C$

1.	$A \vee B$	hyp
2.	$A \to C$	hyp
3.	$B \to C$	hyp
4.	$(A')' \vee B$	1, dn
5.	$A' \to B$	4, imp
6.	$A' \to C$	3, 5, hs
7.	$(A \to C) \wedge (A' \to C)$	2, 5, con
8.	C	Exercise 31

35. $(Y \to Z') \wedge (X' \to Y) \wedge [Y \to (X \to W)] \wedge (Y \to Z) \to (Y \to W)$

1.	$Y \to Z'$	hyp
2.	$Y \to Z$	hyp
3.	Y	hyp
4.	Z'	1, 3, mp
5.	Z	2, 3, mp
6.	W	4, 5, inc

*36. $(A \wedge B)' \wedge (C' \wedge A)' \wedge (C \wedge B')' \to A'$

1.	$(A \wedge B)'$	hyp
2.	$(C' \wedge A)'$	hyp
3.	$(C \wedge B')'$	hyp
4.	$A' \vee B'$	1, De Morgan
5.	$B' \vee A'$	4, comm
6.	$B \to A'$	5, imp
7.	$(C')' \vee A'$	2, De Morgan
8.	$C' \to A'$	7, imp
9.	$C' \vee (B')'$	3, De Morgan
10.	$(B')' \vee C'$	11, comm
11.	$B' \to C'$	10, imp
12.	$B' \to A'$	8, 11, hs
13.	$(B \to A') \wedge (B' \to A')$	6, 12, con
14.	A'	Exercise 31

37. $(P \vee (Q \wedge R)) \wedge (R' \vee S) \wedge (S \to T') \to (T \to P)$

1.	$P \vee (Q \wedge R)$	hyp
2.	$R' \vee S$	hyp
3.	$S \to T'$	hyp
4.	T	hyp
5.	$(T')'$	4, dn
6.	S'	3, 5, mt
7.	$S \vee R'$	2, comm
8.	$(S')' \vee R'$	7, dn
9.	$S' \to R'$	8. imp
10.	R'	6, 9, mp
11.	$R' \vee Q'$	10, add
12.	$Q' \vee R'$	11, comm
13.	$(Q \wedge R)'$	12, De Morgan
14.	$(Q \wedge R) \vee P$	1, comm
15.	P	13, 14, ds

38. The argument is $(E \to Q) \wedge (E \vee B) \wedge Q' \to B$

A proof sequence is:

1.	$E \to Q$	hyp
2.	$E \vee B$	hyp
3.	Q'	hyp
4.	$Q' \to E'$	cont
5.	E'	3, 4, mp
6.	$(E')' \vee B$	2, dn
7.	$E' \to B$	6, imp
8.	B	5, 7, mp

39. The argument is $[(J \to E) \wedge (J \to C)] \to (J \to (E \wedge C))$
A proof sequence is:

1.	$J \to E$	hyp
2.	$J \to C$	hyp
3.	J	hyp
4.	E	1, 3, mp
5.	C	2, 3, mp
6.	$E \wedge C$	4, 5, con

40. The argument is $[(C \wedge W') \wedge ((R \vee S') \to W)] \to (C \wedge S)$
A proof sequence is:

1.	$C \wedge W'$	hyp
2.	$(R \vee S') \to W$	hyp
3.	W'	1, sim
4.	$W' \to (R \vee S')'$	2, cont
5.	$(R \vee S')'$	3, 4, mp
6.	$R' \wedge (S')'$	5, De Morgan
7.	$R' \wedge S$	6, dn
8.	S	7, sim
9.	C	1, sim
10.	$C \wedge S$	8, 9, con

41. The argument is $[(A \to S) \wedge (A \vee C) \wedge S'] \to C$
A proof sequence is:

1.	$A \to S$	hyp
2.	$A \vee C$	hyp
3.	S'	hyp
4.	A'	1, 3, mt
5.	C	2, 4, ds

*42. The argument is $[(R \wedge (F' \vee N)) \wedge N' \wedge (A' \to F)] \to (A \wedge R)$
A proof sequence is:

1.	$R \wedge (F' \vee N)$	hyp
2.	N'	hyp
3.	$A' \to F$	hyp
4.	R	1, sim
5.	$F' \vee N$	1, sim
6.	$N \vee F'$	5, comm
7.	F'	2, 6, ds
8.	$F' \to (A')'$	3, cont
9.	$(A')'$	7, 8, mp
10.	A	9, dn
11.	$A \wedge R$	4, 10, con

43. The argument is $(R \to U)' \wedge (P \vee B')' \to U' \wedge B$

A proof sequence is:

1.	$(R \to U)'$	hyp
2.	$(P \vee B')'$	hyp
3.	$(R' \vee U)'$	1, imp
4.	$(R')' \wedge U'$	3, De Morgan
5.	$R \wedge U'$	4, dn
6.	$P' \wedge (B')'$	2, De Morgan
7.	$P' \wedge B$	6, dn
8.	U'	5, sim
9.	B	7, sim
10.	$U' \wedge B$	8, 9, con

44. The argument is $[((J \vee L) \to C) \wedge T' \wedge (C \to T)] \to J'$

A proof sequence is:

1.	$(J \vee L) \to C$	hyp
2.	T'	hyp
3.	$C \to T$	hyp
4.	$T' \to C'$	3, cont
5.	C'	2, 4, mp
6.	$C' \to (J \vee L)'$	1, cont
7.	$(J \vee L)'$	5, 6, mp
8.	$J' \wedge L'$	7, De Morgan
9.	J'	8, sim

*45. a.

A	B	C	$B \to C$	$A \to (B \to C)$	$A \wedge B$	$(A \wedge B) \to C$	$A \to (B \to C) \leftrightarrow (A \wedge B) \to C$
T	T	T	T	T	T	T	T
T	T	F	F	F	T	F	T
T	F	T	T	T	F	T	T
T	F	F	T	T	F	T	T
F	T	T	T	T	F	T	T
F	T	F	F	T	F	T	T
F	F	T	T	T	F	T	T
F	F	F	T	T	F	T	T

b. $A \to (B \to C) \Leftrightarrow A \to (B' \vee C) \Leftrightarrow A' \vee (B' \vee C) \Leftrightarrow (A' \vee B') \vee C$
$\Leftrightarrow (A \wedge B)' \vee C \Leftrightarrow (A \wedge B) \to C$

c. By part (a) (or (b)),

$$[P_1 \wedge P_2 \wedge \ldots \wedge P_n] \to (R \to S) \Leftrightarrow (P_1 \wedge P_2 \wedge \ldots \wedge P_n \wedge R) \to S$$

which says to take each of $P_1, P_2, \ldots, P_n$, R as hypotheses and deduce S.

46. Let

I = my client is innocent
K = the knife was in the drawer
J = Jason Pritchard saw the knife
O = the knife was there on October 10
H = the hammer was in the barn

Then the argument is

$$[(I' \to K) \wedge (K' \vee J) \wedge (O' \to J') \wedge (O \to (K \wedge H)) \wedge H'] \to I$$

A proof sequence is:

1.	$I' \to K$	hyp
2.	$K' \vee J$	hyp
3.	$O' \to J'$	hyp
4.	$O \to (K \wedge H)$	hyp
5.	H'	hyp
6.	$H' \vee K'$	5, add
7.	$(H \wedge K)'$	6, De Morgan
8.	$(K \wedge H)'$	7, comm
9.	$(K \wedge H)' \to O'$	4, cont
10.	O'	8, 9, mp
11.	J'	3, 10, mp
12.	$J \vee K'$	2, comm
13.	K'	11, 12, ds
14.	$K' \to (I')'$	1, cont
15.	$(I')'$	13, 14, mp
16.	I	15, dn

EXERCISES 1.3

1. a. T b. F c. F d. T

2. *a. true (pick $y = 0$)
   *b. true (pick $y = 0$)
   *c. true (pick $y = -x$)
   *d. false (no one y works for all x's)
   e. false (may have $x = y$)
   f. true (pick $y = -x$)
   g. true (pick $x = 2, y = 4$)
   h. false (may have $x = 0$)

3. a. F b. T c. T d. F e. T

4. *a. true: domain is the integers, $A(x)$ is "x is even", $B(x)$ is "x is odd"
   false: domain is the positive integers, $A(x)$ is "$x > 0$", $B(x)$ is "$x \geq 1$"

b. true: domain is the collection of lines in the plane, P(x, y) is "x is parallel to y"
false: domain is the integers, P(x, y) is "$x < y$"

c. true: domain is the integers, P(x) is "x is even", Q(x, y) is "$y|x$" (y divides x)
false: domain is the collection of all people, P(x) is "x is male", Q(x, y) is "y is a brother of x"

d. true: domain is the nonnegative integers, A(x) is "x is even", B(x, y) is "$x \leq y$"
false: domain is the positive integers, A(x) is "x is even", B(x, y) is "$x \leq y$"

e. true: domain is the integers, A(x) is "$x > 0$", B(x) is "$x \geq 0$"
false: domain is the integers, A(x) is "$x > 0$", B(x) is "x is even"

5. a. scope of $(\forall x)$ is $P(x) \rightarrow Q(y)$; y is a free variable
b. scope of $(\exists x)$ is $A(x) \wedge (\forall y)B(y)$; scope of $(\forall y)$ is $B(y)$; no free variables
c. scope of $(\exists x)$ is $(\forall y)P(x, y) \wedge Q(x, y)$; scope of $(\forall y)$ is $P(x, y)$; y is a free variable
d. scope of $(\exists x)$ is $(\exists y)[A(x, y) \wedge B(y, z) \rightarrow A(a, z)]$; scope of $(\exists y)$ is $A(x, y) \wedge B(y, z) \rightarrow A(a, z)$; z is a free variable

Some parts of Exercises 6-9 have multiple equivalent answers.

6. *a. $(\forall x)(D(x) \rightarrow S(x))$
*b. $(\exists x)[D(x) \wedge (R(x))']$ or $[(\forall x)(D(x) \rightarrow R(x))]'$
*c. $(\forall x)[D(x) \wedge S(x) \rightarrow (R(x))']$
d. $(\exists x)[D(x) \wedge S(x) \wedge R(x)]$
e. $(\forall x)[D(x) \rightarrow (S(x) \wedge R(x))']$ or $(\forall x)[D(x) \wedge S(x) \wedge R(x)]'$
f. $(\forall x)[D(x) \wedge S(x) \rightarrow D(x) \wedge R(x)]$
g. $(\forall x)[D(x) \rightarrow (S(x))']$
h. $S(M) \rightarrow (\forall x)(D(x) \rightarrow S(x))$
i. $R(M) \wedge R(T)$
j. $(\exists x)(D(x) \wedge R(x)) \rightarrow (\forall x)(D(x) \rightarrow S(x))$

7. a. $(\forall x)(B(x) \rightarrow R(x))$
b. $[(\forall x)(B(x) \rightarrow S(x))]'$ or $(\exists x)(B(x) \wedge [S(x)]')$
c. $(\forall x)(S(x) \rightarrow R(x))$
d. $(\exists x)(B(x) \wedge [R(x)]')$
e. $(\exists x)(B(x) \wedge R(x)) \wedge (\forall x)(S(x) \rightarrow [R(x)]')$
f. $(\forall x)(B(x) \wedge R(x) \rightarrow S(x))$
g. $(\forall x)(B(x) \wedge R(x) \rightarrow S(x))$ - this is the same statement as (f)
h. $(\forall x)(S(x) \rightarrow R(x)) \rightarrow (\forall x)(B(x) \rightarrow R(x))$

8. a. $(\exists x)(P(x) \wedge (\forall y)(T(y) \rightarrow F(x, y))$
b. $(\forall x)(P(x) \rightarrow (\exists y)(T(y) \wedge F(x, y))$
c. $[(\forall x)(\forall y)(P(x) \wedge T(y) \rightarrow F(x, y))]'$ or $(\exists x)(\exists y)(P(x) \wedge T(y) \wedge (F(x, y)')$ or $[(\forall x)(P(x) \rightarrow (\forall y)(T(y) \rightarrow F(x, y)))]'$

9. a. $(\exists x)(W(x) \wedge L(x) \wedge C(x))$
*b. $(\forall x)[W(x) \rightarrow (L(x) \wedge C(x))']$
*c. $(\exists x)[L(x) \wedge (\forall y)(A(x, y) \rightarrow J(y))]$ or $(\exists x)(\forall y)[L(x) \wedge (A(x, y) \rightarrow J(y))]$

d. $(\forall x)[J(x) \to (\forall y)(A(x, y) \to J(y))]$ or $(\forall x)(\forall y)[J(x) \to (A(x, y) \to J(y))]$ or $(\forall x)(\forall y)[J(x) \wedge A(x, y) \to J(y)]$
e. $(\forall x)(\forall y)[(J(y) \wedge A(x, y)) \to J(x)]$
f. $(\forall x)([W(x) \wedge L(x)] \to (\exists y)[J(y) \wedge A(x, y)])$ or $(\forall x)(\exists y)([W(x) \wedge L(x)] \to [J(y) \wedge A(x, y)])$
g. $(\exists x)(W(x) \wedge (\forall y)[L(y) \to (A(x, y))'])$ or $(\exists x)(W(x) \wedge (\forall y)[A(x, y) \to (L(y))'])$ or $(\exists x)(\forall y)(W(x) \wedge [L(y) \to (A(x, y))'])$

10.*a. $(\forall x)(C(x) \wedge F(x))'$
*b. $(\exists x)[P(x) \wedge (\forall y)(S(x, y) \to F(y))]$ or $(\exists x)(\forall y)[P(x) \wedge (S(x, y) \to F(y))]$
c. $(\forall x)(\forall y)([P(y) \wedge S(x, y)] \to C(x))$
d. $(\forall x)[F(x) \to (\exists y)(C(y) \wedge S(x, y))]$ or $(\forall x)(\exists y)[F(x) \to (C(y) \wedge S(x, y))]$ or, if there is some *one* Corvette, $(\exists x)[C(x) \wedge (\forall y)(F(y) \to S(y, x))]$
e. $(\exists x)(P(x) \wedge (\forall y)[C(y) \to (S(x, y))'])$ or $(\exists x)(\forall y)(P(x) \wedge [C(y) \to (S(x, y))'])$
f. $(\exists x)(\exists y)(C(x) \wedge F(y) \wedge S(x, y)) \to (\forall x)(\forall y)(C(x) \wedge F(y) \to S(x, y))$

11. a. $(\forall x)[B(x) \to (\forall y)(F(y) \to L(x, y))]$ or $(\forall x)(\forall y)[(B(x) \wedge F(y)) \to L(x, y)]$
b. $(\exists x)[B(x) \wedge (\forall y)(F(y) \to L(x, y))]$
c. $(\forall x)[B(x) \to (\exists y)(F(y) \wedge L(x, y))]$
d. $(\forall x)[B(x) \to (\forall y)((L(x, y))' \to F(y))]$
e. $(\forall y)[F(y) \to (\forall x)(L(x, y) \to B(x))]$ or $(\forall y)(\forall x)[(F(y) \wedge L(x, y)) \to B(x)]$
f. $(\forall x)[B(x) \to (\forall y)(L(x, y) \to F(y))]$
g. $[(\exists x)[B(x) \wedge (\forall y)(L(x, y) \to F(y))]]'$ or $(\forall x)[B(x) \to (\exists y)(L(x, y) \wedge (F(y))']$
h. $(\exists x)[B(x) \wedge (\exists y)(F(y) \wedge L(x, y))]$ or $(\exists x)(\exists y)[B(x) \wedge F(y) \wedge L(x, y)]$
i. $((\exists x)[B(x) \wedge (\forall y)(L(x, y) \to F(y))]$
j. $(\forall x)[B(x) \to (\exists y)(F(y) \wedge (L(x, y))']$
k. $(\forall x)[B(x) \to (\forall y)(F(y) \to (L(x, y))')]$ or $(\forall x)(\forall y)[(B(x) \wedge F(y)) \to (L(x, y))']$
l. $[(\exists x)[B(x) \wedge (\forall y)(F(y) \to (L(x, y))')]]'$ or $(\forall x)[B(x) \to (\exists y)(F(y) \wedge L(x, y))]$

12. a. $(\forall x)(S(x) \to L(x))$
b. $(\exists x)(M(x) \wedge [S(x)]')$
c. $(\forall x)(L(x) \to M(x))$
d. $(\exists x)(S(x) \wedge M(x))$
e. $(\forall x)(\forall y) (S(x) \wedge M(y) \to B(x, y))$
f. $(\exists x)(M(x) \wedge (\forall y)(S(y) \to B(x, y))$
g. $(\forall x)(\forall y)(M(y) \wedge B(x, y) \to S(x))$

13.*a. John is handsome and Kathy loves John
*b. all men are handsome
c. all women love only handsome men
d. a handsome man loves Kathy
e. some pretty woman loves only handsome men
f. John loves all pretty women

14. *a. 2 b. 3 c. 3 d. 1

15 a. No web site features audio.
b. Some web site does not have audio or does not have video.
*c. Some web site has neither audio nor video.
d. Every web site has either audio or video.
e. Some web site does not have text and also either doesn't have audio or doesn't have video.

16. a. There is a nonstudent who eats pizza.
b. Some student does not eat pizza.
c. Every student eats some non-pizza item.

17. a. Every farmer grows something besides corn.
b. Some farmer does not grow corn.
c. Someone besides a farmer grows corn.

18. a. both sides are true exactly when A(x, y) holds for all x, y pairs
b. both sides are true exactly when some x, y pair satisfies the property A(x, y)
*c. if there is a single x that is in relation P to all y, then for every y an x exists (this same x) that is in relation P to y
d. if a has property A, then something in the domain has property A
e. if any member of the domain that has property A also has property B, then if all members of the domain have property A, all have property B

19.*a. domain is the integers, A(x) is "x is even", B(x) is "x is odd"
b. domain is the integers, P(x,y) is "$x + y = 0$"; for every x there is a y ($y = -x$) such that $x + y = 0$ but there is no single integer x that gives 0 when added to every integer y
c. domain is the positive integers, P(x) is "$x > 4$", Q(x) is "$x > 2$". Then every positive integer greater than 4 is greater than 2, so $(\forall x)(P(x) \to Q(x))$ is true. There exists a positive integer greater than 4, but not all positive integers are greater than 2, so $(\exists x)P(x) \to (\forall x)Q(x)$ is false.
d. domain is the integers, A(x) is "x is even". Then $(\forall x)(A(x))'$ is false - it is not the case that every integer is odd (not even) - but $((\forall x)A(x))'$ is true since it is false that every integer is even.

20. a. valid: there is an x in the domain with property A says it is false that everything in the domain fails to have property A.
b. not valid: domain is the integers, P(x) is "x is even", Q(x) is "x is prime". Because there are prime integers, $(\exists x)Q(x)$ and therefore $(\forall x)P(x) \vee (\exists x)Q(x)$ is true. But it is false that every integer is even or prime, so the implication is false.
c. valid: A true for all objects in the domain means it is false that there is some object in the domain for which A is not true.
d. valid: suppose that for every member of the domain, either P(x) or Q(x) is true. If there is some member of the domain for which Q is true, then $(\exists y)Q(y)$ is true. Otherwise all members of the domain have property P and $(\forall x)P(x)$ is true. In either case, $(\forall x)P(x) \vee (\exists y)Q(y)$ is true.

EXERCISES 1.4

1. The conclusion is that pansies are plants. The hypotheses have the form $(\forall x)(F(x) \rightarrow P(x)) \wedge F(p)$. By universal instantiation, $F(p) \rightarrow P(p)$, then by modus ponens, $P(p)$.

*2. The conclusion is that pansies are red. The hypotheses have the form $(\forall x)[F(x) \rightarrow (R(x) \vee P(x))] \wedge F(p) \wedge [P(p)]'$. By universal instantiation, $F(p) \rightarrow (R(p) \vee P(p))$, then by modus ponens, $R(p) \vee P(p)$, and finally by disjunctive syllogism, $R(p)$.

3. The conclusion is that some flowers are small. The hypotheses have the form $(\exists x)(F(x) \wedge P(x)) \wedge (\forall x)(F(x) \wedge P(x) \rightarrow S(x))$. By existential and universal instantiation (in that order), $F(a) \wedge P(a)$ and $F(a) \wedge P(a) \rightarrow S(a)$, so by modus ponens, $S(a)$. Combining $F(a)$ and $S(a)$ and using existential generalization results in $(\exists x)(F(x) \wedge S(x))$.

4. No conclusion is possible. Just because pansies are flowers, it does not make them either red or purple. The hypotheses have the form $(\exists x)(F(x) \wedge R(x))$, $(\exists x)(F(x) \wedge P(x))$, $F(p)$. But existential instantiation does not allow us to use p in removing the existential quantifiers, so we can say nothing further about pansies.

5. The conclusion is that some flowers are weeds. The hypotheses have the form $(\exists x)(F(x) \wedge P(x) \wedge T(x))$, $(\forall x)(F(x) \wedge T(x) \rightarrow B(x))$, and $(\forall x)(F(x) \wedge B(x) \rightarrow W(x))$. By existential and universal instantiation (in that order), $F(a) \wedge P(a) \wedge T(a)$, $F(a) \wedge T(a) \rightarrow B(a)$, and $F(a) \wedge B(a) \rightarrow W(a)$. Simplification gives $F(a) \wedge T(a)$ which, using modus ponens, gives $B(a)$. Combining $F(a)$ with $B(a)$ and using modus ponens results in $W(a)$. Combining $F(a)$ and $W(a)$ and using existential generalization results in $(\exists x)(F(x) \wedge W(x))$

6. 1. Hyp
   2. 1, ei
   3. hyp
   4. 3, ui
   5. 2, 4, mp
   6. 5, eg

7. 1. hyp
   2. hyp
   3. 1, ei
   4. 2, ui
   5. 3, 4, mp
   6. 5, eg

*8. a. The domain is the set of integers, $P(x, y)$ is "$x < y$", and $Q(x, y)$ is "$x > y$"; for every integer x, there is some integer that is larger and there is some integer that is smaller. But it is false that for every integer x there is some one integer that is both larger and smaller than x.

b. To get to step 2, ei was performed on two different existential quantifiers, neither of which was in front with the whole rest of the wff as its scope. Also, both existential quantifiers were removed at once, with the same constant *a* substituted for the variable in each case; this should be done in two steps, and the second would then have to introduce a new constant not previously used in the proof. And at step 3, the existential quantifier was not inserted at the front of the wff.

9. a. domain is the integers, Q(x, y) is "x < y"; for every y there is an x with x < y but there is no single integer x that is less than every integer y.

b. The use of universal generalization at step 4 is illegal because step 3 was deduced by ei from (∃x)Q(x, y) in which y is a free variable.

*10. (∀x)P(x) → (∀x)[P(x) ∨ Q(x)]

1.	(∀x)P(x)	hyp
2.	P(x)	1, ui
3.	P(x) ∨ Q(x)	2, add
4.	(∀x)(P(x) ∨ Q(x))	3, ug (note that P(x) ∨ Q(x) was deduced from (∀x)P(x) in which x is not free

11. (∀x)P(x) ∧ (∃x)Q(x) → (∃x)[P(x) ∧ Q(x)]

1.	(∀x)P(x)	hyp
2.	(∃x)Q(x)	hyp
3.	Q(a)	2, ei
4.	P(a)	1, ui
5.	P(a) ∧ Q(a)	3, 4, con
6.	(∃x)(P(x) ∧ Q(x))	5, eg

12. (∃x)(∃y)P(x, y) → (∃y)(∃x)P(x, y)

1.	(∃x)(∃y)P(x, y)	hyp
2.	(∃y)P(a, y)	1, ei
3.	P(a, b)	2, ei
4.	(∃x)P(x, b)	3, eg
5.	(∃y)(∃x)P(x, y)	4, eg

13. (∀x)(∀y)Q(x, y) → (∀y)(∀x)Q(x, y)

1.	(∀x)(∀y)Q(x,y)	hyp
2.	(∀y)Q(x, y)	1, ui
3.	Q(x, y)	2, ui
4.	(∀x)Q(x, y)	3, ug (x not free in (∀x)(∀y)Q(x, y))
5.	(∀y)(∀x)Q(x, y)	4, ug (y not free in (∀x)(∀y)Q(x, y))

14. (∀x)P(x) ∧ (∃x)[P(x)]' → (∃x)Q(x)

1.	(∀x)P(x)	hyp
2.	(∃x)[P(x)]'	hyp
3.	[P(a)]'	2, ei
4.	P(a)	1, ui
5.	Q(a)	3, 4, inc
6.	(∃x)Q(x)	5, eg

*15. $(\exists x)[A(x) \wedge B(x)] \rightarrow (\exists x)A(x) \wedge (\exists x)B(x)$

1.	$(\exists x)(A(x) \wedge B(x))$	hyp
2.	$A(a) \wedge B(a)$	1, ei
3.	$A(a)$	2, sim
4.	$B(a)$	2, sim
5.	$(\exists x)A(x)$	3, eg
6.	$(\exists x)B(x)$	4, eg
7.	$(\exists x)A(x) \wedge (\exists x)B(x)$	5, 6, con

16. Using an equivalence to rewrite the conclusion, we want to prove
$(\exists x)(R(x) \vee S(x)) \rightarrow [((\exists x)R(x))' \rightarrow (\exists x)S(x)]$

1.	$(\exists x)(R(x) \vee S(x))$	hyp
2.	$R(a) \vee S(a)$	1, ei
3.	$((\exists x)R(x))'$	hyp
4.	$(\forall x)(R(x))'$	3, neg
5.	$(R(a))'$	4, ui
6.	$S(a)$	2, 5, ds
7.	$(\exists x)S(x)$	6, eg

17. $(\forall x)[P(x) \rightarrow Q(x)] \rightarrow [(\forall x)P(x) \rightarrow (\forall x)Q(x)]$

1.	$(\forall x)[P(x) \rightarrow Q(x)]$	hyp
2.	$(\forall x)P(x)$	hyp
3.	$P(x) \rightarrow Q(x)$	1, ui
4.	$P(x)$	2, ui
5.	$Q(x)$	3, 4, mp
6.	$(\forall x)Q(x)$	5, ug

18. $[(\forall x)P(x) \rightarrow (\forall x)Q(x)] \rightarrow (\forall x)[P(x) \rightarrow Q(x)]$
Domain is the integers, P(x) is "$x < 3$" and Q(x) is "$x < 2$." (The left side is true because $(\forall x)P(x)$ is false, but the right side is false.)

*19. $(\exists x)(\forall y)Q(x, y) \rightarrow (\forall y)(\exists x)Q(x, y)$

1.	$(\exists x)(\forall y)Q(x, y)$	hyp
2.	$(\forall y)Q(a, y)$	1, ei
3.	$Q(a, y)$	2, ui
4.	$(\exists x)Q(x, y)$	3, eg
5.	$(\forall y)(\exists x)Q(x, y)$	4, ug

20. $(\forall x)P(x) \vee (\exists x)Q(x) \rightarrow (\forall x)[P(x) \vee Q(x)]$
Domain is the integers, P(x) is "$x < 5$" and Q(x) is "x is even."

21. $(\forall x)[A(x) \rightarrow B(x)] \rightarrow [(\exists x)A(x) \rightarrow (\exists x)B(x)]$

1.	$(\forall x)(A(x) \rightarrow B(x))$	hyp
2.	$(\exists x)A(x)$	hyp
3.	$A(a)$	2, ei
4.	$A(a) \rightarrow B(a)$	1, ui
5.	$B(a)$	3,4, mp
6.	$(\exists x)B(x)$	5, eg

22. $(\forall y)[Q(x, y) \to P(x)] \to [(\exists y)Q(x, y) \to P(x)]$

1. $(\forall y)(Q(x, y) \to P(x))$ hyp
2. $(\exists y)Q(x, y)$ hyp
3. $Q(x, a)$ 2, ei
4. $Q(x, a) \to P(x)$ 1, ui
5. $P(x)$ 3, 4, mp

*23. $[P(x) \to (\exists y)Q(x, y)] \to (\exists y)[P(x) \to Q(x, y)]$

1. $P(x) \to (\exists y)Q(x, y)$ hyp
2. $P(x)$ temporary hyp
3. $(\exists y)Q(x, y)$ 1, 2, mp
4. $Q(x, a)$ 3, ei
5. $P(x) \to Q(x, a)$ temporary hyp discharged
6. $(\exists y)(P(x) \to Q(x, y))$ 5, eg

24. $(\exists x)[P(x) \to Q(x)] \wedge (\forall y)[Q(y) \to R(y)] \wedge (\forall x)P(x) \to (\exists x)R(x)$

1. $(\exists x)[P(x) \to Q(x)]$ hyp
2. $(\forall y)[Q(y) \to R(y)]$ hyp
3. $(\forall x)P(x)$ hyp
4. $P(a) \to Q(a)$ 1, ei
5. $P(a)$ 3, ui
6. $Q(a)$ 4, 5, mp
7. $Q(a) \to R(a)$ 2, ui
8. $R(a)$ 6, 7, mp
9. $(\exists x)R(x)$ 8, eg

25. a. $(\forall x)(M(x) \to P(x)) \wedge (\forall x)(S(x) \to M(x)) \to (\forall x)(S(x) \to P(x))$

A proof sequence is:

1. $(\forall x)(M(x) \to P(x))$ hyp
2. $(\forall x)(S(x) \to M(x))$ hyp
3. $M(x) \to P(x)$ 1, ui
4. $S(x) \to M(x)$ 2, ui
5. $S(x) \to P(x)$ 3, 4 hs
6. $(\forall x)(S(x) \to P(x))$ 5, ug

b. $(\forall x)(M(x) \to [P(x)]') \wedge (\forall x)(S(x) \to M(x)) \to (\forall x)(S(x) \to [P(x)]')$

1. $(\forall x)(M(x) \to [P(x)]')$ hyp
2. $(\forall x)(S(x) \to M(x))$ hyp
3. $M(x) \to [P(x)]'$ 1, ui
4, $S(x) \to M(x)$ 2, ui
5. $S(x) \to [P(x)]'$ 3, 4, hs
6. $(\forall x)(S(x) \to [P(x)]')$ 5, ug

c. $(\forall x)(M(x) \rightarrow P(x)) \wedge (\exists x)(S(x) \wedge M(x)) \rightarrow (\exists x)(S(x) \wedge P(x))$

1. $(\forall x)(M(x) \rightarrow P(x))$	hyp
2. $(\exists x)(S(x) \wedge M(x))$	hyp
3. $S(a) \wedge M(a)$	2, ei
4. $M(a)$	3, sim
5. $M(a) \rightarrow P(a)$	1, ui
6. $P(a)$	4, 5, mp
7. $S(a)$	3, sim
8. $S(a) \wedge P(a)$	6, 7, con
9. $(\exists x)(S(x) \wedge P(x))$	8, eg

d. $(\forall x)(M(x) \rightarrow [P(x)]') \wedge (\exists x)(S(x) \wedge M(x)) \rightarrow (\exists x)(S(x) \wedge [P(x)]')$

1. $(\forall x)(M(x) \rightarrow [P(x)]')$	hyp
2. $(\exists x)(S(x) \wedge M(x))$	hyp
3. $S(a) \wedge M(a)$	2, ei
4. $M(a) \rightarrow [P(a)]'$	1, ui
5. $M(a)$	3, sim
6. $[P(a)]'$	4, 5, mp
7. $S(a)$	3, sim
8. $S(a) \wedge [P(a)]'$	6, 7, con
9. $(\exists x)(S(x) \wedge [P(x)]')$	8, eg

26. The argument is:

$(\exists x)[P(x) \wedge F(x)] \wedge (\forall y)[F(y) \rightarrow S(y)] \rightarrow (\exists x)[P(x) \wedge S(x)]$

A proof sequence is:

1. $(\exists x)[P(x) \wedge F(x)]$	hyp
2. $(\forall y)[F(y) \rightarrow S(y)]$	hyp
3. $P(a) \wedge F(a)$	1, ei
4. $F(a) \rightarrow S(a)$	2, ui
5. $F(a)$	3, sim
6. $S(a)$	4, 5, mp
7. $P(a)$	3, sim
8. $P(a) \wedge S(a)$	6, 7, con
9. $(\exists x)[P(x) \wedge S(x)]$	8. eg

*27 The argument is:

$(\forall x)(\forall y)[C(x) \wedge A(y) \rightarrow B(x, y)] \wedge C(s) \wedge (\exists x)(S(x) \wedge \{B(s, x)]') \rightarrow (\exists x)[A(x)]'$

A proof sequence is:

1. $(\forall x)(\forall y)[C(x) \wedge A(y) \rightarrow B(x, y)]$	hyp
2. $C(s)$	hyp
3. $(\forall y)[C(s) \wedge A(y) \rightarrow B(s, y)]$	1, ui
4. $(\exists x)(S(x) \wedge \{B(s, x)]')$	hyp
5. $S(a) \wedge \{B(s, a)]'$	4, ei
6. $C(s) \wedge A(a) \rightarrow B(s, a)$	3, ui
7. $[B(s, a)]'$	6, sim

8. $[C(s) \wedge A(a)]'$	6, 7, mt
9. $[C(s)]' \vee \{A(a)]'$	8, De Morgan
10. $[[C(s)]']'$	2, dn
11. $[A(a)]'$	9, 10, ds
12. $(\exists x)[A(x)]'$	11, eg

28. The argument is:
$(\exists x)(A(x) \wedge (N(x))') \wedge (\forall x)(G(x) \rightarrow N(x)) \wedge (\forall x)(G(x) \vee C(x)) \rightarrow (\exists x)(A(x) \wedge C(x))$

A proof sequence is:

1. $(\exists x)(A(x) \wedge (N(x))')$	hyp
2. $A(a) \wedge (N(a))'$	1, ei
3. $(\forall x)(G(x) \rightarrow N(x))$	hyp
4. $G(a) \rightarrow N(a)$	3, ui
5. $(N(a))' \rightarrow (G(a))'$	4, cont
6. $(N(a))'$	2, sim
7. $(G(a))'$	5, 6, mp
8. $(\forall x)(G(x) \vee C(x))$	hyp
9. $G(a) \vee C(a)$	8, ui
11. $C(a)$	7, 9, ds
12. $A(a)$	2, sim
13. $A(a) \wedge C(a)$	11, 12, con
14. $(\exists x)(A(x) \wedge C(x))$	13, eg

*29. The argument is:
$(\forall x)(M(x) \rightarrow I(x) \vee G(x)) \wedge (\forall x)(G(x) \wedge L(x) \rightarrow F(x)) \wedge (I(j))' \wedge L(j) \rightarrow [(M(j) \rightarrow F(j))]$

A proof sequence is:

1. $(\forall x)(M(x) \rightarrow I(x) \vee G(x))$	hyp
2. $(\forall x)(G(x) \wedge L(x) \rightarrow F(x))$	hyp
3. $M(j) \rightarrow I(j) \vee G(j)$	1, ui
4. $G(j) \wedge L(j) \rightarrow F(j)$	2, ui
5. $M(j)$	hyp
6. $I(j) \vee G(j)$	3, 5, mp
7. $(I(j))'$	hyp
8. $G(j)$	6, 7, ds
9. $L(j)$	hyp
10. $G(j) \wedge L(j)$	8, 9, con
11. $F(j)$	4,10, mp

30. The argument is:

$(\exists x)(M(x) \wedge (\forall y)R(x, y)) \wedge (\forall x)(\forall y)(R(x, y) \rightarrow T(x, y)) \rightarrow (\exists x)(M(x) \wedge (\forall y)T(x, y))$

A proof sequence is:

1.	$(\exists x)(M(x) \wedge (\forall y)R(x, y))$	hyp
2.	$M(a) \wedge (\forall y)R(a, y)$	1, ei
3.	$M(a)$	2, sim
4.	$(\forall x)(\forall y)(R(x, y) \rightarrow T(x, y))$	hyp
5.	$(\forall y)(R(a, y) \rightarrow T(a, y))$	4, ui
6.	$(R(a, y) \rightarrow T(a, y))$	5, ui
7.	$(\forall y)R(a, y))$	2, sim
8.	$R(a, y)$	7, ui
8.	$T(a, y))$	6, 8, mp
9.	$(\forall y)T(a, y)$	8, ug
10.	$M(a) \wedge (\forall y)T(a, y)$	3, 9, con
11.	$(\exists x)(M(x) \wedge (\forall y)T(x, y))$	10, eg

31. The argument is:

$(\forall x)(C(x) \rightarrow (\exists y)W(x, y)) \wedge (\forall x)(\forall y)(W(x, y) \rightarrow S(x, y)) \wedge C(m) \rightarrow (\exists y)S(m. y)$

A proof sequence is:

1.	$(\forall x)(C(x) \rightarrow (\exists y)W(x, y))$	hyp
2.	$C(m) \rightarrow (\exists y)W(m, y)$	1, ui
3.	$C(m)$	hyp
4.	$(\exists y)W(m, y)$	2, 3, mp
5.	$(\forall x)(\forall y)(W(x, y) \rightarrow S(x, y))$	hyp
6.	$(\forall y)(W(m, y) \rightarrow S(m, y))$	5, ui
7.	$W(m, a)$	4, ei
8.	$W(m, a) \rightarrow S(m, a)$	6, ui
9.	$S(m, a)$	7, 8, mp
10.	$(\exists y)S(m, y)$	9, eg

32. The argument is:

$(\forall x)(\forall y)((A(x) \wedge S(x, y)) \rightarrow D(y)) \wedge (\exists x)(\exists y)(A(x) \wedge S(x, y)) \rightarrow (\exists x)D(x)$

A proof sequence is:

1.	$(\forall x)(\forall y)((A(x) \wedge S(x, y)) \rightarrow D(y))$	hyp
2.	$(\exists x)(\exists y)(A(x) \wedge S(x, y))$	hyp
3.	$(\exists y)(A(a) \wedge S(a, y))$	2, ei
4.	$A(a) \wedge S(a, b)$	3, ei
5.	$(\forall y)((A(a) \wedge S(a, y)) \rightarrow D(y))$	1, ui
6.	$(A(a) \wedge S(a, b)) \rightarrow D(b)$	5, ui
7.	$D(b)$	4, 6, mp
8.	$(\exists x)D(x)$	7, eg

33. $[(\exists x)[P(x)]']' \leftrightarrow (\forall x)[[P(x)]']'$ neg, using $[P(x)]'$ for A(x)
$[(\exists x)[P(x)]']' \leftrightarrow (\forall x)P(x)$ dn
$[(\forall x)P(x)]' \leftrightarrow ([(\exists x)[P(x)]']')'$ cont (each direction)
$[(\forall x)P(x)]' \leftrightarrow (\exists x)[P(x)]'$ dn

34. a. 3 b. 5 c. 2 (an even prime number) d. 2 $((-1)^2 = 1)$
e. 11 $(2^{11} - 1 = 2047 = 23*89)$

EXERCISES 1.5

1. no 2. yes *3. fish

4. rabbit 5. fox
deer

6. fish
little-fish
fish
raccoon
fox
deer
deer

*7. *herbivore*(x) **if** *eat*(x, y) **and** *plant*(y)

8. little-fish
rabbit
deer

9. a. Anita
b. Mike
Kim
c. Judith
Sam
Mike
Kim
Joan
Hamal
Enrique
Jefferson

10. For example,
capitol(north-dakota, bismark)
capitol(california, sacramento)
capitol(hawaii, honolulu)
capitol(pennsylvania, harrisburg)

capitol(florida, tallahassee)
capitol(new-york, albany)

big(sacramento)
big(honolulu)
big(albany)

small(bismark)
small(harrisburg)
small(tallahassee)

eastern(pennsylvania)
eastern(florida)
eastern(new-york)

western(north-dakota)
western(california)
western(hawaii)

a. **which**(x: *small*(x))
b. **which**(x: *capitol*(x, y) **and** *small*(y))
c. **which**(x: *eastern*(x) **and** *capitol*(x, y) **and** *big*(y))
d. *cosmopolitan*(x) **if** *big*(x) **and** *capitol*(y, x) **and** *western*(y)
e. **which**(x: *cosmopolitan*(x))

*11. a. **is**(*author-of*(mark-twain, hound-of-the-baskervilles))
b. **which**(x: *author-of*(faulkner, x))
c. *nonfiction-author*(x) **if** *author-of*(x, y) **and** **not**(*fiction*(y))
d. **which**(x: *nonfiction-author*(x))

12. a. *father-of*(x, y) **if** *parent-of*(x, y) **and** *male*(x)
b. *daughter-of*(x, y) **if** *parent-of*(y, x) **and** female(x)
*c. *ancestor-of*(x, y) **if** *parent-of*(x, y)
ancestor-of(x, y) **if** *parent-of*(x, z) **and** *ancestor-of*(z, y)

13. a. **which**(x: *small*(x) **and** *part-of*(x, y))
b. **which**(y: *small*(x) **and** *part-of*(x, y) **and** *big*(y))
c. *component-of*(x, y) **if** *part-of*(x, y)
component-of(x, y) **if** *part-of*(x, z) **and** *component-of*(z, y)

EXERCISES 1.6

*1. $x + 1 = y - 1$, or $x = y - 2$

2. $2x > y$, or $x > y/2$

3. Working backwards from the postcondition using the assignment rule,

$\{x + 3 = 4\}$
$y = x + 3$
$\{2y = 8 \quad \text{or} \quad y = 4\}$
$y = 2*y$
$\{y = 8\}$

The first assertion, $x + 3 = 4$, is equivalent to the precondition $x = 1$. The assignment rule, applied twice, proves the program segment correct.

4. Working backwards from the postcondition using the assignment rule,

$\{x + 2 > 2\}$
$y = x + 2$
$\{y + 1 > 3 \quad \text{or} \quad y > 2\}$
$z = y + 1$
$\{z > 3\}$

The first assertion, $x + 2 > 2$, is equivalent to the precondition $x > 0$. The assignment rule, applied twice, proves the program segment correct.

*5. The desired postcondition is $y = x(x - 1)$. Working back from the postcondition, using the assignment rule, gives

$\{x(x - 1) = x(x - 1)\}$
$y = x - 1$
$\{xy = x(x - 1)\}$
$y = x * y$
$\{y = x(x - 1)\}$

Because the precondition is always true, so is each subsequent assertion, including the postcondition.

6. The desired postcondition is $y = 2x + 1$. Working back from the postcondition, using the assignment rule, gives

$\{x + x = 2x\}$
$y = x$
$\{y + y = 2x\}$
$y = y + y$
$\{y + 1 = 2x + 1 \quad \text{or} \quad y = 2x\}$
$y = y + 1$
$\{y = 2x + 1\}$

Because the precondition is always true, so is each subsequent assertion, including the postcondition.

*7. The two implications to prove are

$$\{y = 0 \text{ and } y < 5\}\ y = y + 1\ \{y = 1\}$$

and

$$\{y = 0 \text{ and } y \geq 5\}\ y = 5\ \{y = 1\}$$

The first implication holds because

$$\{y = 0\}\ y = y + 1\ \{y = 1\}$$

is true by the assignment rule, and the second is true because the antecedent is false. The program segment is correct by the conditional rule.

8. The two implications to prove are

$$\{x = 7 \text{ and } x \leq 0\}\ y = x\ \{y = 14\}$$

and

$$\{x = 7 \text{ and } x > 0\}\ y = 2{*}x\ \{y = 14\}$$

The first is true because the antecedent is false, and the second is true because

$$\{x = 7\}\ y = 2{*}x\ \{y = 14\}$$

holds by the assignment rule. The program segment is correct by the conditional rule.

9. The desired postcondition follows from the definition of minimum: $(x < y \text{ and } min = x)$ or $(x > y \text{ and } min = y)$. The two implications to prove are

$$\{x \neq y \text{ and } x \leq y\}\ min = x\ \{(x < y \text{ and } min = x) \text{ or } (x > y \text{ and } min = y)\}$$
$$\{x \neq y \text{ and } x > y\}\ min = y\ \{(x < y \text{ and } min = x) \text{ or } (x > y \text{ and } min = y)\}$$

Using the assignment rule on the first implication gives the precondition

$$(x < y \text{ and } x = x) \text{ or } (x > y \text{ and } x = y)$$

Because the second disjunct is false, this is equivalent to $(x < y \text{ and } x = x)$ or $(x < y)$ or $(x \neq y \text{ and } x \leq y)$.

Using the assignment rule on the second implication gives the precondition

$$(x < y \text{ and } y = x) \text{ or } (x > y \text{ and } y = y)$$

which is equivalent to $(x > y)$ or $(x \neq y \text{ and } x > y)$.

The program segment is correct by the conditional rule.

10. The desired postcondition follows from the definition of absolute value for a nonzero number: $(x > 0 \text{ and } abs = x)$ or $(x < 0 \text{ and } abs = -x)$. The two implications to prove are

$\{x \neq 0 \text{ and } x \geq 0\}$ abs = x $\{(x > 0 \text{ and abs} = x) \text{ or } (x < 0 \text{ and abs} = -x)\}$
$\{x \neq 0 \text{ and } x < 0\}$ abs = -x $\{(x > 0 \text{ and abs} = x) \text{ or } (x < 0 \text{ and abs} = -x)\}$

Using the assignment rule on the first implication gives the precondition

$$(x > 0 \text{ and } x = x) \text{ or } (x < 0 \text{ and } x = -x)$$

Because the second disjunct is false, this is equivalent to $(x > 0 \text{ and } x = x)$ or $(x > 0)$ or $(x \neq 0 \text{ and } x \geq 0)$.

Using the assignment rule on the second implication gives the precondition

$$(x > 0 \text{ and } -x = x) \text{ or } (x < 0 \text{ and } -x = -x)$$

which is equivalent to $(x < 0)$ or $(x \neq 0 \text{ and } x < 0)$.

The program segment is correct by the conditional rule.

11. For the top section of the program, we can work backwards from the postcondition using the assignment rule twice:

```
{z + 1 = 4   or   z = 3}
    x = z + 1;
{x + 2 = 6   or   x = 4}
    y = x + 2;
{y = 6}
```

which agrees with the given precondition. For the bottom section of the program, we use the conditional rule to prove

```
{y = 6}
    if (y > 0)
        z = y + 1;
    else
        z = 2*y;
{z = 7}
```

One implication is

$$\{y = 6 \wedge y > 0\} \ z = y + 1 \ \{z = 7\}$$

which is true by the assignment rule. The other implication is

$$\{y = 6 \wedge y \leq 0\} \ z = 2*y \ \{z = 7\}$$

which is true because the antecedent is false.

CHAPTER 2: Proofs, Recursion, and Analysis of Algorithms

The most oft-heard refrain in teaching this class is "I can't do proofs." What this means is not that the student cannot "do" a proof but that he or she lacks the organizational skills to go about setting up the attempt. Are the definitions well understood? Is the given hypothesis in mind? Is the desired conclusion in mind? As I try to point out, if you don't know what you have and you don't know where you're going, you can't hope to get from here to there (think of a proof as building a bridge from one side of a river to the other). I sometimes have the student write the hypothesis at the top of the page, and the conclusion at the bottom. Then we list all the properties (ammunition) we can draw from the hypothesis - this is where truly understanding the definitions comes in. This process alone often allows the student to complete the proof without any further help, using a direct proof. If not, we repeat the process looking at the contrapositive. If the desired result is to show that something is *not* true, I point out that this is a 99% clue to use a proof by contradiction. The first section of this chapter is intended to catalog the options - exhaustive proof (which only applies under certain circumstances), direct proof, proof by contraposition, and proof by contradiction.

Proof by mathematical induction (Section 2.2) is a special case. Even though students have usually seen induction before, they view an inductive proof with extreme suspicion. It is difficult to get them to accept the inductive assumption, as they tend to think that this is equivalent to assuming what they want to prove. Here is where the emphasis on formal logic in the previous chapter is helpful; looking at the structure of the principle of mathematical induction as a (major) implication one of whose antecedents *is also a (minor) implication* seems to help. They are then more willing to accept the inductive assumption as merely the antecedent of this minor implication. Once students are comfortable with the first principle of induction, the second principle of induction goes pretty well. One must take pains to point out conditions under which the second principle is more useful, even though the two principles are provably equivalent. At any rate, students need to do lots of inductive proofs, and not always just summation formulas, so I've provided many exercises in this section. Proof of correctness using loop invariants (Section 2.3) is a nice application of induction that ties in well with Section 1.6.

Recursion requires the same leap of faith as induction. An interesting exercise to introduce recursion is to ask a group of students to line up facing the class, and ask the student on the left end to tell how many students are standing to his or her right. This person almost always steps out of line and counts (an iterative solution). Then ask the same student what information he or she would need from the person on the right to figure out the answer without stepping out of line, and ask the student on the right end if he or she has a simple way to answer this question. Then start from the right end (zero persons on my right - all I had to do was look) and progress to the left (if person on my right tells me x persons are on his or her right, then x + 1 are on my right) until the problem is once again solved. Section 2.4 emphasizes the variety of things that are recursively defined - sequences, sets, operations, and algorithms.

Recurrence relations, introduced in Section 2.4, are also discussed in Section 2.5 in the context of analysis of algorithms (order of magnitude discussions will come in Chapter 4.)

EXERCISES 2.1

*1. a. Converse: Healthy plant growth implies sufficient water.
Contrapositive: If there is not healthy plant growth, then there is not sufficient water.
b. Converse: Increased availability of information implies further technological advances.
Contrapositive: If there is not increased availability of information, then there are no further technological advances.
c. Converse: If there is a modification of the program then errors will be introduced.
Contrapositive: No modification of the program implies that errors will not be introduced.
d. Converse: Good insulation or storm windows throughout implies fuel savings.
Contrapositive: Poor insulation and some windows not storm windows implies no fuel savings.

2. $P' \rightarrow Q'$ is the contrapositive of the converse $Q \rightarrow P$, so the inverse and the converse are equivalent.

3. For example:
a. a nonsquare rectangle
b. 0
c. a short, blue-eyed redhead
d. a redhead who is short

4 a. Half of this statement is true. If n is an odd integer, n = 2k + 1 for some integer k. Then 3n + 5 = 3(2k + 1) + 5 = 6k + 8 = 2(3k + 4), which is an even integer. However, the converse is false. Consider the even integer 6. If 3n + 5 = 6, then 3n = 1 and n = 1/3, which is not an integer at all, much less an odd integer. See Exercise 18.

b. Half of this statement is true. If n is an even integer, n = 2k for some integer k. Then 3n + 2 = 3(2k) + 2 = 6k + 2 = 2(3k + 1), which is an even integer. However, the converse is false. Consider the even integer 10. If 3n + 2 = 10, then 3n = 8 and n = 8/3, which is not an integer at all, much less an even integer. See Exercise 19.

*5. $25 = 5^2 = 9 + 16 = 3^2 + 4^2$
$100 = (10)^2 = 36 + 64 = 6^2 + 8^2$
$169 = (13)^2 = 25 + 144 = 5^2 + (12)^2$

6. 4 = 2 + 2
6 = 3 + 3
8 = 3 + 5
10 = 5 + 5
12 = 5 + 7

7.

n	n!	2^n
1	1	2
2	2	4
3	6	8

8.

n	n^2	2^n
2	4	4
3	9	8
4	16	16

9. Let $x = 2m$, $y = 2n$, where m and n are integers. Then $x + y = 2m + 2n = 2(m + n)$, where $m + n$ is an integer, so $x + y$ is even.

10. Let $x = 2m$, $y = 2n$ for integers m and n, and assume that $x + y$ is odd. Then $x + y = 2m + 2n = 2k + 1$ for some integer k or $2(m + n - k) = 1$ where $m + n - k$ is an integer. This is a contradiction since 1 is not even.

*11. Let $x = 2m + 1$, $y = 2n + 1$, where m and n are integers. Then $x + y = (2m + 1) + (2n + 1) = 2m + 2n + 2 = 2(m + n + 1)$, where $m + n + 1$ is an integer, so $x + y$ is even.

12. Let $x = 2m$ and $y = 2n + 1$, where m and n are integers. Then $x + y = (2m) + (2n + 1) = (2m + 2n) + 1 = 2(m + n) + 1$, where $m + n$ is an integer, so $x + y$ is odd.

13. For two consecutive integers, one is even and one is odd. The product of an even integer and an odd integer is even by the proof of Example 9.

14. Let n be an integer. Then $n + n^2 = n^2 + n = n(n + 1)$, which is even by Exercise 13.

*15. Let $x = 2m$ where m is an integer. Then $x^2 = (2m)^2 = 4m^2$, where m^2 is an integer, so x^2 is divisible by 4.

16. $3(n^2 + 2n + 3) - 2n^2 = 3n^2 + 6n + 9 - 2n^2 = n^2 + 6n + 9 = (n + 3)^2$

17. The contrapositive is: if $x + 1 \leq 0$, then $x \leq 0$. If $x + 1 \leq 0$, then $x \leq -1 < 0$, so $x < 0$ and therefore $x \leq 0$.

18. If n is odd, then $n = 2k + 1$ for some integer k. Then $3n + 5 = 3(2k + 1) + 5 = 6k + 8$. For the converse, if $3n + 5 = 6k + 8$ for some integer k, then $3n = 6k + 3$ or $3n = 3(2k + 1)$ and $n = 2k + 1$ for some integer k, so n is an odd integer.

19. If n is even, then $n = 2k$ for some integer k. Then $3n + 2 = 3(2k) + 2 = 6k + 2$. For the converse, if $3n + 2 = 6k + 2$ for some integer k, then $3n = 6k$ and $n = 2k$ for some integer k, so n is an even integer.

*20. If $x < y$ then multiplying both sides of the inequality by the positive numbers x and y in turn gives $x^2 < xy$ and $xy < y^2$ and therefore $x^2 < xy < y^2$ or $x^2 < y^2$. For the other direction, if $x^2 < y^2$ then

$y^2 - x^2 > 0$	(definition of $<$)
$(y + x)(y - x) > 0$	(factoring)
$(y + x) < 0$ and $(y - x) < 0$	(a positive number is the product
or	of two negatives or two positives)
$(y + x) > 0$ and $(y - x) > 0$	

But it cannot be that $(y + x) < 0$ because y and x are both positive, therefore $(y + x) > 0$ and $y - x > 0$ and $y > x$.

21. Let $x^2 + 2x - 3 = 0$ and assume that $x = 2$. Then

$$2^2 + 2(2) - 3 = 0$$

or

$$4 + 4 - 3 = 0$$

or

$$5 = 0$$

This is a contradiction, so $x \neq 2$.

22. Let x be a prime number with $x = 2k$, where k is an integer. Then both 2 and k divide x. Because x is prime, x is divisible only by itself and 1, so $x = 2$ and $k = 1$. Therefore $x = 2$.

*23. Let x and y be divisible by n. Then $x = k_1n$ and $y = k_2n$, where k_1 and k_2 are integers, and $x + y = k_1n + k_2n = (k_1 + k_2)n$, where $k_1 + k_2$ is an integer. Therefore $x + y$ is divisible by n.

24. Proof by contraposition: if one of two integers is divisible by an integer n, then so is their product. Let $x = kn$ where k is an integer. Then $xy = (kn)y = (ky)n$ where ky is an integer. Therefore the product xy is divisible by n.

25. $n + (n + 1) + (n + 2) = 3n + 3 = 3(n + 1)$

*26. Let $x = 2n + 1$. Then $x^2 = (2n + 1)^2 = 4n^2 + 4n + 1 = 4n(n + 1) + 1$. But $n(n + 1)$ is even (Exercise 13), so $n(n + 1) = 2k$ for some integer k. Therefore $x^2 = 4(2k) + 1 = 8k + 1$.

27. $n^3 - (n - 1)^3 = n^3 - [n^3 - 3n^2 + 3n - 1] = 3n^2 - 3n + 1 = 3n(n - 1) + 1$. Then $n(n - 1)$ is even by Exercise 11, and $3n(n - 1)$ is even by the proof of Example 9, thus $3n(n - 1) + 1$ is odd.

28. Proof by contradiction: assume that $m^2 + n^2 = k^2$ where m and n are odd integers and k is an integer. By Exercise 26, $m^2 = 8k_1 + 1$ and $n^2 = 8k_2 + 1$ for integers k_1 and k_2. Therefore $(8k_1 + 1) + (8k_2 + 1) = k^2$, or $2[4k_1 + 4k_2 + 1] = k^2$. Then 2 divides k^2, so 2 divides k, hence 4 is a factor of k^2, which can be written as 4x. Therefore $2[4k_1 + 4k_2 + 1] = 4x$ or $4k_1 + 4k_2 + 1 = 2x$. This is a contradiction because $4k_1 + 4k_2 + 1$ is odd while 2x is even.

*29. $m^2n^2 = (mn)^2$

30. Proof by cases, depending on whether x and y are positive, negative, or zero (it turns out to be a little neater to dispose of the zero case separately.)

Case 1: $x = 0$ or $y = 0$.
Subcase a: $x = 0$. Then $|x| = 0$, $xy = 0$, and $|xy| = 0$. Therefore $|xy| = 0 = 0 \cdot |y| = |x||y|$.
Subcase b: $y = 0$. Similar to Subcase a.

Case 2: $x > 0, y > 0$. Then $|x| = x$, $|y| = y$. Also, $xy > 0$ and $|xy| = xy$. Therefore $|xy| = xy = |x||y|$.

Case 3: $x < 0, y > 0$. Then $|x| = -x$, $|y| = y$. Also, $xy < 0$ and $|xy| = -xy$. Therefore $|xy| = -xy = (-x)y = |x||y|$.

Case 4: $x > 0, y < 0$. Then $|x| = x$, $|y| = -y$. Also, $xy < 0$ and $|xy| = -xy$. Therefore $|xy| = -xy = x(-y) = |x||y|$.

Case 5: $x < 0, y < 0$. Then $|x| = -x$, $|y| = -y$. Also, $xy > 0$ and $|xy| = xy$. Therefore $|xy| = xy = (-x)(-y) = |x||y|$.

31. Proof by cases, depending on whether x and y are negative.

Case 1: $x \geq 0, y \geq 0$. Then $|x| = x$, $|y| = y$. Also, $x + y \geq 0$ and $|x + y| = x + y$. Therefore $|x + y| = x + y = |x| + |y|$.

Case 2: $x \geq 0, y < 0$. Then $|x| = x$, $|y| = -y$.
Subcase a: $x + y \geq 0$. Then $|x + y| = x + y$. Therefore $|x + y| = x + y < x + (-y)$ (remember that y is negative, so -y is positive) $= |x| + |y|$.
Subcase b: $x + y < 0$. Then $|x + y| = -(x + y)$. Therefore $|x + y| = -(x + y) = (-x) + (-y) \leq x + (-y)$ (remember that $x \geq 0$, so $-x \leq 0$) $= |x| + |y|$.

Case 3: $x < 0, y \geq 0$. Similar to Case 2 with the roles of x and y reversed.

Case 4: $x < 0, y < 0$. Then $|x| = -x$, $|y| = -y$. Also, $x + y < 0$ and $|x + y| = -(x + y)$. Therefore $|x + y| = -(x + y) = (-x) + (-y) = |x| + |y|$.

32. Proof by contradiction. If $x_1 < A$, $x_2 < A$, ..., and $x_n < A$, then $x_1 + x_2 + \dots + x_n < A + A + \dots + A = nA$, and $(x_1 + x_2 + \dots + x_n)/n < A$, which contradicts the definition of A as the average of $x_1, \dots, x_n$.

33. Following the proof of Example 11, $\sqrt{4} = p/q$, $4 = p^2/q^2$, $4q^2 = p^2$, and 4 divides p^2. But we cannot now conclude that 4 divides p; for example 4 divides 36 but 4 does not divide 6. In Example 11, 2 divides p^2 implies 2 divides p because 2 is prime.

34. Assume that $\sqrt{3}$ is rational. Then $\sqrt{3} = p/q$ where p and q are integers, $q \neq 0$, and p and q have no common factors (other than ± 1). If $\sqrt{3} = p/q$ then $3 = p^2/q^2$ or $3q^2 = p^2$. Then 3 divides p^2 so 3 divides p. Thus 3 is a factor of p or 9 is a factor of p^2, and the equation $3q^2 = p^2$ can be written $3q^2 = 9x$ or $q^2 = 3x$. Then 3 divides q^2 so 3 divides q. Therefore 3 is a common factor of p and q, a contradiction.

35. Assume that $\sqrt{5}$ is rational. Then $\sqrt{5} = p/q$ where p and q are integers, $q \neq 0$, and p and q have no common factors (other than ± 1). If $\sqrt{5} = p/q$ then $5 = p^2/q^2$ or $5q^2 = p^2$. Then 5 divides p^2 so 5 divides p. Thus 5 is a factor of p or 25 is a factor of p^2, and the equation $5q^2 = p^2$ can be written $5q^2 = 25x$ or $q^2 = 5x$. Then 5 divides q^2 so 5 divides q. Therefore 5 is a common factor of p and q, a contradiction.

36. Assume that $\sqrt[3]{2}$ is rational. Then $\sqrt[3]{2} = p/q$ where p and q are integers, $q \neq 0$, and p and q have no common factors (other than ± 1). If $\sqrt[3]{2} = p/q$ then $2 = p^3/q^3$ or $2q^3 = p^3$. Then 2 divides p^3 so 2 divides p. Thus 2 is a factor of p, or 8 is a factor of p^3, and the equation $2q^3 = p^3$ can be written $2q^3 = 8x$ or $q^3 = 4x$. Then 2 divides 4x so 2 divides q^3 or 2 divides q. Therefore 2 is a common factor of p and q, a contradiction.

*37. Proof:
If x is even, then x = 2n and
$$x(x+1)(x+2) = (2n)(2n+1)(2n+2) = 2[(n)(2n+1)(2n+2)]$$
which is even.
If x is odd, then x = 2n + 1 and
$$x(x+1)(x+2) = (2n+1)(2n+2)(2n+3) = 2[(2n+1)(n+1)(2n+3)]$$
which is even.

38. Counterexample: 2 + 3 + 4 = 9

39. Counterexample: $3 \times 9 = 27$

*40. Proof:
If x is even, then x = 2n and
$$2n + (2n)^3 = 2n + 8n^3 = 2(n + 4n^3)$$
which is even.
If x is odd, then x = 2n + 1 and
$$(2n+1) + (2n+1)^3 = (2n+1) + (8n^3 + 12n^2 + 6n + 1) = 8n^3 + 12n^2 + 8n + 2 = 2(4n^3 + 6n^2 + 4n + 1)$$
which is even.

41. Counterexample: 3 is a positive integer that cannot be written as the sum of two squares.

42. Proof by contradiction: Suppose x is positive but $x + \frac{1}{x} < 2$. Then $\frac{x^2+1}{x} < 2$. Multiplying both sides by x > 0, $x^2 + 1 < 2x$ or $x^2 - 2x + 1 < 0$ or $(x - 1)^2 < 0$, a contradiction because the square of an integer cannot be negative. Therefore $x + \frac{1}{x} \geq 2$.

43. Counterexample: 5 is prime, but 5 + 4 = 9 is not prime.

44. Proof: $n^2 - 1 = (n + 1)(n - 1)$ where $n - 1 > 1$, which is a non-trivial factorization, so the number is not prime.

*45. Counterexample: $4^2 + 4 + 1 = 21 = 3(7)$, not prime.

46. Counterexample: $9 = 2^3 + 1$, and 9 = (3)(3), so is not prime.

47. Proof: let n = 2k with k > 1. Then $2^n - 1 = 2^{2k} - 1 = (2^k)^2 - 1 = (2^k + 1)(2^k - 1)$. Because k > 1, $2^k - 1 > 1$ and this is a non-trivial factorization, so $2^n - 1$ is not prime.

48. Proof: Let x and y be rational numbers, x = p/q, y = r/s with p, q, r, s integers and q, s, ≠ 0. Then xy = (p/q)(r/s) = pr/qs, where pr and qs are integers with qs ≠ 0. Thus xy is rational.

*49. Proof: Let x and y be rational numbers, x = p/q, y = r/s with p, q, r, s integers and q, s ≠ 0. Then x + y = p/q + r/s = (ps + rq)/qs, where ps + rq and qs are integers with qs ≠ 0. Thus x + y is rational.

50. Counterexample: $\sqrt{2}$ is irrational but $\sqrt{2} \times \sqrt{2} = 2$, which is rational.

51. Proof by contradiction: Let x be a rational number and y be an irrational number. Assume that x + y = r where r is a rational number. Then y = (-x) + r, which is the sum of two rational numbers. The sum of two rational numbers is rational [Exercise 44], so this is a contradiction.

52. Angle 3 plus Angle 4 sum to 180° by the third fact. Angle 3 plus (Angle 1 and Angle 2) sum to 180° by the first fact. Therefore Angle 4 = Angle 1 plus Angle 2 in size.

*53. Angle 6 plus Angle 5 plus the right angle sum to 180° by the first fact. The right angle is 90° by the fourth fact. Therefore Angle 6 plus Angle 5 sum to 90°. Angle 6 is the same size as Angle 3 by the second fact. Therefore Angle 3 plus Angle 5 sum to 90°.

54. Assume that Angle 1 and Angle 5 are the same size. As in Exercise 48, Angle 3 plus Angle 5 sum to 90°. Because Angle 1 and Angle 5 are the same size, Angle 3 plus Angle 1 sum to 90°. Also, Angle 3 plus Angle 1 plus Angle 2 sum to 180° by the first

fact. Therefore 90° plus Angle 2 sum to 180°, or Angle 2 = 90°. Angle 2 is a right angle by the fourth fact.

55. The pairs are

$1 + 100 = 101$
$2 + 99 = 101$
$\vdots$
$50 + 51 = 101$

There are 50 such pairs, so the sum is 50(101) = 5050.

EXERCISES 2.2

*1. P(1): $4(1) - 2 = 2(1)^2$ or $2 = 2$ true

Assume P(k): $2 + 6 + 10 + \ldots + (4k - 2) = 2k^2$

Show P(k + 1): $2 + 6 + 10 + \ldots + [4(k + 1) - 2] = 2(k + 1)^2$

	$2 + 6 + 10 + \ldots + [4(k + 1) - 2]$	left side of P(k + 1)
=	$2 + 6 + 10 + \ldots + (4k - 2) + [4(k + 1) - 2]$	
=	$2k^2 + 4(k + 1) - 2$	using P(k)
=	$2k^2 + 4k + 2$	
=	$2(k^2 + 2k + 1)$	
=	$2(k + 1)^2$	right side of P(k + 1)

2. P(1): $2 = 1(1 + 1)$ true

Assume P(k): $2 + 4 + 6 + \ldots + 2k = k(k + 1)$

Show P(k + 1): $2 + 4 + 6 + \ldots + 2(k + 1) = (k + 1)[(k + 1) + 1]$

	$2 + 4 + 6 + \ldots + 2(k + 1)$	left side of P(k + 1)
=	$2 + 4 + 6 + \ldots + 2k + 2(k + 1)$	
=	$k(k + 1) + 2(k + 1)$	using P(k)
=	$(k + 1)(k + 2)$	factoring
=	$(k + 1)[(k + 1) + 1]$	right side of P(k + 1)

*3. P(1): $1 = 1(2(1) - 1)$ true

Assume P(k): $1 + 5 + 9 + \ldots + (4k - 3) = k(2k - 1)$

Show P(k + 1): $1 + 5 + 9 + \ldots + [4(k + 1) - 3] = (k + 1)[2(k + 1) - 1]$

	$1 + 5 + 9 + \ldots + [4(k + 1) - 3]$	left side of P(k + 1)
=	$1 + 5 + 9 + \ldots + (4k - 3) + [4(k + 1) - 3]$	
=	$k(2k - 1) + 4(k + 1) - 3$	using P(k)
=	$2k^2 - k + 4k + 1$	
=	$2k^2 + 3k + 1$	
=	$(k + 1)(2k + 1)$	
=	$(k + 1)[2(k + 1) - 1]$	right side of P(k + 1)

4. P(1): $1 = 1(1 + 1)(1 + 2)/6 = 2(3)/6$ true

Assume P(k): $1 + 3 + 6 + \dots + \frac{k(k+1)}{2} = k(k + 1)(k + 2)/6$

Show P(k + 1): $1 + 3 + 6 + \dots + \frac{(k+1)(k+2)}{2} = (k + 1)(k + 2)(k + 3)/6$

$1 + 3 + 6 + \dots + \frac{(k+1)(k+2)}{2}$ left side of P(k + 1)

$= 1 + 3 + 6 + \dots + \frac{k(k+1)}{2} + \frac{(k+1)(k+2)}{2}$

$= k(k + 1)(k + 2)/6 + \frac{(k+1)(k+2)}{2}$ using P(k)

$= (k + 1)(k + 2)\left[\frac{k}{6} + \frac{1}{2}\right]$ factoring

$= (k + 1)(k + 2)\left[\frac{k+3}{6}\right]$

$= (k + 1)(k + 2)(k + 3)/6$ right side of P(k + 1)

*5. P(1): $6 - 2 = 1[3(1) + 1]$ true

Assume P(k): $4 + 10 + 16 + \dots + (6k - 2) = k(3k + 1)$

Show P(k + 1): $4 + 10 + 16 + \dots + [6(k + 1) - 2] = (k + 1)[3(k + 1) + 1]$

$4 + 10 + 16 + \dots + [6(k + 1) - 2]$ left side of P(k + 1)

$= 4 + 10 + 16 + \dots + (6k - 2) + [6(k + 1) - 2]$

$= k(3k + 1) + 6(k + 1) - 2$ using P(k)

$= 3k^2 + k + 6k + 4$

$= 3k^2 + 7k + 4$

$= (k + 1)(3k + 4)$

$= (k + 1)[3(k + 1) + 1]$ right side of P(k + 1)

6. P(1): $5 = 5(1 + 1)/2$ true

Assume P(k): $5 + 10 + 15 + \dots + 5k = 5k(k + 1)/2$

Show P(k + 1): $5 + 10 + 15 + \dots + 5(k + 1) = 5(k + 1)(k + 2)/2$

$5 + 10 + 15 + \dots + 5(k + 1)$ left side of P(k + 1)

$= 5 + 10 + 15 + \dots + 5k + 5(k + 1)$

$= 5k(k + 1)/2 + 5(k + 1)$ using P(k)

$= 5(k + 1)\left[\frac{k}{2} + 1\right]$ factoring

$= 5(k + 1)(k + 2)/2$ right side of P(k + 1)

or write $5 + 10 + 15 + \dots + 5n$ as $5(1 + 2 + 3 + \dots + n)$ and use the result of Practice 7.

7. $P(1)$: $1^2 = 1(1 + 1)(2 + 1)/6$ true
Assume $P(k)$: $1^2 + 2^2 + \ldots + k^2 = k(k + 1)(2k + 1)/6$
Show $P(k + 1)$: $1^2 + 2^2 + \ldots + (k + 1)^2 = (k + 1)(k + 2)(2(k + 1) + 1)/6$

$1^2 + 2^2 + \ldots + (k + 1)^2$ left side of $P(k + 1)$
$= 1^2 + 2^2 + \ldots + k^2 + (k + 1)^2$
$= k(k + 1)(2k + 1)/6 + (k + 1)^2$ using $P(k)$
$= (k + 1)\left[\dfrac{k(2k+1)}{6} + k + 1\right]$ factoring
$= (k + 1)\left[\dfrac{2k^2 + k + 6k + 6}{6}\right]$
$= (k + 1)\left[\dfrac{2k^2 + 7k + 6}{6}\right]$
$= \dfrac{(k+1)(k+2)(2k+3)}{6}$
$= (k + 1)(k + 2)(2(k + 1) + 1)/6$ right side of $P(k + 1)$

8. $P(1)$: $1^3 = 1^2(1 + 1)^2/4$ true
Assume $P(k)$: $1^3 + 2^3 + \ldots + k^3 = k^2(k + 1)^2/4$
Show $P(k + 1)$: $1^3 + 2^3 + \ldots + (k + 1)^3 = (k + 1)^2(k + 2)^2/4$

$1^3 + 2^3 + \ldots + (k + 1)^3$ left side of $P(k + 1)$
$= 1^3 + 2^3 + \ldots + k^3 + (k + 1)^3$
$= k^2(k + 1)^2/4 + (k + 1)^3$ using $P(k)$
$= (k + 1)^2\left[\dfrac{k^2}{4} + k + 1\right]$ factoring
$= (k + 1)^2\left[\dfrac{k^2 + 4k + 4}{4}\right]$
$= (k + 1)^2(k + 2)^2/4$ right side of $P(k + 1)$

*9. $P(1)$: $1^2 = 1(2 - 1)(2 + 1)/3$ true
Assume $P(k)$: $1^2 + 3^2 + \ldots + (2k - 1)^2 = k(2k - 1)(2k + 1)/3$
Show $P(k + 1)$: $1^2 + 3^2 + \ldots + [2(k + 1) - 1]^2 = (k + 1)(2(k + 1) - 1)(2(k + 1) + 1)/3$

$1^2 + 3^2 +. \ldots + [2(k + 1) - 1]^2$ left side of $P(k + 1)$
$= 1^2 + 3^2 + \ldots + (2k - 1)^2 + [2(k + 1) - 1]^2$
$= \dfrac{k(2k-1)(2k+1)}{3} + [2(k+1) - 1]^2$ using $P(k)$

$= \dfrac{k(2k-1)(2k+1)}{3} + (2k+1)^2$

$= (2k+1)\left[\dfrac{k(2k-1)}{3} + 2k+1\right]$ factoring

$= (2k+1)\left[\dfrac{2k^2 - k + 6k + 3}{3}\right]$

$= \dfrac{(2k+1)(2k^2+5k+3)}{3}$

$= (2k+1)(k+1)(2k+3)/3$

$= (k+1)(2(k+1)-1)(2(k+1)+1)/3$ right side of P(k+1)

10. P(1): $1^4 = 1(1+1)(2+1)(3+3-1)/30 = 2\cdot3\cdot5/30$ true

Assume P(k): $1^4 + 2^4 + \ldots + k^4 = k(k+1)(2k+1)(3k^2+3k-1)/30$

Show P(k + 1): $1^4 + 2^4 + \ldots + (k+1)^4$

$= (k+1)(k+2)(2(k+1)+1)(3(k+1)^2+3(k+1)-1)/30$

$= (k+1)(k+2)(2k+3)(3k^2+9k+5)/30$

$= \dfrac{k+1}{30}(2k^2+7k+6)(3k^2+9k+5)$

$= \dfrac{k+1}{30}(6k^4+39k^3+91k^2+89k+30)$

$1^4 + 2^4 + \ldots + (k+1)^4$ left side of P(k + 1)

$= 1^4 + 2^4 + \ldots + k^4 + (k+1)^4$

$= k(k+1)(2k+1)(3k^2+3k-1)/30 + (k+1)^4$ using P(k)

$= (k+1)\left[\dfrac{k(2k+1)(3k^2+3k-1)}{30} + (k+1)^3\right]$ factoring

$= \dfrac{k+1}{30}[(2k^2+k)(3k^2+3k-1) + 30(k+1)^3]$

$= \dfrac{k+1}{30}(6k^4+6k^3-2k^2+3k^3+3k^2-k+30(k^3+3k^2+3k+1))$

$= \dfrac{k+1}{30}(6k^4+39k^3+91k^2+89k+30)$ right side of P(k + 1)

11. P(1): $1\cdot3 = 1(2)(9)/6$ true

Assume P(k): $1\cdot3 + 2\cdot4 + \ldots + k(k+2) = k(k+1)(2k+7)/6$

Show P(k + 1): $1\cdot3 + 2\cdot4 + \quad + (k+1)(k+3) = (k+1)(k+2)(2(k+1)+7)/6$

$1\cdot3 + 2\cdot4 + \quad + (k+1)(k+3)$ left side of P(k + 1)

$= 1\cdot3 + 2\cdot4 + \quad + k(k+2) + (k+1)(k+3)$

$= k(k+1)(2k+7)/6 + (k+1)(k+3)$ using P(k)

$= (k+1)[k(2k+7) + 6(k+3)]/6$

$= (k+1)(2k^2+13k+18)/6$

$= (k+1)(k+2)(2k+9)/6$

$= (k+1)(k+2)(2(k+1)+7)/6$ right side of P(k + 1)

12. P(1): $a^0 = \frac{a^1 - 1}{a - 1}$ or $1 = \frac{a-1}{a-1}$ true

Assume P(k): $1 + a + \ldots + a^{k-1} = \frac{a^k - 1}{a - 1}$

Show P(k + 1): $1 + a + \ldots + a^k = \frac{a^{k+1} - 1}{a - 1}$

$$
\begin{aligned}
& 1 + a + \ldots + a^k && \text{left side of P(k + 1)} \\
= \; & 1 + a + \ldots + a^{k-1} + a^k \\
= \; & \frac{a^k - 1}{a - 1} + a^k && \text{using P(k)} \\
= \; & \frac{a^k - 1 + a^k(a-1)}{a - 1} \\
= \; & \frac{a^k - 1 + a^{k+1} - a^k}{a - 1} \\
= \; & \frac{a^{k+1} - 1}{a - 1} && \text{right side of P(k + 1)}
\end{aligned}
$$

*13. P(1): $\frac{1}{1\cdot 2} = \frac{1}{1+1}$ true

Assume P(k): $\frac{1}{1\cdot 2} + \frac{1}{2\cdot 3} + \ldots + \frac{1}{k(k+1)} = \frac{k}{k+1}$

Show P(k + 1): $\frac{1}{1\cdot 2} + \frac{1}{2\cdot 3} + \ldots + \frac{1}{(k+1)(k+2)} = \frac{k+1}{k+2}$

$$
\begin{aligned}
& \frac{1}{1\cdot 2} + \frac{1}{2\cdot 3} + \ldots + \frac{1}{(k+1)(k+2)} && \text{left side of P(k + 1)} \\
= \; & \frac{1}{1\cdot 2} + \frac{1}{2\cdot 3} + \ldots + \frac{1}{k(k+1)} + \frac{1}{(k+1)(k+2)} \\
= \; & \frac{k}{k+1} + \frac{1}{(k+1)(k+2)} && \text{using P(k)} \\
= \; & \frac{k(k+2)+1}{(k+1)(k+2)} \\
= \; & \frac{k^2 + 2k + 1}{(k+1)(k+2)} \\
= \; & \frac{(k+1)^2}{(k+1)(k+2)} \\
= \; & \frac{k+1}{k+2} && \text{right side of P(k + 1)}
\end{aligned}
$$

14. P(1): $\dfrac{1}{1\cdot 3}=\dfrac{1}{2+1}$ true

Assume P(k): $\dfrac{1}{1\cdot 3}+\dfrac{1}{3\cdot 5}+\ldots+\dfrac{1}{(2k-1)(2k+1)}=\dfrac{k}{2k+1}$

Show P(k + 1): $\dfrac{1}{1\cdot 3}+\dfrac{1}{3\cdot 5}+\ldots+\dfrac{1}{(2(k+1)-1)(2(k+1)+1}=\dfrac{k+1}{2(k+1)+1}$

$\dfrac{1}{1\cdot 3}+\dfrac{1}{3\cdot 5}+\ldots+\dfrac{1}{(2(k+1)-1)(2(k+1)+1}$ left side of P(k + 1)

$= \dfrac{1}{1\cdot 3}+\dfrac{1}{3\cdot 5}+\ldots+\dfrac{1}{(2k-1)(2k+1)}+\dfrac{1}{(2(k+1)-1)(2(k+1)+1)}$

$= \dfrac{k}{2k+1}+\dfrac{1}{(2k+1)(2k+3)}$ using P(k)

$= \dfrac{k(2k+3)+1}{(2k+1)(2k+3)}$

$= \dfrac{2k^2+3k+1}{(2k+1)(2k+3)}$

$= \dfrac{(k+1)(2k+1)}{(2k+1)(2k+3)}$

$= \dfrac{k+1}{2k+3}$ right side of P(k + 1)

*15. P(1): $1^2 = (-1)^2(1)(2)/2$ true

Assume P(k): $1^2 - 2^2 + \ldots + (-1)^{k+1}k^2 = (-1)^{k+1}(k)(k+1)/2$

Show P(k + 1): $1^2 - 2^2 + \ldots + (-1)^{k+2}(k+1)^2 = (-1)^{k+2}(k+1)(k+2)/2$

$1^2 - 2^2 + \ldots + (-1)^{k+2}(k+1)^2$ left side of P(k + 1)

$= 1^2 - 2^2 + \ldots + (-1)^{k+1}k^2 + (-1)^{k+2}(k+1)^2$

$= (-1)^{k+1}(k)(k+1)/2 + (-1)^{k+2}(k+1)^2$ using P(k)

$= [(-1)^{k+1}(k)(k+1) + 2(-1)^{k+2}(k+1)^2]/2$

$= (-1)^{k+2}(k+1)[k(-1)^{-1} + 2(k+1)]/2$

$= (-1)^{k+2}(k+1)[-k + 2k + 2]/2$

$= (-1)^{k+2}(k+1)(k+2)/2$ right side of P(k + 1)

16. P(1): $2 \cdot 3^0 = 3^1 - 1$ or $2 = 3 - 1$ true

Assume P(k): $2 + 6 + 18 + \ldots + 2 \cdot 3^{k-1} = 3^k - 1$

Show P(k + 1): $2 + 6 + 18 + \ldots + 2 \cdot 3^k = 3^{k+1} - 1$

$2 + 6 + 18 + \ldots + 2 \cdot 3^k$ left side of P(k + 1)

$= 2 + 6 + 18 + \ldots + 2 \cdot 3^{k-1} + 2 \cdot 3^k$

$= 3^k - 1 + 2 \cdot 3^k$ using P(k)

$= 3 \cdot 3^k - 1$

$= 3^{k+1} - 1$ right side of P(k + 1)

17. P(1): $2^2 = (2)(1)(2)(2 + 1)/3$ or $4 = (2)(6)/3$ true
Assume P(k): $2^2 + 4^2 + \dots + (2k)^2 = 2k(k + 1)(2k + 1)/3$
Show P(k + 1): $2^2 + 4^2 + \dots + [2(k + 1)]^2 = 2(k + 1)(k + 2)[2(k + 1) + 1]/3$

$$\begin{aligned}
& 2^2 + 4^2 + \dots + [2(k+1)]^2 && \text{left side of } P(k+1)\\
= {} & 2^2 + 4^2 + \dots + (2k)^2 + [2(k+1)]^2 &&\\
= {} & 2k(k+1)(2k+1)/3 + [2(k+1)]^2 && \text{using } P(k)\\
= {} & 2(k+1)[k(2k+1)/3 + 2(k+1)] &&\\
= {} & 2(k+1)[k(2k+1) + 6(k+1)]/3 &&\\
= {} & 2(k+1)[2k^2 + 7k + 6]/3 &&\\
= {} & 2(k+1)(k+2)(2k+3)/3 &&\\
= {} & 2(k+1)(k+2)[2(k+1)+1]/3 && \text{right side of } P(k+1)
\end{aligned}$$

18. P(1): $1 \cdot 2^1 = (1 - 1)2^{1+1} + 2$ true
Assume P(k): $1 \cdot 2^1 + 2 \cdot 2^2 + 3 \cdot 2^3 + \cdots + k \cdot 2^k = (k - 1)2^{k+1} + 2$
Show P(k + 1): $1 \cdot 2^1 + 2 \cdot 2^2 + 3 \cdot 2^3 + \cdots + (k + 1) \cdot 2^{k+1} = (k)2^{k+2} + 2$

$$\begin{aligned}
& 1 \cdot 2^1 + 2 \cdot 2^2 + 3 \cdot 2^3 + \cdots + (k+1) \cdot 2^{k+1} && \text{left side of } P(k+1)\\
= {} & 1 \cdot 2^1 + 2 \cdot 2^2 + 3 \cdot 2^3 + \cdots + k \cdot 2^k + (k+1) \cdot 2^{k+1} &&\\
= {} & (k-1)2^{k+1} + 2 + (k+1) \cdot 2^{k+1} && \text{using } P(k)\\
= {} & 2^{k+1}[(k-1) + (k+1)] + 2 && \text{factoring out } 2^{k+1}\\
= {} & 2^{k+1}[2k] + 2 &&\\
= {} & 2^{k+2}(k) + 2 && \text{right side of } P(k+1)
\end{aligned}$$

19. P(1): $1 \cdot 2 = (1)(2)(3)/3$ true
Assume P(k): $1 \cdot 2 + 2 \cdot 3 + 3 \cdot 4 + \dots + k(k + 1) = \dfrac{k(k+1)(k+2)}{3}$
Show P(k + 1): $1 \cdot 2 + 2 \cdot 3 + 3 \cdot 4 + \dots + (k + 1)(k + 2) = \dfrac{(k+1)(k+2)(k+3)}{3}$

$$\begin{aligned}
& 1 \cdot 2 + 2 \cdot 3 + 3 \cdot 4 + \dots + (k+1)(k+2) = && \text{left side of } P(k+1)\\
= {} & 1 \cdot 2 + 2 \cdot 3 + 3 \cdot 4 + \dots + k(k+1) + (k+1)(k+2) &&\\
= {} & \frac{k(k+1)(k+2)}{3} + (k+1)(k+2) && \text{using } P(k)\\
= {} & (k+1)(k+2)[k/3 + 1] &&\\
= {} & (k+1)(k+2)\left(\frac{k+3}{3}\right) && \text{right side of } P(k+1)
\end{aligned}$$

20. P(1): $1 \cdot 2 \cdot 3 = \dfrac{1(2)(3)(4)}{4}$ true
Assume P(k): $1 \cdot 2 \cdot 3 + 2 \cdot 3 \cdot 4 + \cdots + k(k + 1)(k + 2) = \dfrac{k(k+1)(k+2)(k+3)}{4}$

Show P(k + 1): $1 \cdot 2 \cdot 3 + 2 \cdot 3 \cdot 4 + \cdots + (k+1)(k+2)(k+3) = \dfrac{(k+1)(k+2)(k+3)(k+4)}{4}$

$$
\begin{aligned}
& 1 \cdot 2 \cdot 3 + 2 \cdot 3 \cdot 4 + \cdots + (k+1)(k+2)(k+3) && \text{left side of } P(k+1) \\
= \; & 1 \cdot 2 \cdot 3 + 2 \cdot 3 \cdot 4 + \cdots + (k)(k+1)(k+2) + (k+1)(k+2)(k+3) \\
= \; & \frac{k(k+1)(k+2)(k+3)}{4} + (k+1)(k+2)(k+3) && \text{using } P(k) \\
= \; & (k+1)(k+2)(k+3)\left[\frac{k}{4} + 1\right] \\
= \; & (k+1)(k+2)(k+3)\left[\frac{k+4}{4}\right] \\
= \; & \frac{(k+1)(k+2)(k+3)(k+4)}{4} && \text{right side of } P(k+1)
\end{aligned}
$$

21. P(1): $\dfrac{1}{1 \cdot 4} = \dfrac{1}{3 \cdot 1 + 1}$ true

Assume P(k): $\dfrac{1}{1 \cdot 4} + \dfrac{1}{4 \cdot 7} + \dfrac{1}{7 \cdot 10} + \ldots + \dfrac{1}{(3k-2)(3k+1)} = \dfrac{k}{3k+1}$

Show P(k + 1): $\dfrac{1}{1 \cdot 4} + \dfrac{1}{4 \cdot 7} + \dfrac{1}{7 \cdot 10} + \ldots + \dfrac{1}{(3(k+1)-2)(3(k+1)+1)} = \dfrac{k+1}{3(k+1)+1}$

$$
\begin{aligned}
& \frac{1}{1 \cdot 4} + \frac{1}{4 \cdot 7} + \frac{1}{7 \cdot 10} + \ldots + \frac{1}{(3(k+1)-2)(3(k+1)+1)} && \text{left side of } P(k+1) \\
= \; & \frac{1}{1 \cdot 4} + \frac{1}{4 \cdot 7} + \frac{1}{7 \cdot 10} + \ldots + \frac{1}{(3k-2)(3k+1)} + \frac{1}{(3(k+1)-2)(3(k+1)+1)} \\
= \; & \frac{k}{3k+1} + \frac{1}{(3(k+1)-2)(3(k+1)+1)} && \text{using } P(k) \\
= \; & \frac{k}{3k+1} + \frac{1}{(3k+1)(3k+4)} \\
= \; & \frac{k(3k+4)+1}{(3k+1)(3k+4)} \\
= \; & \frac{3k^2+4k+1}{(3k+1)(3kk+4)} \\
= \; & \frac{(3k+1)(k+1)}{(3k+1)(3k+4)} = \frac{k+1}{3(k+1)+1} && \text{right side of } P(k+1)
\end{aligned}
$$

22. P(1): $1 \cdot 1! = 2! - 1$ or $1 = 2 - 1$ true

Assume P(k): $1 \cdot 1! + \ldots + k \cdot k! = (k+1)! - 1$

Show P(k + 1): $1 \cdot 1! + \ldots + (k+1)(k+1)! = (k+2)! - 1$

	$1 \cdot 1! + ... + (k+1)(k+1)!$	left side of P(k + 1)
=	$1 \cdot 1! + ... + k \cdot k! + (k+1)(k+1)!$	
=	$(k+1)! - 1 + (k+1)(k+1)!$	using P(k)
=	$(k+1)!(1 + k + 1) - 1$	
=	$(k+1)!(k+2) - 1$	
=	$(k+2)! - 1$	right side of P(k + 1)

*23. P(1): $a = \dfrac{a - ar}{1-r} = \dfrac{a(1-r)}{1-r}$ true

Assume P(k): $a + ar + ... + ar^{k-1} = \dfrac{a - ar^k}{1-r}$

Show P(k + 1): $a + ar + ... + ar^k = \dfrac{a - ar^{k+1}}{1-r}$

	$a + ar + \quad + ar^k$	left side of P(k + 1)
=	$a + ar + \quad + ar^{k-1} + ar^k$	
=	$\dfrac{a - ar^k}{1-r} + ar^k$	using P(k)
=	$\dfrac{a - ar^k + ar^k(1-r)}{1-r}$	
=	$\dfrac{a - ar^{k+1}}{1-r}$	right side of P(k + 1)

24. P(1): $a = \dfrac{1}{2}(2a)$ true

Assume P(k): $a + (a + d) + \quad + [a + (k - 1)d] = \left[\dfrac{k}{2}\right][2a + (k - 1)d]$

Show P(k + 1): $a + (a + d) + \quad + [a + kd] = \left[\dfrac{k+1}{2}\right][2a + kd]$

	$a + (a + d) + ... + [a + kd]$	left side of P(k + 1)
=	$a + (a + d) + \quad + [a + (k - 1)d] + [a + kd]$	
=	$\left[\dfrac{k}{2}\right][2a + (k - 1)d] + a + kd$	using P(k)
=	$\left[\dfrac{k}{2}\right][2a + (k - 1)d] + \dfrac{2a + 2kd}{2}$	
=	$\dfrac{2ka + k^2d - kd + 2a + 2kd}{2}$	
=	$\dfrac{k(2a + kd) + (2a + kd)}{2}$	
=	$\dfrac{(k+1)}{2}(2a + kd)$	right side of P(k + 1)

25. a. Geometric progression with a = 2, r = 5, n = 10. Sum = $(2 - 2(5^{10}))/(1-5)$ = 4,882,812

b. Rewrite with an additional term of 4 in the front: $4 + 4 \cdot 7 + 4 \cdot 7^2 + 4 \cdot 7^3 + \dots + 4 \cdot 7^{12}$
Geometric progression with a = 4, r = 7, n = 13. Sum $=[(4 - 4(7)^{13})/(1 - 7)] - 4$ (subtract the 4 we added) = 64,592,673,600

c. Rewrite as $1 + (1 + 6) + (1 + 2 \cdot 6) + \dots + (1 + 8 \cdot 6)$. Arithmetic progression with a = 1, d = 6, n = 9. Sum = (9/2)[2(1) + 8(6)] = 225

d. Rewrite as $12 + (12 + 5) + (12 + 2 \cdot 5) + (12 + 3 \cdot 5) + \dots + (12 + 16 \cdot 5)$.
Arithmetic progression with a = 12, d = 5, n = 17. Sum = (17/2)[24 + 16(5)] = 884

26. The base case is still P(1) because 1 is the smallest positive odd integer.

$P(1)$: $(-2)^0 + (-2)^1 = \dfrac{1-2^2}{3}$ or $1 - 2 = \dfrac{1-4}{3}$ true

Assume P(k) where k is odd: $(-2)^0 + (-2)^1 + (-2)^2 + \dots + (-2)^k = \dfrac{1-2^{k+1}}{3}$

The next odd integer is k + 2.

Show P(k + 2): $(-2)^0 + (-2)^1 + (-2)^2 + \dots + (-2)^{k+2} = \dfrac{1-2^{k+3}}{3}$

$(-2)^0 + (-2)^1 + (-2)^2 + \dots + (-2)^{k+2}$ left side of P(k + 2)

$= (-2)^0 + (-2)^1 + (-2)^2 + \dots + (-2)^k + (-2)^{k+1} + (-2)^{k+2}$

$= \dfrac{1-2^{k+1}}{3} + (-2)^{k+1} + (-2)^{k+2}$ using P(k)

$= \dfrac{1-2^{k+1}}{3} + 2^{k+1} - 2^{k+2}$ because, since k is odd, k + 1 is even and k + 2 is odd

$= \dfrac{1-2^{k+1} + 3 \cdot 2^{k+1} - 3 \cdot 2^{k+2}}{3}$

$= \dfrac{1+2 \cdot 2^{k+1} - 3 \cdot 2^{k+2}}{3}$

$= \dfrac{1+2^{k+2} - 3 \cdot 2^{k+2}}{3}$

$= \dfrac{1-2 \cdot 2^{k+2}}{3}$

$= \dfrac{1-2^{k+3}}{3}$ right side of P(k + 2)

27. P(3): $3^2 \geq 2(3) + 3$ or $9 \geq 9$ true

Assume P(k): $k^2 \geq 2k + 3$

Show P(k + 1): $(k + 1)^2 \geq 2(k + 1) + 3$

$(k + 1)^2$ left side of P(k + 1)

$= k^2 + 2k + 1$

$\geq (2k + 3) + 2k + 1$ using P(k)

$= 2k + 2k + 4$

$\geq 2k + 6 + 4$ because $k \geq 3$

$>$	$2k + 5$	
$=$	$2(k + 1) + 3$	right side of P(k + 1)

*28. P(2): $2^2 > 2 + 1$ true

Assume P(k): $k^2 > k + 1$

Show P(k + 1): $(k + 1)^2 > k + 2$

	$(k + 1)^2$	left side of P(k + 1)
$=$	$k^2 + 2k + 1$	
$>$	$(k + 1) + 2k + 1$	using P(k)
$=$	$3k + 2$	
$>$	$k + 2$	right side of P(k + 1)

29. P(7): $7^2 > 5 \cdot 7 + 10$ or $49 > 45$ true

Assume P(k): $k^2 > 5k + 10$

Show P(k + 1): $(k + 1)^2 > 5(k + 1) + 10 = 5k + 15$

	$(k + 1)^2$	left side of P(k + 1)
$=$	$k^2 + 2k + 1$	
$>$	$(5k + 10) + 2k + 1$	using P(k)
$=$	$7k + 11$	
$=$	$6k + k + 11$	
$>$	$6k + 17$	because $k > 6$
$>$	$5k + 15$	right side of P(k + 1)

30. P(5): $32 > 25$ true

Assume P(k): $2^k > k^2$

Show P(k + 1): $2^{k+1} > (k + 1)^2$

	2^{k+1}	left side of P(k + 1)
$=$	$2 \cdot 2^k$	
$>$	$2k^2$	using P(k)
$=$	$k^2 + k^2$	
$>$	$k^2 + 3k$	by Example 18
$=$	$k^2 + 2k + k$	
$>$	$k^2 + 2k + 1$	because $k > 1$
$=$	$(k + 1)^2$	right side of P(k + 1)

31. P(4): $4! > 4^2$ or $1 \cdot 2 \cdot 3 \cdot 4 = 24 > 16$ true

Assume P(k): $k! > k^2$

Show P(k + 1): $(k + 1)! > (k + 1)^2$

$(k + 1)!$	
$= k!(k + 1)$	
$> k^2(k + 1)$	using P(k)
$> (k + 1)(k + 1)$	by Exercise 25 since $k \geq 4$
$= (k + 1)^2$	right side of P(k + 1)

32. P(7): $7! > 3^7$ or $5040 > 2187$ true
Assume P(k): $k! > 3^k$
Show P(k + 1): $(k + 1)! > 3^{k+1}$

$(k + 1)!$	left side of P(k + 1)
$= k!(k + 1)$	
$> 3^k(k + 1)$	using P(k)
$> 3^k(3)$	because $k \geq 7$
$= 3^{k+1}$	right side of P(k + 1)

*33. P(4): $2^4 < 4!$ or $16 < 24$ true
Assume P(k): $2^k < k!$
Show P(k + 1): $2^{k+1} < (k + 1)!$

2^{k+1}	left side of P(k + 1)
$= 2 \cdot 2^k$	
$< 2 \cdot k!$	using P(k)
$< (k + 1)k!$	because $k \geq 4$
$= (k + 1)!$	right side of P(k + 1)

34. P(1): $2^0 \leq 1!$ true
Assume P(k): $2^{k-1} \leq k!$
Show P(k + 1): $2^k \leq (k + 1)!$

2^k	left side of P(k + 1)
$= 2 \cdot 2^{k-1}$	
$\leq 2 \cdot k!$	using P(k)
$\leq (k+1)k!$	because $k \geq 1$ or $k + 1 \geq 2$
$= (k + 1)!$	right side of P(k + 1)

35. P(2): $2! < 2^2$ or $2 < 4$ true
Assume P(k): $k! < k^k$
Show P(k + 1): $(k + 1)! < (k + 1)^{k+1}$

$(k + 1)!$	left side of P(k + 1)
$= (k + 1)k!$	
$< (k + 1)k^k$	using P(k)

$< \ (k+1)(k+1)^k$
$= \ (k+1)^{k+1}$ right side of P(k + 1)

An alternate proof is:
$n(n-1)(n-2) \cdots 3 \cdot 2 \cdot 1 < n \cdot n \cdot n \cdots n \cdot n \cdot n$ (n terms) $= n^n$

36. P(2): $(1+x)^2 > 1 + x^2$ or $1 + 2x + x^2 > 1 + x^2$ true because $x > 0$ implies $2x > 0$
Assume P(k): $(1+x)^k > 1 + x^k$
Show P(k + 1): $(1+x)^{k+1} > 1 + x^{k+1}$

$(1+x)^{k+1}$ left side of P(k + 1)
$= \ (1+x)^k(1+x)$
$> \ (1+x^k)(1+x)$ using P(k)
$= \ 1 + x^k + x + x^{k+1}$
$> \ 1 + x^{k+1}$ because $x^k + x > 0$

37. P(1): $\left(\frac{a}{b}\right)^2 < \frac{a}{b}$ or $\frac{a^2}{b^2} < \frac{a}{b}$
From a < b and a > 0, b > 0, we get aa < ab and then aab < abb. Dividing by the positive numbers b and b^2, we get
$a^2 < \frac{ab^2}{b}$ and $\frac{a^2}{b^2} < \frac{a}{b}$
Assume P(k): $\left(\frac{a}{b}\right)^{k+1} < \left(\frac{a}{b}\right)^k$
Show P(k + 1): $\left(\frac{a}{b}\right)^{k+2} < \left(\frac{a}{b}\right)^{k+1}$

$\left(\frac{a}{b}\right)^{k+2}$ left side of P(k + 1)
$= \ \left(\frac{a}{b}\right)^{k+1}\left(\frac{a}{b}\right) < \left(\frac{a}{b}\right)^{k}\left(\frac{a}{b}\right)$ using P(k)
$= \ \left(\frac{a}{b}\right)^{k+1}$ right side of P(k + 1)

*38. P(2): $1 + 2 < 2^2$ or $3 < 4$ true
Assume P(k): $1 + 2 + \ldots + k < k^2$
Show P(k + 1): $1 + 2 + \ldots + (k+1) < (k+1)^2$

$1 + 2 + \ldots + (k+1)$ left side of P(k + 1)
$= \ 1 + 2 + \ldots + k + (k+1)$
$< \ k^2 + k + 1$ using P(k)
$< \ k^2 + 2k + 1 = (k+1)^2$ right side of P(k + 1)

39. a. P(1): $1 + \frac{1}{2} < 2$ true

Assume P(k): $1 + \frac{1}{2} + \ldots + \frac{1}{2^k} < 2$

Show P(k + 1): $1 + \frac{1}{2} + \ldots + \frac{1}{2^{k+1}} < 2$

$1 + \frac{1}{2} + \ldots + \frac{1}{2^{k+1}}$ left side of P(k + 1)

$= 1 + \frac{1}{2} + \ldots + \frac{1}{2^k} + \frac{1}{2^{k+1}}$

$< 2 + \frac{1}{2^{k+1}}$ using P(k)

but $2 + \frac{1}{2^{k+1}}$ is not less than 2.

b. P(1): $1 + \frac{1}{2} = 2 - \frac{1}{2}$ true

Assume P(k): $1 + \frac{1}{2} + \ldots + \frac{1}{2^k} = 2 - \frac{1}{2^k}$

Show P(k + 1): $1 + \frac{1}{2} + \ldots + \frac{1}{2^{k+1}} = 2 - \frac{1}{2^{k+1}}$

$1 + \frac{1}{2} + \ldots + \frac{1}{2^{k+1}}$

$= 1 + \frac{1}{2} + \ldots + \frac{1}{2^k} + \frac{1}{2^{k+1}}$

$= 2 - \frac{1}{2^k} + \frac{1}{2^{k+1}}$ using P(k)

$= 2 - \frac{2}{2^{k+1}} + \frac{1}{2^{k+1}} = 2 - \frac{1}{2^{k+1}}$

40. Prove that $1 + \frac{1}{2} + \frac{1}{3} + \cdots + \frac{1}{2^n} \geq 1 + \frac{n}{2}$ for $n \geq 1$

P(1): $1 + \frac{1}{2} \geq 1 + \frac{1}{2}$ true

Assume P(k): $1 + \frac{1}{2} + \frac{1}{3} + \cdots + \frac{1}{2^k} \geq 1 + \frac{k}{2}$

Show P(k + 1): $1 + \frac{1}{2} + \frac{1}{3} + \cdots + \frac{1}{2^{k+1}} \geq 1 + \frac{k+1}{2}$

$1 + \frac{1}{2} + \frac{1}{3} + \cdots + \frac{1}{2^{k+1}}$ left side of P(k + 1)

$$= 1 + \frac{1}{2} + \frac{1}{3} + \cdots + \frac{1}{2^k} + \underbrace{\frac{1}{2^k + 1} + \frac{1}{2^k + 2} \cdots + \frac{1}{2^{k+1}}}$$

these denominators count from 2^k up to 2^{k+1}, a total of $2^{k+1} - 2^k = 2^k(2 - 1) = 2^k$ terms

$$\geq 1+\frac{k}{2}+\frac{1}{2^k+1}+\frac{1}{2^k+2}\cdots+\frac{1}{2^{k+1}}$$ using P(k)

$$\geq 1+\frac{k}{2}+\underbrace{\frac{1}{2^{k+1}}+\frac{1}{2^{k=1}}\cdots+\frac{1}{2^{k+1}}}_{2^k \text{ terms}}$$ because each denominator is $\leq 2^{k+1}$

$$= 1+\frac{k}{2}+2^k\left(\frac{1}{2^{k+1}}\right)$$

$$= 1+\frac{k}{2}+\frac{1}{2} = 1+\frac{k+1}{2}$$ right side of P(k+1)

*41. P(1): $2^3 - 1 = 8 - 1 = 7$ and $7 \mid 7$ true

Assume P(k): $7 \mid 2^{3k} - 1$ so $2^{3k} - 1 = 7m$ or $2^{3k} = 7m + 1$ for some integer m

Show P(k + 1): $7 \mid 2^{3(k+1)} - 1$

$2^{3(k+1)} - 1 = 2^{3k+3} - 1 = 2^{3k} \cdot 2^3 - 1$

$= (7m + 1)2^3 - 1$ using P(k)

$= 7(2^3m) + 8 - 1$

$= 7(2^3m + 1)$ where $2^3m + 1$ is an integer, so $7 \mid 2^{3(k+1)} - 1$

42. P(1): $3^2 + 7 = 9 + 7 = 16$ and $8 \mid 16$ true

Assume P(k): $8 \mid 3^{2k} + 7$ so $3^{2k} + 7 = 8m$ or $3^{2k} = 8m - 7$ for some integer m

Show P(k + 1): $8 \mid 3^{2(k+1)} + 7$

$3^{2(k+1)} + 7 = 3^{2k+2} + 7 = 3^{2k} \cdot 3^2 + 7$

$= (8m - 7)9 + 7$ using P(k)

$= 8(9m) - 63 + 7 = 8(9m) - 56$

$= 8(9m - 7)$ where $9m - 7$ is an integer, so $8 \mid 3^{2(k+1)} + 7$

43. P(1): $7 - 2 = 5$ and $5 \mid 5$ true

Assume P(k): $5 \mid 7^k - 2^k$ so $7^k - 2^k = 5m$ or $7^k = 5m + 2^k$ for some integer m

Show P(k + 1): $5 \mid 7^{k+1} - 2^{k+1}$

$7^{k+1} - 2^{k+1} = 7 \cdot 7^k - 2^{k+1}$

$= 7(5m + 2^k) - 2^{k+1}$ using P(k)

$= 5(7m) + 2^k(7 - 2)$

$= 5(7m + 2^k)$ where $7m + 2^k$ is an integer, so $5 \mid 7^{k+1} - 2^{k+1}$

44. P(1): $13 - 6 = 7$ and $7 \mid 7$ true

Assume P(k): $7 \mid 13^k - 6^k$ so $13^k - 6^k = 7m$ or $13^k = 7m + 6^k$ for some integer m

Show P(k + 1): $7 \mid 13^{k+1} - 6^{k+1}$

$13^{k+1} - 6^{k+1} = 13(13^k) - 6^{k+1}$

$= 13(7m + 6^k) - 6^{k+1}$ using P(k)

$= 7(13m) + 6^k(13 - 6)$

$= 7(13m + 6^k)$ where $13m + 6^k$ is an integer, so $7 \mid 13^{k+1} - 6^{k+1}$

*45. P(1): $2 + (-1)^2 = 2 + 1 = 3$ and $3 \mid 3$ true

Assume P(k): $3 \mid 2^k + (-1)^{k+1}$ so $2^k + (-1)^{k+1} = 3m$ or $2^k = 3m - (-1)^{k+1}$ for some integer m

Show P(k + 1): $3 \mid 2^{k+1} + (-1)^{k+2}$

$2^{k+1} + (-1)^{k+2} = 2 \cdot 2^k + (-1)^{k+2}$

$= 2(3m - (-1)^{k+1}) + (-1)^{k+2}$ using P(k)

$= 3(2m) - 2(-1)^{k+1} + (-1)^{k+2}$

$= 3(2m) + (-1)^{k+1}(-2 + (-1))$

$= 3(2m) + (-1)^{k+1} (-3)$

$= 3(2m - (-1)^{k+1})$ where $2m - (-1)^{k+1}$ is an integer, so $3 \mid 2^{k+1} + (-1)^{k+2}$

46. P(1): $2^{5+1} + 5^{1+2} = 2^6 + 5^3 = 64 + 125 = 189 = 27 \cdot 7$ and $27 \mid 27 \cdot 7$ true

Assume P(k): $27 \mid 2^{5k+1} + 5^{k+2}$ so $2^{5k+1} + 5^{k+2} = 27m$ or $2^{5k+1} = 27m - 5^{k+2}$ for some integer m

Show P(k + 1): $27 \mid 2^{5(k+1)+1} + 5^{(k+1)+2}$

$2^{5(k+1)+1} + 5^{(k+1)+2} = 2^{5k+1+5} + 5^{k+3} = 2^{5k+1} \cdot 2^5 + 5^{k+3}$

$= (27m - 5^{k+2})2^5 + 5^{k+3}$ using P(k)

$= 27(m2^5) - 5^{k+2} \cdot 2^5 + 5^{k+2} \cdot 5$

$= 27(m2^5) + 5^{k+2}(5 - 2^5)$

$= 27(m2^5) + 5^{k+2}(-27)$

$= 27(m2^5 - 5^{k+2})$ where $m2^5 - 5^{k+2}$ is an integer, so $27 \mid 2^{5(k+1)+1} + 5^{(k+1)+2}$

47. P(1): $3^{4+2} + 5^{2+1} = 3^6 + 5^3 = 729 + 125 = 854 = 61 \cdot 14$ and $14 \mid 61 \cdot 14$ true

Assume P(k): $14 \mid 3^{4k+2} + 5^{2k+1}$ so $3^{4k+2} + 5^{2k+1} = 14m$ or $3^{4k+2} = 14m - 5^{2k+1}$ for some integer m

Show P(k + 1): $14 \mid 3^{4(k+1)+2} + 5^{2(k+1)+1}$

$3^{4(k+1)+2} + 5^{2(k+1)+1} = 3^{4k+2} \cdot 3^4 + 5^{2k+3}$

$= (14m - 5^{2k+1})3^4 + 5^{2k+3}$ using P(k)

$= 14(m3^4) - 5^{2k+1} \cdot 3^4 + 5^{2k+1} \cdot 5^2$

$= 14(m3^4) - 5^{2k+1} (3^4 - 5^2)$

$= 14(m3^4) - 5^{2k+1}(81 - 25)$

$= 14(m3^4) - 5^{2k+1}(56)$

$= 14(m3^4 - 4 \cdot 5^{2k+1})$ where $m3^4 - 4 \cdot 5^{2k+1}$ is an integer, so $14 \mid 3^{4(k+1)+2} + 5^{2(k+1)+1}$

48. P(1): $7^2 + 16 - 1 = 49 + 15 = 64$ and $64 \mid 64$ true

Assume P(k): $64 \mid 7^{2k} + 16k - 1$ so $7^{2k} + 16k - 1 = 64m$ or $7^{2k} = 64m - 16k + 1$ for some integer m

Show P(k + 1): $64 \mid 7^{2(k+1)} + 16(k + 1) - 1$

$7^{2(k+1)} + 16(k + 1) - 1 = 7^{2k} \cdot 7^2 + 16(k + 1) - 1$

$= (64m - 16k + 1) \cdot 7^2 + 16(k + 1) - 1$ using P(k)

$= 64(m7^2) - 16k \cdot 7^2 + 16k + 7^2 + 16 - 1$

$= 64(m7^2) + 16k(1 - 7^2) + 64$

$= 64(m7^2 + 1) + 16k(-48)$

$= 64(m7^2 + 1 - 12k)$ where $m7^2 + 1 - 12k$ is an integer, so $64 \mid 7^{2(k+1)} + 16(k + 1) - 1$

*49. P(1): $10 + 3 \cdot 4^3 + 5 = 10 + 192 + 5 = 207 = 9 \cdot 23$ true

Assume P(k): $9 \mid 10^k + 3 \cdot 4^{k+2} + 5$ so $10^k + 3 \cdot 4^{k+2} + 5 = 9m$ or $10^k = 9m - 3 \cdot 4^{k+2} - 5$ for some integer m

Show P(k + 1): $9 \mid 10^{k+1} + 3 \cdot 4^{k+3} + 5$

$10^{k+1} + 3 \cdot 4^{k+3} + 5 = 10 \cdot 10^k + 3 \cdot 4^{k+3} + 5$

$= 10(9m - 3 \cdot 4^{k+2} - 5) + 3 \cdot 4^{k+3} + 5$ using P(k)

$= 9(10m) - 30 \cdot 4^{k+2} - 50 + 3 \cdot 4^{k+2} \cdot 4 + 5$

$= 9(10m) - 45 - 3 \cdot 4^{k+2}(10 - 4) = 9(10m - 5) - 18 \cdot 4^{k+2}$

$= 9(10m - 5 - 2 \cdot 4^{k+2})$ where $10m - 5 - 2 \cdot 4^{k+2}$ is an integer, so $9 \mid 10^{k+1} + 3 \cdot 4^{k+3} + 5$

50. P(1): $1^3 - 1 = 0$ and $3|0$ true

Assume P(k): $3|k^3 - k$ or $k^3 - k = 3m$ for some integer n

Show P(k + 1): $3|(k + 1)^3 - (k + 1)$

$(k + 1)^3 - (k + 1) = (k^3 + 3k^2 + 3k + 1) - (k + 1)$

$= k^3 - k + 3(k^2 + k)$

$= 3m + 3(k^2 + k)$ using P(k)

$= 3(m + k^2 + k)$ where $m + k^2 + k$ is an integer, so $3|(k + 1)^3 - (k + 1)$

51. P(1): $1^3 + 2(1) = 3$ and $3|3$ true

Assume P(k): $3|k^3 + 2k$ so $k^3 + 2k = 3m$ for some integer m

Show P(k + 1): $3|(k + 1)^3 + 2(k + 1)$

$(k + 1)^3 + 2(k + 1) = k^3 + 3k^2 + 3k + 1 + 2k + 2$

$= k^3 + 2k + 3(k^2 + k + 1)$

$= 3m + 3(k^2 + k + 1)$ using P(k)

$= 3(m + k^2 + k + 1)$ where $m + k^2 + k + 1$ is an integer, so $3|(k + 1)^3 + 2(k + 1)$

This result also follows directly from Exercise 45:
$n^3 + 2n = n^3 - n + 3n = 3m + 3n$ (by Exercise 45) $= 3(m + n)$

52. P(1): $x^1 - 1$ is divisible by $x - 1$ true
Assume P(k): $x^k - 1$ is divisible by $x - 1$, so $x^k - 1 = (x - 1)q(x)$ for some polynomial $q(x)$, or $x^k = (x - 1)q(x) + 1$
Show P(k + 1): $x^{k+1} - 1$ is divisible by $x - 1$

$x^{k+1} - 1 = x \cdot x^k - 1 = x((x - 1)q(x) + 1) - 1$ using P(k)
$= (x - 1)[xq(x)] + x - 1$
$= (x - 1)[xq(x) + 1]$ where $xq(x) + 1$ is a polynomial, so $x - 1$ divides $x^{k+1} - 1$

*53. P(1): $\cos\theta + i\sin\theta = \cos\theta + i\sin\theta$ true
Assume P(k): $(\cos\theta + i\sin\theta)^k = \cos k\theta + i\sin k\theta$
Show P(k + 1): $(\cos\theta + i\sin\theta)^{k+1} = \cos(k+1)\theta + i\sin(k+1)\theta$

$(\cos\theta + i\sin\theta)^{k+1} = (\cos\theta + i\sin\theta)^k(\cos\theta + i\sin\theta)$
$= (\cos k\theta + i\sin k\theta)(\cos\theta + i\sin\theta)$ using P(k)
$= \cos k\theta\cos\theta + i\sin k\theta\cos\theta + i\cos k\theta\sin\theta + i^2\sin k\theta\sin\theta$
$= \cos k\theta\cos\theta - \sin k\theta\sin\theta + i(\sin k\theta\cos\theta + \cos k\theta\sin\theta)$
$= \cos(k\theta + \theta) + i\sin(k\theta + \theta)$

54. P(1): $\sin\theta = \dfrac{\sin^2\theta}{\sin\theta}$ true

Assume P(k): $\sin\theta + \sin 3\theta + \ldots + \sin(2k - 1)\theta = \dfrac{\sin^2 k\theta}{\sin\theta}$

Show P(k + 1): $\sin\theta + \ldots + \sin(2k + 1)\theta = \dfrac{\sin^2(k+1)\theta}{\sin\theta}$

$\sin\theta + \ldots + \sin(2k+1)\theta$
$= \sin\theta + \quad + \sin(2k - 1)\theta + \sin(2k + 1)\theta$
$= \dfrac{\sin^2 k\theta}{\sin\theta} + \sin(2k + 1)\theta$ using P(k)
$= \dfrac{\sin^2 k\theta + \sin\theta\sin(2k+1)\theta}{\sin\theta}$
$= \dfrac{\sin^2 k\theta + \sin\theta\sin(2k\theta+\theta)}{\sin\theta}$
$= \dfrac{\sin^2 k\theta + \sin\theta(\sin 2k\theta\cos\theta + \cos 2k\theta\sin\theta)}{\sin\theta}$
$= \dfrac{\sin^2 k\theta + 2\sin\theta\sin k\theta\cos k\theta\cos\theta + \cos 2k\theta\sin^2\theta}{\sin\theta}$
$= \dfrac{\sin^2 k\theta + (\cos^2 k\theta - \sin^2 k\theta)\sin^2\theta + 2\sin\theta\sin k\theta\cos k\theta\cos\theta}{\sin\theta}$

$$= \frac{\sin^2 k\theta + \cos^2 k\theta \sin^2\theta - \sin^2 k\theta \sin^2\theta + 2\sin\theta \sin k\theta \cos k\theta \cos\theta}{\sin\theta}$$

$$= \frac{\sin^2 k\theta(1 - \sin^2\theta) + \cos^2 k\theta \sin^2\theta + 2\sin\theta \sin k\theta \cos k\theta \cos\theta}{\sin\theta}$$

$$= \frac{\sin^2 k\theta \cos^2\theta + \cos^2 k\theta \sin^2\theta + 2\sin\theta \sin k\theta \cos k\theta \cos\theta}{\sin\theta}$$

$$= \frac{[\sin k\theta \cos\theta + \cos k\theta \sin\theta]^2}{\sin\theta}$$

$$= \frac{[\sin(k\theta + \theta)]^2}{\sin\theta}$$

$$= \frac{\sin^2(k+1)\theta}{\sin\theta} \qquad \text{right side of } P(k+1)$$

*55. The statement to be proved is that $n(n + 1)(n + 2)$ is divisible by 3 for $n \geq 1$.
$P(1)$: $1(1 + 1)(1 + 2) = 6$ is divisible by 3 true
Assume $P(k)$: $k(k + 1)(k + 2) = 3m$ for some integer m.
Show $P(k + 1)$: $(k + 1)(k + 2)(k + 3)$ is divisible by 3

$$\begin{aligned} & (k+1)(k+2)(k+3) = (k+1)(k+2)k + (k+1)(k+2)3 \\ = \; & 3m + (k+1)(k+2)3 \qquad \text{using } P(k) \\ = \; & 3[m + (k+1)(k+2)] \end{aligned}$$

56. Let n be fixed but arbitrary and show $x^n \cdot x^m = x^{n+m}$ for all $m \geq 1$.
Let $x^j \cdot x = x^{j+1}$ be equation (1).
$m = 1$: $x^n \cdot x^1 = x^{n+1}$ by (1)
Assume $x^n \cdot x^k = x^{n+k}$
Then $x^n \cdot x^{k+1} = x^n \cdot x^k \cdot x$ by (1)
$= x^{n+k} \cdot x$ by the inductive assumption
$= x^{n+k+1}$ by (1)

57.

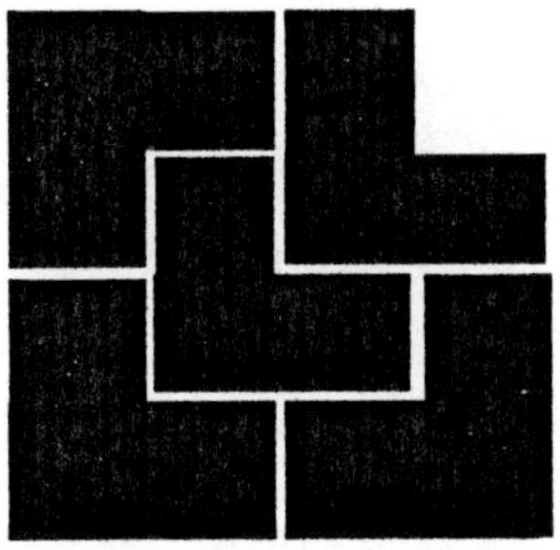

58. Trial and error shows that it is not possible to tile a 3×3 board.

59. Proof by induction on n.
$P(1)$ is true because 1 line divides the plane into 2 regions, and $(1^2 + 1 + 2)/2 = 2$.
Assume that $P(k)$ is true: k lines divide the plane into $(k^2 + k + 2)/2$ regions.
Show $P(k + 1)$, that $k + 1$ lines divide the plane into $[(k + 1)^2 + (k + 1) + 2]/2$ regions.

As an illustration, consider what happens when we go from 2 lines to 3. Initially there are 4 regions in the plane (figure a). As we draw the third line, by the time it intersects one of the two original lines, it has split one of the original regions into 2, thus adding one new region (figure b). By the time it hits the second and last line, another region has been added (figure c). And after it leaves the last line, a final new region is added (figure d).

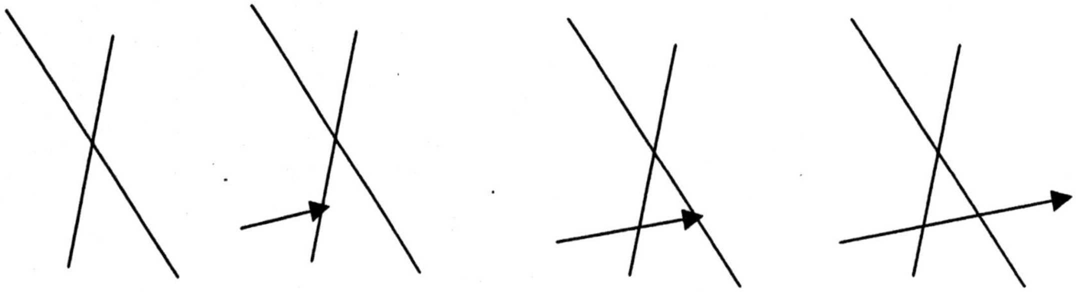

(a) (b) (c) (d)

A new line therefore creates one more region than the number of lines it crosses. When line k + 1 is added, it will cross k lines (since no two lines are parallel and have no common intersection points). Therefore k + 1 new regions are created. The total number of regions is therefore k + 1 more than the number present with k lines, or

$$\frac{k^2+k+2}{2}+k+1=\frac{k^2+k+2+2(k+1)}{2}=\frac{k^2+3k+4}{2}=\frac{(k+1)^2+(k+1)+2}{2}$$

60. Proof is by induction on the length of the string. For a string of two characters, the single processed character could be 0 or 1. If it is 0, the parity bit stays 0, and the total number of 1s is zero, an even number. If it is 1, the parity bit switches from 0 to 1, and the total number of 1s is two, an even number. Now assume that a string of k characters contains an even number of 1s. A k + 1-length string is a k-length string with one new processed character. There are four cases:

	New character	Old parity bit	New parity bit
1.	0	0	0
2.	0	1	1
3.	1	0	1
4.	1	1	0

In cases (1) and (2) no 1s are added, so the total number of 1s remains the same as before, an even number. In case (3), the total number of 1s is increased by two, so it is still an even number. In case (4), a 1 is added and a 1 is removed, so the total number of 1s remains the same as before, an even number.

*61. P(1) is 1 = 1 + 1 which is not true.

62. The brief explanation is that P(k) $\rightarrow$ P(k + 1) fails when k = 1.
Students find this problem quite difficult, so here is a more detailed discussion. P(1) is obviously true; any set consisting of just one machine has the property that all machines in that set have the same manufacturer. Assume P(k): in ANY set of k computers, all have the same manufacturer. Show P(k + 1): in ANY set of k + 1 computers, all have the same manufacturer.

The fallacious argument goes like this:

Assume P(k). Now consider any set of k + 1 machines

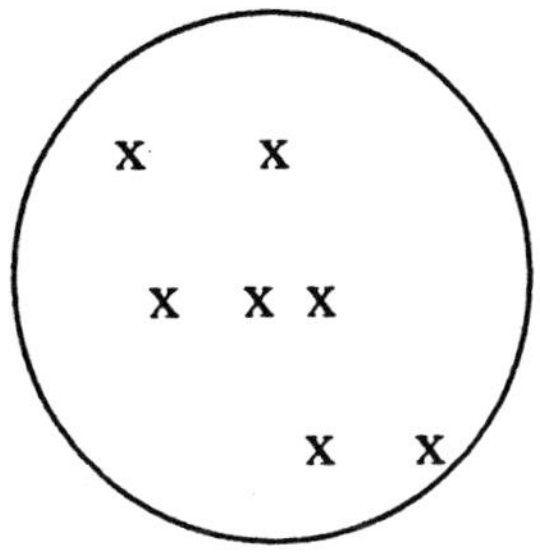

Remove one, leaving a set of k. Then by P(k), all have the same manufacturer (so make them all uppercase X):

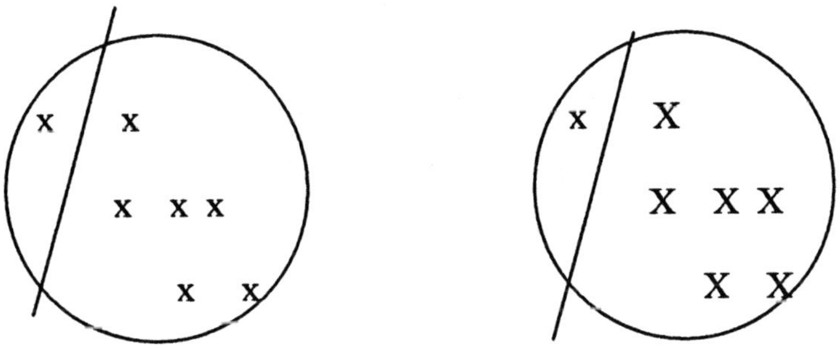

Now put the one back in and remove a different one, again leaving a set of k machines:

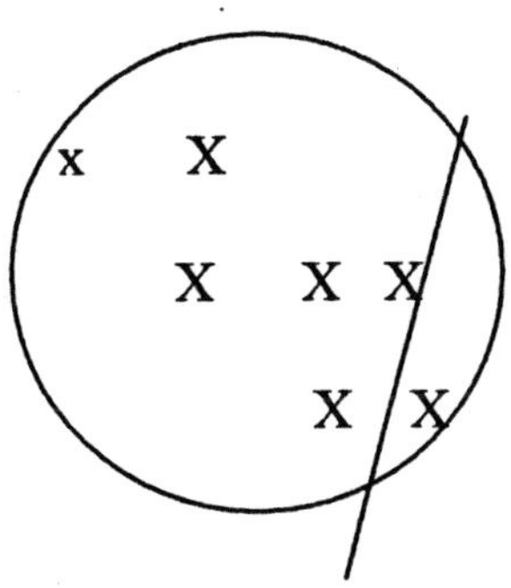

By P(k), these k machines must all have the same manufacturer. Therefore (****) the original Hal (lowercase above) must be uppercase like the rest:

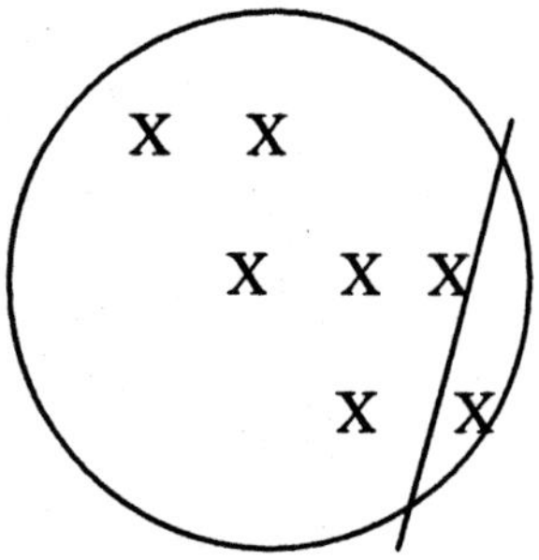

and therefore all are made by the same manufacturer.

OK, where does this fail? Look at the sentence above with ****. What forces Hal to turn uppercase is the "overlap" – those machines that are not Hal and not the other singled-out machine:

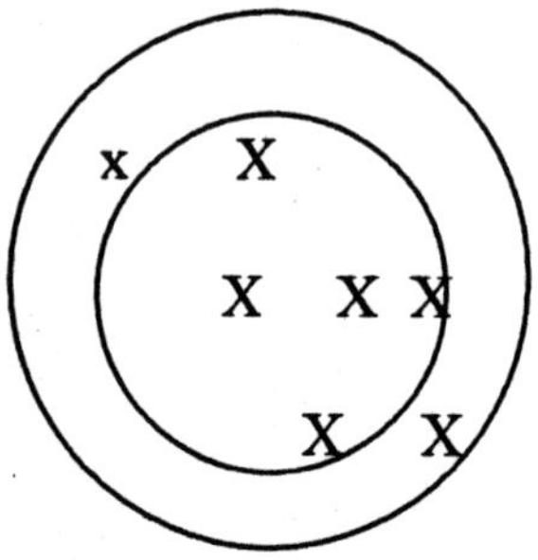

What if there is no overlap, i.e., what if there is just Hal and the other machine?

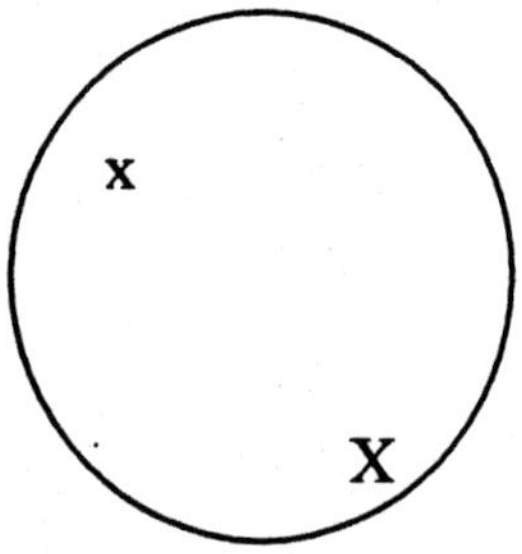

Then when you take the other machine out, Hal can be anything at all.

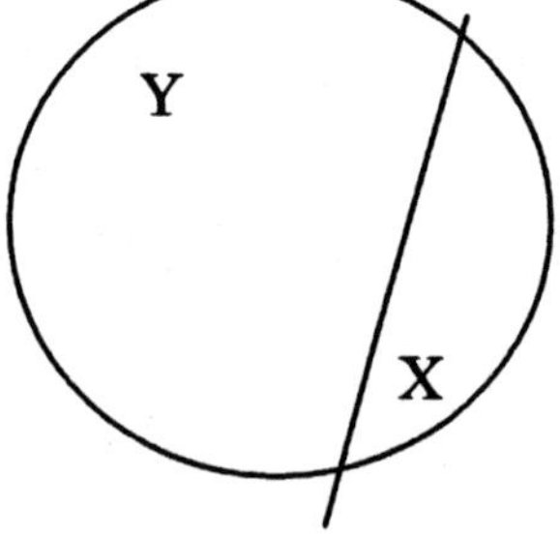

So this "proof" fails when k + 1 equals 2. In other words, the implication

$P(1) \to P(2)$

does not work. But FOR ALL OTHER CASES, the implication is true. That is, $P(2) \to P(3)$, $P(3) \to P(4)$, etc. – these are all true. Using the ladder-climbing analogy, we can get to the first rung (P(1) is true) and we can get from the 2nd rung to the 3rd, from the 3rd rung to the 4th, etc., but we can't get from the 1st rung to the second, so that's the hole.

63. a. Let $P(n)$ be the property that any word composed of a juxtaposition of n subwords has an even number of o's. Then $P(1)$ is true because the only words with 1 subword are the words moon, noon, and soon, all of which have 2 o's. Assume that $P(k)$ is true and consider $P(k+1)$. For any word composed of $k + 1$ subwords, break the word into two parts compoased of k subwords and 1 subword. By the inductive hypothesis, the part with k subwords has an even number m of o's. The part with 1 subword has 2 o's. The total number of o's is therefore $m + 2$, an even number. This verifies $P(k + 1)$ and completes the proof.

b. Let $P(n)$ be the property that any word composed of a juxtaposition of n subwords has an even number of o's. Then $P(1)$ is true because the only words with 1 subword are the words *moon*, *noon*, and *soon*, all of which have 2 o's. Assume that $P(r)$ is true for all r, $1 \leq r \leq k$ and consider $P(k+1)$. For any word composed of $k + 1$ subwords, break the word into two parts compoased of r_1 and r_2 subwords, with $1 \leq r_1 \leq k$, $1 \leq r_2 \leq k$, and $r_1 + r_2 = k + 1$. By the inductive hypothesis, r_1 contains m_1 o's, an even number, and r_2 contains m_2 o's, an even number. Then the original word contains $m_1 + m_2$ o's, an even number. This verifies $P(k + 1)$ and completes the proof.

64. a. For the base case, $n = 3$, the polygon is a triangle, and the sum of the interior angles of a triangle is 180°, which is (3 - 2)180°. Assume that for a k-sided simple closed polygon, $k \geq 3$, the sum of the interior angles is $(k - 2)180°$. Now consider a simple closed ploygon with $k + 1$ sides. Such a polygon can be formed by joining a triangle with a simple closed polygon of k sides (see figure). By the inductive hypothesis the k-sided polygon has interior angles adding to $(k - 2)180°$; the triangle has interior angles adding to 180°. The sum of the interior angles of the smaller polygon and the triangle equals the sum of the angles of the original polygon, and that sum is $(k - 2)180° + 180° = (k - 1)180°$.

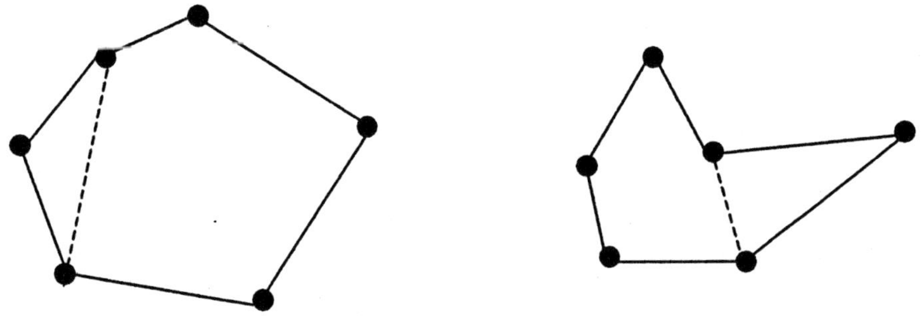

b. For the base case, $n = 3$, the polygon is a triangle, and the sum of the interior angles of a triangle is 180°, which is (3 - 2)180°. Assume that for any r-sided simple closed polygon, $3 \leq r \leq k$, the sum of the interior angles is $(r - 2)180°$. Now consider a simple closed polygon with $k + 1$ sides. Such a polygon can be formed by joining two such polygons with r_1 and r_2 sides (see figure). Then $3 \leq r_1 \leq k$ and $3 \leq r_2 \leq k$ with $r_1 + r_2 - 2 = k + 1$ (the common side of the two smaller polygons is counted as part of r_1 and as part of r_2 but is not a side of the original polygon). By the inductive hypothesis the interior angles of the two smaller polygons add up to

$(r_1 - 2)180° + (r_2 - 2)180°$. The sum of the interior angles of the two smaller polygons equals the sum of the angles of the original polygon, and that sum is $(r_1 + r_2 - 4)180° = (k - 1)180°$.

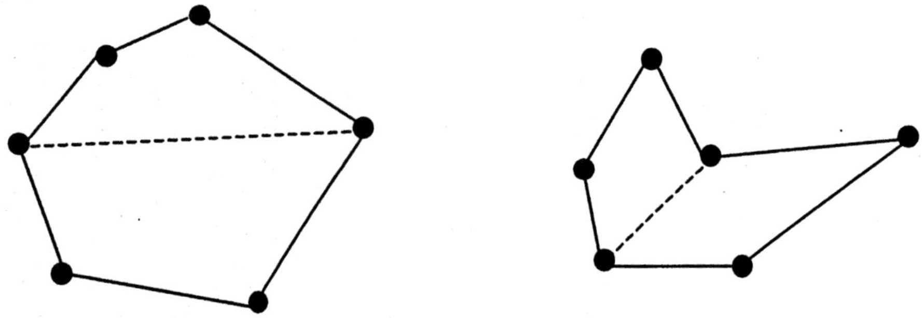

*65. For the base case, the simplest such wff is a single statement letter, which has 1 symbol; 1 is odd. Assume that for any such wff with r symbols, $1 \le r \le k$, r is odd. Consider a wff with k + 1 symbols. It must have the form $(P) \wedge (Q)$, $(P) \vee (Q)$, or $(P) \rightarrow (Q)$, where P has r_1 symbols, $1 \le r_1 < k$, and Q has r_2 symbols, $1 \le r_2 < k$. By the inductive hypothesis, both r_1 and r_2 are odd. The number of symbols in the original wff is then $r_1 + r_2 + 5$ (four parentheses plus one connective), which is odd.

*66. P(2) and P(3) are true by the equations 2 = 2 and 3 = 3. Now assume that P(r) is true for any r, $2 \le r \le k$, and consider P(k + 1). We may assume that $k + 1 \ge 4$, so that $(k + 1) - 2 \ge 2$ and by the inductive hypothesis can be written as a sum of 2's and 3's. Adding an additional 2 gives k + 1 as a sum of 2's and 3's.

67. P(12), P(13), P(14), and P(15) are true by the equations
$12 = 4 + 4 + 4$
$13 = 4 + 4 + 5$
$14 = 5 + 5 + 4$
$15 = 5 + 5 + 5$
Now assume that P(r) is true for any r, $12 \le r \le k$, and consider P(k + 1). We may assume that $k + 1 \ge 16$, so that $(k + 1) - 4 \ge 12$ and by the inductive hypothesis can be written as a sum of 4's and 5's. Adding an additional 4 gives k + 1 as a sum of 4's and 5's.

68. P(14), P(15), and P(16) are true by the equations
$14 = 8 + 3 + 3$
$15 = 3 + 3 + 3 + 3 + 3$
$16 = 8 + 8$
Now assume that P(r) is true for any r, $14 \le r \le k$, and consider P(k + 1). We may assume that $k + 1 \ge 17$, so that $(k + 1) - 3 \ge 14$ and by the inductive hypothesis can be written as a sum of 3's and 8's. Adding an additional 3 gives k + 1 as a sum of 3's and 8's.

*69. P(64), P(65), P(66), P(67), and P(68) are true by the equations

$64 = 6(5) + 2(17)$
$65 = 13(5)$
$66 = 3(5) + 3(17)$
$67 = 10(5) + 17$
$68 = 4(17)$

Now assume that P(r) is true for any r, $64 \le r \le k$, and consider P(k + 1). We may assume that $k + 1 \ge 69$, so that $(k + 1) - 5 \ge 64$ and by the inductive hypothesis can be written as a sum of 5's and 17's. Adding an additional 5 gives k + 1 as a sum of 5's and 17's.

70. Clearly the \$20 can be generated. Let P(n) be the statement that \$n(10) can be generated using only \$20 and \$50 bills, and prove P(n) for $n \ge 4$. P(4) and P(5) are true by the equations

P(4): \$40 = \$20 + \$20
P(5): \$50

Now assume that P(r) is true for any r, $4 \le r \le k$ and consider P(k + 1). We may assume that $k + 1 \ge 6$, so that $(k + 1) - 2 \ge 4$ and by the inductive hypothesis \$[(k + 1) – 2](10) can be generated using only \$20 and \$50 bills. Therefore \$(k + 1)(10) can be generated by adding another \$20 bill.

71. The formula is P(k) = k(k - 1)/2. When k = 1, zero handshakes are performed, and P(1) = 1(1 - 1)/2 = 0. Assume P(k) = k(k - 1)/2, and consider a group of k + 1 people. Identify person x and remove him from the group. The remaining group of k people performs (by inductive hypothesis) k(k - 1)/2 handshakes. If x shakes hands with each of these k people, all of the handshaking will be done. The number of handshakes is therefore

$$\begin{aligned} k(k-1)/2 + k &= k(k-1)/2 + 2k/2 \\ &= (k^2 - k + 2k)/2 \\ &= (k^2 + k)/2 \\ &= (k+1)k/2 = P(k+1) \end{aligned}$$

EXERCISES 2.3

1. Q(0): $j_0 = i_0^2$ true since i = 1, j = 1 before loop is entered

Assume Q(k): $j_k = i_k^2$

Show Q(k + 1): $j_{k+1} = (i_{k+1})^2$

$j_{k+1} = j_k + 2i_k + 1 = i_k^2 + 2i_k + 1 = (i_k + 1)^2 = (i_{k+1})^2$

At loop termination, $j = i^2$ and i = x so $j = x^2$

*2. Q(0): $j_0 = (i_0 - 1)!$ true since j = 1, i = 2 before loop is entered and 1 = 1!
Assume Q(k): $j_k = (i_k - 1)!$
Show Q(k + 1): $j_{k+1} = (i_{k+1} - 1)!$

$j_{k+1} = j_k \cdot i_k = (i_k - 1)!\,i_k = (i_k)! = (i_{k+1} - 1)!$
At loop termination, j = (i -1)! and i = x + 1 so j = x!

3. Q(0): $j_0 = x^{i_0}$ true since j = x, i = 1 before loop is entered
Assume Q(k): $j_k = x^{i_k}$
Show Q(k + 1): $j_{k+1} = x^{i_{k+1}}$

$j_{k+1} = j_k \cdot x = x^{i_k} \cdot x = x^{i_k+1} = x^{i_{k+1}}$
At loop termination, $j = x^i$ and i = y so $j = x^y$

4. Q(0): $x = q_0 y + r_0$ true since q = 0, r = x before loop is entered
Assume Q(k): $x = q_k y + r_k$
Show Q(k + 1): $x = q_{k+1} y + r_{k+1}$

$q_{k+1} y + r_{k+1} = (q_k + 1)y + r_k - y = q_k y + r_k = x$
At loop termination, x = qy + r and $0 \le r < y$

5. 2420 = 34*70 + 40, 70 = 1*40 + 30, 40 = 1*30 + 10, 30 = 3*10 + 0, so gcd(2420, 70) = 10

*6. 735 = 8*90 + 15, 90 = 6*15 + 0, so gcd(735, 90) = 15

7. 1326 = 5*252 + 66, 252 = 3*66 + 54, 66 = 1*54 + 12, 54 = 4*12 + 6, 12 = 2*6 + 0, so gcd(1326, 252) = 6

8. 1018215 = 377*2695 + 2200, 2695 = 1*2200 + 495, 2200 = 4*495 + 220, 495 = 2*220 + 55, 220 = 4*55 + 0, so gcd(1018215, 2695) = 55

9. Q: $j = x - i$
Q(0): $j_0 = x - i_0$ true since j = x, i = 0 before loop is entered
Assume Q(k): $j_k = x - i_k$
Show Q(k + 1): $j_{k+1} = x - i_{k+1}$

$j_{k+1} = j_k - 1 = x - i_k - 1 = x - (i_k + 1) = x - i_{k+1}$
At loop termination, j = x - i and i = y so j = x - y

*10. Q: $j = x * y^i$
Q(0): $j_0 = x * y^{i_0}$ true since j = x, i = 0 before loop is entered
Assume Q(k): $j_k = x * y^{i_k}$
Show Q(k + 1): $j_{k+1} = x * y^{i_{k+1}}$

$j_{k+1} = j_k * y = x * y^{i_k} * y = x * y^{i_k+1} = x * y^{i_{k+1}}$
At loop termination, $j = x * y^i$ and i = n, so $j = x * y^n$

11. Q: $j = (i + 1)^2$
 Q(0): $j_0 = (i_0 + 1)^2$ true since j = 4, i = 1 before loop is entered
 Assume Q(k): $j_k = (i_k + 1)^2$
 Show Q(k + 1): $j_{k+1} = (i_{k+1} + 1)^2$

 $j_{k+1} = j_k + 2i_k + 3 = (i_k + 1)^2 + 2i_k + 3 = i_k^2 + 2i_k + 1 + 2i_k + 3 = i_k^2 + 4i_k + 4 = (i_k + 2)^2 = (i_k + 1 + 1)^2 = (i_{k+1} + 1)^2$
 At loop termination, $j = (i + 1)^2$ and i = x, so $j = (x + 1)^2$

12. Q: $j = 2^i$
 Q(0): $j_0 = 2^{i_0}$ true since j = 2, i = 1 before loop is entered
 Assume Q(k): $j_k = 2^{i_k}$
 Show Q(k + 1): $j_{k+1} = 2^{i_{k+1}}$

 $j_{k+1} = 2 * j_k = 2 * 2^{i_k} = 2^{i_k+1} = 2^{i_{k+1}}$
 At loop termination, $j = 2^i$ and i = n, so $j = 2^n$

*13. Q: $j = x*i!$
 Q(0): $j_0 = x*i_0!$ true since j = x, i = 1 before loop is entered
 Assume Q(k): $j_k = x*(i_k)!$
 Show Q(k + 1): $j_{k+1} = x*(i_{k+1})!$

 $j_{k+1} = j_k*(i_k + 1) = x*(i_k)!(i_k + 1) = x*(i_k + 1)! = x*(i_{k+1})!$
 At loop termination, $j = x*i!$ and i = n so $j = x*n!$

14. Q: $j = a_n x^{n-i} + a_{n-1}x^{n-i-1} + \ldots + a_i x^0$
 Q(0): $j_0 = a_n x^{n-i_0} + a_{n-1}x^{n-i_0-1} + \ldots + a_{i_0}x^0$ true since $j = a_n$ and i = n before loop is entered, so equation becomes $a_n = a_n x^{n-n}$, or $a_n = a_n$

 Assume Q(k): $j_k = a_n x^{n-i_k} + a_{n-1}x^{n-i_k-1} + \ldots + a_{i_k}x^0$
 Show Q(k + 1): $j_{k+1} = a_n x^{n-i_{k+1}} + a_{n-1}x^{n-i_{k+1}-1} + \ldots + a_{i_{k+1}}x^0$

 $j_{k+1} = j_k * x + a_{i_k-1} = (a_n x^{n-i_k} + a_{n-1}x^{n-i_k-1} + \ldots + a_{i_k}x^0)x + a_{i_k-1}$
 $= a_n x^{n-i_k+1} + a_{n-1}x^{n-i_k} + \ldots + a_{i_k}x^1 + a_{i_k-1}$
 $= a_n x^{n-i_{k+1}} + a_{n-1}x^{n-i_{k+1}-1} + \ldots + a_{i_k}x^1 + a_{i_{k+1}}$
 At termination,
 $j = a_n x^{n-i} + a_{n-1}x^{n-i-1} + \ldots + a_i x^0$ and $i = 0$ so
 $j = a_n x^n + a_{n-1}x^{n-1} + \ldots + a_0 x^0$

15. (a) makes one too many passes through the loop and adds a[n + 1] to the sum.
(b) is correct:

Q: $j = a[1] + \dots + a[i-1]$
Q(0): $j_0 = a[1] + \dots + a[i_0 - 1]$ — true since j = 0, i = 1 before loop is entered, so equation becomes 0 = a[1] + ... + a[0]; the right side has no terms, so has the value 0

Assume Q(k): $j_k = a[1] + \dots + a[i_k - 1]$
Show Q(k + 1): $j_{k+1} = a[1] + \dots + a[i_{k+1} - 1]$

Note that $j_{k+1} = j_k + a[i_k]$ and $i_{k+1} = i_k + 1$

$$j_{k+1} = j_k + a[i_k] = a[1] + \dots + a[i_k - 1] + a[i_k]$$
$$= a[1] + \dots + a[i_k - 1] + a[i_{k+1} - 1]$$

At loop termination, $j = a[1] + \dots + a[i-1]$ and $i = n + 1$, so $j = a[1] + \dots + a[n]$
(c) begins the sum with a[0]
(d) adds a[n + 1] to the sum

EXERCISES 2.4

*1. 10, 20, 30, 40, 50

2. 2, 1/2, 2, 1/2, 2

3. 1, 5, 14, 30, 55

*4. $1, 1+\frac{1}{2}, 1+\frac{1}{2}+\frac{1}{3}, 1+\frac{1}{2}+\frac{1}{3}+\frac{1}{4}, 1+\frac{1}{2}+\frac{1}{3}+\frac{1}{4}+\frac{1}{5}$

5. 1, 2, 6, 24, 120

6. 1, 5, 47, 755, 18879

*7. 2, 2, 6, 14, 34

8. 3, 5, 13, 49, 235

9. 2, 3, 6, 18, 108

10. 1, 2, 3, 10, 22

*11.
$$F(n + 1) + F(n - 2) = F(n - 1) + F(n) + F(n - 2)$$
$$= [F(n - 2) + F(n - 1)] + F(n)$$
$$= F(n) + F(n) = 2F(n)$$

12. $F(n) = F(n - 2) + F(n - 1)$
$= F(n - 3) + F(n - 4) + F(n - 2) + F(n - 3)$
$= 2F(n - 3) + F(n - 4) + F(n - 2)$
$= 2[F(n - 4) + F(n - 5)] + F(n - 4) + [F(n - 3) + F(n - 4)]$
$= 4F(n - 4) + 2F(n - 5) + F(n - 3)$
$= 4F(n - 4) + 2F(n - 5) + [F(n - 4) + F(n - 5)]$
$= 5F(n - 4) + 3F(n - 5)$

13. $[F(n + 1)]^2 = [F(n - 1) + F(n)]^2$
$= [F(n - 1)]^2 + 2F(n - 1)F(n) + [F(n)]^2$
$= F(n - 1)[F(n - 1) + F(n) + F(n)] + [F(n)]^2$
$= F(n - 1)[F(n + 1) + F(n)] + [F(n)]^2$
$= F(n - 1)F(n + 2) + [F(n)]^2$

14. $F(n + 3) = F(n + 2) + F(n + 1)$
$= F(n + 1) + F(n) + F(n + 1)$
$= 2F(n + 1) + F(n)$

15. $F(n + 6) = F(n + 5) + F(n + 4)$
$= F(n + 4) + F(n + 3) + F(n + 4)$
$= [F(n + 3) + F(n + 2)] + F(n + 3) + [F(n + 3) + F(n + 2)]$
$= 3F(n + 3) + F(n + 2) + F(n + 2)$
$= 3F(n + 3) + [F(n + 3) - F(n + 1)] + F(n + 2)$
$= 4F(n + 3) - F(n + 1) + [F(n + 1) + F(n)]$
$= 4F(n + 3) + F(n)$

*16. $n = 1$: $F(1) = F(3) - 1$ or $1 = 2 - 1$ true
Assume true for $n = k$: $F(1) + ... + F(k) = F(k + 2) - 1$
Show true for $n = k + 1$: $F(1) + ... + F(k + 1) = F(k + 3) - 1$

$F(1) + ... + F(k + 1)$
$= F(1) + ... + F(k) + F(k + 1)$
$= F(k + 2) - 1 + F(k + 1)$ inductive hypothesis
$= F(k + 3) - 1$ recurrence relation

17. $n = 1$: $F(2) = F(3) - 1$ or $1 = 2 - 1$ true
Assume true for $n = k$: $F(2) + ... + F(2k) = F(2k + 1) - 1$
Show true for $n = k + 1$: $F(2) + ... + F(2(k + 1)) = F(2(k + 1) + 1) - 1$
$F(2) + ... + F(2(k + 1))$
$= F(2) + ... + F(2k) + F(2(k + 1))$
$= F(2k + 1) - 1 + F(2(k + 1))$ inductive hypothesis
$= F(2k + 1) + F(2k + 2) - 1$
$= F(2k + 3) - 1$ recurrence relation
$= F(2(k + 1) + 1) - 1$

18. $n = 1$: $F(1) = F(2)$ or $1 = 1$ true

Assume true for $n = k$: $F(1) + F(3) + \ldots + F(2k - 1) = F(2k)$

Show true for $n = k + 1$: $F(1) + F(3) + \ldots + F(2(k+1) - 1) = F(2(k + 1))$

$$\begin{aligned} & F(1) + F(3) + \ldots + F(2(k+1) - 1) \\ = {} & F(1) + F(3) + \ldots + F(2k - 1) + F(2(k+1) - 1) \\ = {} & F(2k) + F(2(k + 1) - 1) && \text{inductive hypothesis} \\ = {} & F(2k) + F(2k + 1) \\ = {} & F(2k + 2) && \text{recurrence relation} \\ = {} & F(2(k + 1)) \end{aligned}$$

19. $n = 1$: $[F(1)]^2 = F(1)F(2)$ or $1^2 = (1)(1)$ true

Assume true for $n = k$: $[F(1)]^2 + [F(2)]^2 + \ldots + [F(k)]^2 = F(k)F(k + 1)$

Show true for $n = k + 1$: $[F(1)]^2 + [F(2)]^2 + \ldots + [F(k + 1)]^2 = F(k + 1)F(k + 2)$

$$\begin{aligned} & [F(1)]^2 + [F(2)]^2 + \ldots + [F(k + 1)]^2 \\ = {} & [F(1)]^2 + [F(2)]^2 + \ldots + [F(k)]^2 + [F(k + 1)]^2 \\ = {} & F(k)F(k + 1) + [F(k + 1)]^2 && \text{inductive hypothesis} \\ = {} & F(k + 1)[F(k) + F(k + 1)] \\ = {} & F(k + 1)F(k + 2) && \text{recurrence relation} \end{aligned}$$

*20. $n = 1$: $F(4) = 2F(2) + F(1)$ or $3 = 2(1) + 1$ true

$n = 2$: $F(5) = 2F(3) + F(2)$ or $5 = 2(2) + 1$ true

Assume for all r, $1 \le r \le k$, $F(r + 3) = 2F(r + 1) + F(r)$

Show that $F(k + 4) = 2F(k + 2) + F(k + 1)$ where $k + 1 \ge 3$.

$$\begin{aligned} & F(k + 4) = F(k + 2) + F(k + 3) && \text{recurrence relation} \\ = {} & 2F(k) + F(k - 1) + 2F(k + 1) + F(k) && \text{inductive hypothesis} \\ = {} & 2[F(k) + F(k + 1)] + [F(k - 1) + F(k)] \\ = {} & 2F(k + 2) + F(k + 1) && \text{recurrence relation} \end{aligned}$$

21. $n = 1$: $F(7) = 4F(4) + F(1)$ or $13 = 4(3) + 1$ true

$n = 2$: $F(8) = 4F(5) + F(2)$ or $21 = 4(5) + 1$ true

Assume for all r, $1 \le r \le k$, $F(r + 6) = 4F(r + 3) + F(r)$

Show that $F(k + 7) = 4F(k + 4) + F(k + 1)$ where $k + 1 \ge 3$.

$$\begin{aligned} & F(k + 7) = F(k + 5) + F(k + 6) && \text{recurrence relation} \\ = {} & 4F(k + 2) + F(k - 1) + 4F(k + 3) + F(k) && \text{inductive hypothesis} \\ = {} & 4[F(k + 2) + F(k + 3)] + [F(k - 1) + F(k)] \\ = {} & 4F(k + 4) + F(k + 1) && \text{recurrence relation} \end{aligned}$$

22. $n = 1$: $F(1) < 2$ or $1 < 2$ true

$n = 2$: $F(2) < 2^2$ or $1 < 4$ true

Assume for all r, $1 \le r \le k$, $F(r) < 2^r$

Show that $F(k + 1) < 2^{k+1}$ where $k + 1 \ge 3$.

$$F(k + 1) = F(k - 1) + F(k) \qquad \text{recurrence relation}$$

$< \ 2^{k-1} + 2^k$ inductive hypothesis

$= \ 2^{k-1}(1 + 2)$

$= \ 3(2^{k-1}) < 4(2^{k-1}) = 2^2(2^{k-1}) = 2^{k+1}$

23. n = 6: $F(6) > \left(\frac{3}{2}\right)^5$ or 8 > 7.594, true

n = 7: $F(7) > \left(\frac{3}{2}\right)^6$ or 13 > 11.39, true

Assume for all r, $6 \le r \le k$, $F(r) > \left(\frac{3}{2}\right)^{r-1}$

Show that $F(k + 1) > \left(\frac{3}{2}\right)^k$ where $k + 1 \ge 8$ so $k - 1 \ge 6$

F(k + 1) = F(k) + F(k - 1) >

$$\left(\frac{3}{2}\right)^{k-1} + \left(\frac{3}{2}\right)^{k-2} = \left(\frac{3}{2}\right)^{k-2}\left(1+\frac{3}{2}\right) = \left(\frac{3}{2}\right)^{k-2}\left(\frac{5}{2}\right) = \left(\frac{3}{2}\right)^{k-2}\left(\frac{10}{4}\right) > \left(\frac{3}{2}\right)^{k-2}\left(\frac{9}{4}\right) = \left(\frac{3}{2}\right)^{k-2}\left(\frac{3}{2}\right)^2 = \left(\frac{3}{2}\right)^k$$

24. a. $p^2 = \frac{(1+\sqrt{5})^2}{2^2}$

$$= \frac{1+2\sqrt{5}+5}{2} = \frac{6+2\sqrt{5}}{4} = \frac{3+\sqrt{5}}{2} = \frac{2}{2} + \frac{1+\sqrt{5}}{2} = 1 + p$$

The proof that $1 + q = q^2$ is similar.

b. n = 1: $F(1) = \frac{p-q}{p-q} = 1$ true

n = 2: $F(2) = \frac{p^2-q^2}{p-q} = \frac{(p-q)(p+q)}{p-q} = p + q = \frac{1+\sqrt{5}}{2} + \frac{1-\sqrt{5}}{2} = \frac{2}{2} = 1$ true

Assume for all r, $1 \le r \le k$, $F(r) = \frac{p^r - q^r}{p-q}$

Show that $F(k + 1) = \frac{p^{k+1} - q^{k+1}}{p-q}$ where $k + 1 \ge 3$.

F(k + 1) = F(k - 1) + F(k) =

$$\frac{p^{k-1}-q^{k-1}}{p-q} + \frac{p^k - q^k}{p-q} = \frac{p^{k-1}(1+p) - q^{k-1}(1+q)}{p-q} = \frac{p^{k-1}p^2 - q^{k-1}q^2}{p-q} = \frac{p^{k+1} - q^{k+1}}{p-q}$$

c. Substitute the expressions for p and q into the formula in part b and carry out algebraic manipulations.

25. a. $n = 0$: $S(0) = 1$ is odd
$n = 1$: $S(1) = 1$ is odd
Assume for all r, $0 \le r \le k$, S(r) is odd
Show that $S(k + 1)$ is odd where $k + 1 \ge 2$

$S(k + 1) = 2S(k) + S(k - 1)$ recurrence relation
$= 2(\text{odd}) + \text{odd}$ inductive hypothesis
$= \text{even} + \text{odd} = \text{odd}$

b. The first few terms of this sequence are: $S(0) = 1$, $S(1) = 1$, $S(2) = 3$, $S(3) = 7$, $S(4) = 17$, $S(5) = 41$.

$n = 4$: $S(4) < 6S(2)$ or $17 < 6(3) = 18$ true
$n = 5$: $S(5) < 6S(3)$ or $41 < 6(7) = 42$ true
Assume for all r, $4 \le r \le k$, $S(r) < 6S(r - 2)$
Show that $S(k + 1) < 6S(k - 1)$ where $k + 1 \ge 6$

$S(k + 1) = 2S(k) + S(k - 1)$ recurrence relation
$< 2[6S(k - 2)] + 6S(k - 3)$ inductive hypothesis
$= 6[2S(k - 2) + S(k - 3)]$
$= 6S(k - 1)$

26. $n = 0$: $T(0) \le \left(\frac{5}{2}\right)^0$ or $1 \le 1$ true

$n = 1$: $T(1) \le \left(\frac{5}{2}\right)^1$ or $2 \le \frac{5}{2}$ true

Assume for all r, $0 \le r \le k$, $T(r) \le \left(\frac{5}{2}\right)^r$

Show that $T(k + 1) \le \left(\frac{5}{2}\right)^{k+1}$ where $k + 1 \ge 2$

$T(k + 1) = 2T(k) + T(k - 1)$ recurrence relation

$\le 2\left(\frac{5}{2}\right)^k + \left(\frac{5}{2}\right)^{k-1}$ inductive hypothesis

$= \left(\frac{5}{2}\right)^{k-1} [2\left(\frac{5}{2}\right) + 1]$

$= \left(\frac{5}{2}\right)^{k-1}\left(\frac{12}{2}\right) < \left(\frac{5}{2}\right)^{k-1}\left(\frac{25}{4}\right) = \left(\frac{5}{2}\right)^{k-1}\left(\frac{5}{2}\right)^2 = \left(\frac{5}{2}\right)^{k+1}$

27. a. 1, 3, 4, 7, 11
b. $n = 2$: $L(2) = F(2) + F(0)$ or $3 = 2 + 1$ true
$n = 3$: $L(3) = F(3) + F(1)$ or $4 = 3 + 1$ true
Assume for all r, $2 \le r \le k$, $L(r) = F(r) + F(r - 2)$
Show that $L(k + 1) = F(k + 1) + F(k - 1)$ where $k + 1 \ge 4$

$L(k + 1) = L(k) + L(k - 1)$ recurrence relation
$= F(k) + F(k - 2) + F(k - 1) + F(k - 3)$ inductive hypothesis
$= F(k) + F(k - 1) + F(k - 2) + F(k - 3)$
$= F(k + 1) + F(k - 1)$

28. No; the next value in the sequence, A(5), equals $4*2^3 + 1 = 33$, which is not a Fibonacci number.

*29. Yes, this sequence continues to generate all of the Fibonacci numbers; this is true because of Exercise 16: F(1) + F(2) + ...+ F(n) = F(n + 2) - 1, or
1 + F(1) + F(2) + ...+ F(n) = F(n + 2).

30. No. The next value in the sequence, C(6), is 8, which is in the Fibonacci sequence.

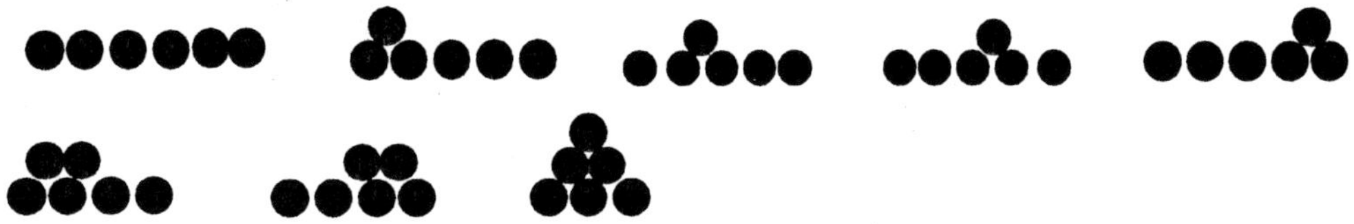

But the next value in this sequence, C(7), is 12, which is not in the Fibonacci sequence:

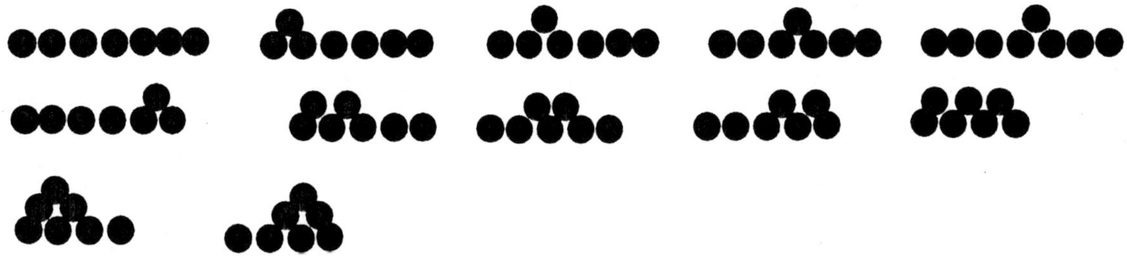

31. Yes, this sequence continues to generate all the Fibonacci numbers. To paint an (n + 1)-story building, there are two cases. If floor n + 1 is painted yellow, then the remaining n floors can be painted in any of D(n) ways. If floor n + 1 is pointed blue, then floor n must be painted yellow; the remaining n - 1 floors can be painted in any of D(n - 1) ways. Hence D(n + 1) = D(n) + D(n - 1), which is the formula for the Fibonacci sequence.

32. F(1) = F(2) = 1 because the rabbits do not breed until they are two months old.
F(n) = the number of pairs at the end of n months
= (the number of pairs at the end of n - 1 months) +
(the number of offspring pairs produced during month n, born to the pairs existing at month n - 2)
= F(n - 1) + F(n - 2)

33. $C(2) = \sum_{k=1}^{2} C(k-1)C(2-k) = C(0)C(1) + C(1)C(0) = 1*1 + 1*1 = 2$

$$C(3) = \sum_{k=1}^{3} C(k-1)C(3-k) = C(0)C(2) + C(1)C(1) + C(2)C(0) = 1*2 + 1*1 + 2*1 = 5$$

$$C(4) = \sum_{k=1}^{4} C(k-1)C(4-k) = C(0)C(3) + C(1)C(2) + C(2)C(1) + C(3)C(0)$$
$$= 1*5 + 1*2 + 2*1 + 5*1 = 14$$

*34. S(1) = a
S(n) = rS(n - 1) for $n \geq 2$

35. S(1) = a
S(n) = S(n - 1) + d for $n \geq 2$

*36. a A(l) = 50,000
A(n) = 3A(n - 1) for $n \geq 2$
b. 4

37. a. P(l) = 500
P(n) = (l.1)P(n - 1) for $n \geq 2$
b. 4

38. b and c

39. a, b, c, f, and g

*40 a, b, and e

41. a, b, and d

*42. 1. Any unary predicate in x is a wff.
2. If P and Q are unary predicate wffs in x, so are $(P \wedge Q)$, $(P \vee Q)$, $(P \rightarrow Q)$, (P'), $(P \leftrightarrow Q)$, $(\forall x)P$, and $(\exists x)P$.

Note that this allows expressions such as $(\forall x)(\exists x)P(x)$ to be wffs, meaning that such expressions are syntactically correct. Within any interpretation, the truth value is unaffected by the outermost quantifiers. For example, if the domain is the integers and P(x) is the predicate x = 0, then $(\forall x)(\exists x)P(x)$ is true because $(\exists x)P(x)$ is true.

43. 1. Any integer is a well-formed formula.
2. If P and Q are well-formed formulas, so are (P + Q), (P- Q), (P*Q), and (P/Q).

44. 1. λ is well-balanced.
2. If A and B are strings of well-balanced parentheses, so are (A) and AB.

45. 1. The string 0 belongs to the set.
2. If x belongs to the set, so do 1x, x1, 0x0, 00x, and x00.

*46. 1. $\lambda^R = \lambda$
2. If x is a string with a single character, $x^R = x$.
3. If x = yz, then $x^R = z^R y^R$

47. 1. $|\lambda| = 0$
2. If x is a single character, $|x| = 1$.
3. If $x = yz$, then $|x| = |y| + |z|$.

*48. <positive digit> ::= 1|2|3|4|5|6|7|8|9
<digit> ::= 0|<positive digit>
<positive integer> ::= <positive digit>|<positive integer><digit>

49. <digit> ::= 0|1|2|3|4|5|6|7|8|9
<number> ::= <digit>|<number><digit>
<point> ::= .
<sign> ::= + | -
<decimal number> ::= <number><point>|<sign><number><point>|
<decimal number><digit>

*50. $1! = 1$
$n! = n(n-1)!$ for $n \geq 2$

51. $m + 0 = m$
$m + n = m + (n-1) + 1$ for $n \geq 1$

52. a. $\max(a_1, a_2) = \begin{cases} a_1 & \text{if } a_1 \geq a_2 \\ a_2 & \text{if } a_1 < a_2 \end{cases}$

$\max(a_1, \ldots, a_n) = \max(\max(a_1, \ldots, a_{n-1}), a_n)$ for $n > 2$

b. $\min(a_1, a_2) = \begin{cases} a_1 & \text{if } a_1 \leq a_2 \\ a_2 & \text{if } a_1 > a_2 \end{cases}$

$\min(a_1, \ldots, a_n) = \min(\min(a_1, \ldots, a_{n-1}), a_n)$ for $n > 2$

53. a. $A_1 \wedge A_n$ defined as in Section 1.1
$A_1 \wedge \ldots \wedge A_n = (A_1 \wedge \ldots \wedge A_{n-1}) \wedge A_n$ for $n > 2$ (1)

b. For any n with $n \geq 3$ and any p with $1 \leq p \leq n-1$,

$(A_1 \wedge \ldots \wedge A_p) \wedge (A_{p+1} \wedge \ldots \wedge A_n) \Leftrightarrow A_1 \wedge \ldots \wedge A_n$

Proof: For $n = 3$,

$(A_1) \wedge (A_2 \wedge A_3) \Leftrightarrow (A_1 \wedge A_2) \wedge A_3$ by equivalence 2b
$= A_1 \wedge A_2 \wedge A_3$ by equation (1)

Assume that for $n = k$ and $1 \le p \le k - 1$,

$(A_1 \wedge ... \wedge A_p) \wedge (A_{p+1} \wedge ... \wedge A_k) \Leftrightarrow A_1 \wedge ... \wedge A_k$

Then for $1 \le p \le k$,

$(A_1 \wedge ... \wedge A_p) \wedge (A_{p+1} \wedge ... \wedge A_{k+1})$	
$= (A_1 \wedge ... \wedge A_p) \wedge [(A_{p+1} \wedge ... \wedge A_k) \wedge A_{k+1}]$	by equation (1)
$\Leftrightarrow [(A_1 \wedge ... \wedge A_p) \wedge (A_{p+1} \wedge ... \wedge A_k)] \wedge A_{k+1}$	by equivalence 2b
$\Leftrightarrow (A_1 \wedge ... \wedge A_k) \wedge A_{k+1}$	by inductive hypothesis
$= A_1 \wedge ... \wedge A_{k+1}$	by equation (1)

*54. $A \vee (B_1 \wedge B_2) \Leftrightarrow (A \vee B_1) \wedge (A \vee B_2)$ by equivalence 3a

Assume that

$A \vee (B_1 \wedge ... \wedge B_k) \Leftrightarrow (A \vee B_1) \wedge ... \wedge (A \vee B_k)$

Then

$A \vee (B_1 \wedge ... \wedge B_{k+1})$	
$= A \vee [(B_1 \wedge ... \wedge B_k) \wedge B_{k+1}]$	by Exercise 52
$\Leftrightarrow (A \vee (B_1 \wedge ... \wedge B_k)) \wedge (A \vee B_{k+1})$	by equivalence 3a
$\Leftrightarrow [(A \vee B_1) \wedge ... \wedge (A \vee B_k)] \wedge (A \vee B_{k+1})$	by inductive hypothesis
$\Leftrightarrow (A \vee B_1) \wedge ... \wedge (A \vee B_{k+1})$	by Exercise 52

The proof of the other statement is similar.

55. $(B_1 \vee B_2)' \Leftrightarrow B_1' \wedge B_2'$ by DeMorgan's Laws

Assume that

$(B_1 \vee ... \vee B_k)' \Leftrightarrow B_1' \wedge ... \wedge B_k'$.

Then

$(B_1 \vee \quad \vee B_{k+1})' = [(B_1 \vee ... \vee B_k) \vee B_{k+1}]'$	by the recursive definition of disjunction
$\Leftrightarrow (B_1 \vee ... \vee B_k)' \wedge B_{k+1}'$	by DeMorgan's Law
$\Leftrightarrow (B_1' \wedge ... \wedge B_k') \wedge B_{k+1}'$	by inductive hypothesis
$= B_1' \wedge ... \wedge B_{k+1}'$	by Exercise 41

The proof of the other statement is similar.

56. **if** n = 1 **then**
    return 1
else
    return 3*S(n - 1)
end if

57.
```
if n = 1 then
    return 2
else
    return S(n - 1)/2
end if
```

*58.
```
if n = 1 then
    return 1
else
    return S(n - 1) + (n - 1)
end if
```

59.
```
if n = 1 then
    return 2
else
    return S(n - 1)*S(n - 1)
end if
```

60.
```
if n = 1 then
    return a
else
    if n = 2 then
        return b
    else
        return S(n - 2) + S(n - 1)
    end if
end if
```

*61.
```
if n = 1 then
    return p
else
    if odd(n) then
        return S(n - 1) + (n - 1)*q
    else
        return S(n - 1) - (n - 1)*q
    end if
end if
```

62. Mystery(n) = n

63. g returns the value 0 if x is not in the list, and the value 10 if x is in the list

*64. If the list has 1 element or 0 elements, then we are done, else exchange the first and last element in the list and invoke the algorithm on the list minus its first and last elements.

65. If the number is less than 10, the sum is the single digit, else divide the number by 10, and add the remainder to the result of invoking the algorithm on the quotient.

66. Divide a by b. If the remainder r is 0, then gcd(a, b) = b, else invoke the algorithm on b and r instead of a and b.

67. If the stack contains one disk, move it to the target peg. Otherwise, invoke the algorithm on the top n – 1 disks to stack them on order on the third peg, move the bottom disk to the target peg, then invoke the algorithm on the n – 1 disks to stack them on the target peg.

*68. 4 10 -6 2 5
4 5 -6 2 10
4 2 -6 5 10
-6 2 4 5 10

69. 9 0 2 6 4
4 0 2 6 9
2 0 4 6 9
0 2 4 6 9

70. New Orleans, Charlotte, Indianapolis

71. eggs, shortening, flour

72. Q: $CurrentValue = 2^{i-1}$

$Q(0)$: $CurrentValue_0 = 2^{i_0-1}$ true since $CurrentValue_0 = 2$ and $i_0 = 2$ and $2 = 2^{2-1}$

Assume $Q(k)$: $CurrentValue_k = 2^{i_k-1}$

Show $Q(k + 1)$: $CurrentValue_{k+1} = 2^{i_{k+1}-1}$

$CurrentValue_{k+1} = 2*CurrentValue_k = 2(2^{i_k-1}) = 2^{i_k} = 2^{i_{k+1}-1}$

At termination, $CurrentValue = 2^{i-1}$ and $i = n + 1$, so $CurrentValue = 2^{n+1-1} = 2^n$

*73. The recurrence relation matches Equation (6) with c = 1 and g(n) = 5. From Equation (8), the solution is

$$S(n) = 5 + \sum_{i=2}^{n} 5 = 5 + (n-1)5 = n(5)$$

74. The recurrence relation matches Equation (6) with c = 2 and $g(n) = 2^n$. From Equation (8), the solution is

$$F(n) = 2^{n-1}(2) + \sum_{i=2}^{n} 2^{n-i}2^i = 2^n + \sum_{i=2}^{n} 2^n = 2^n + (n-1)2^n = n2^n$$

75. The recurrence relation matches Equation (6) with c = 2 and g(n) = 1. From Equation (8), the solution is

$$T(n) = 2^{n-1}(1) + \sum_{i=2}^{n} 2^{n-i}(1) = 2^{n-1} + 2^{n-2} + \ldots + 2 + 1 = 2^n - 1$$

*76. The recurrence relation matches Equation (6) with c = 1 and g(n) = n. From Equation (8), the solution is

$$A(n) = 1 + \sum_{i=2}^{n} i = 1 + 2 + 3 + \ldots + n = n(n+1)/2$$

77. The recurrence relation matches Equation (6) with c = 1 and g(n) = 2n - 1. From Equation (8), the solution is

$$S(n) = 1 + \sum_{i=2}^{n}(2i-1) = 1+3+\ldots+(2n-1) = n^2$$

78. The recurrence relation matches Equation (6) with c = 2 and $g(n) = n2^n$. From Equation (8), the solution is

$$P(n) = 2^{n-1}(2) + \sum_{i=2}^{n} 2^{n-i}(i)(2^i) = 2^n + \sum_{i=2}^{n}(i)2^n = 2^n + 2^n[n(n+1)/2 - 1] = 2^n[n(n+1)/2]$$

*79. F(n) = nF(n - 1)
= n[(n - 1)F(n - 2)] = n(n - 1)F(n - 2)
= n(n - 1)[(n - 2)F(n - 3)] = n(n - 1)(n - 2)F(n - 3)
In general, F(n) = n(n - 1)(n - 2)...(n - (k - 1))F(n - k)
When n - k = 1, k = n - 1,
F(n) = n(n - 1)(n - 2)...(2)F(1) = n(n - 1)(n - 2)...(2)(1) = n!

Now prove by induction that F(n) = n!
F(1): F(1) = 1! = 1 true
Assume F(k): F(k) = k!
Show F(k + 1): F(k + 1) = (k + 1)!
F(k + 1) = (k + 1)F(k) = (k + 1)k! = (k + 1)!

80. S(n) = nS(n - 1) + n!
= n[(n - 1)S(n - 2) + (n - 1)!] + n!
= n(n - 1)S(n - 2) + 2n!
= n(n - 1)[(n - 2)S(n - 3) + (n - 2)!] + 2n!
= n(n - 1)(n - 2)S(n - 3) + 3n!
In general, S(n) = n(n - 1)(n - 2)...(n - (k - 1))S(n - k) + (k)n!
When n - k = 1, k = n - 1,
S(n) = n(n - 1)(n - 2)...(2)S(1) + (n - 1)n!
= n(n - 1)(n - 2)...(2)(1) + (n - 1)n!
= n! + (n - 1)n! = n(n!)

Now prove by induction that S(n) = n(n!).
S(1): S(1) = 1(1!) = 1 true
Assume S(k) = k(k!)
Show S(k + 1) = (k + 1)(k + 1)!
S(k + 1) = (k + 1)S(k) + (k + 1)! = (k + 1)(k)(k!) + (k + 1)!
= (k)(k + 1)! + (k + 1)! = (k + 1)(k + 1)!

81. The recurrence relation is A(n) = 1.5A(n - 1) with a base case A(1) = 1200. This is a linear, first-order recurrence relation with constant coefficients, so Equation (8) applies and gives the solution $A(n) = (1.5)^{n-1}(1200)$. $A(12) = (1.5)^{11}(1200) = 103,797$.

*82. The recurrence relation is $T(n) = 0.95T(n - 1)$ with a base case $T(1) = X$. This is a linear, first-order recurrence relation with constant coefficients, so Equation (8) applies and gives the solution $T(n) = (0.95)^{n-1}(X)$. At the end of 20 years (the beginning of the 21st year), the amount is $T(21) = (0.95)^{20}(X) = 0.358(X)$, which is slightly more than one-third the original amount X.

83. The recurrence relation is $A(n) = (1.08)A(n - 1) + 100$ with a base case of $A(1) = 1000$. This is a linear, first-order recurrence relation with constant coefficients, so Equation (8) applies and gives the solution $A(n) = (1.08)^{n-1}(1000) + 100[1.08^{n-2} + 1.08^{n-3} + \ldots + 1.08 + 1] = (1.08)^{n-1}(1000) + 100\,[1 - (1.08)^{n-1}]/[1 - (1.08)]$ from the formula for the sum of a geometric sequence. At the end of 7 years (the beginning of the 8th year), the account is worth $(1.08)^7(1000) + 100[1 - (1.08)^7]/(1-1.08)$ which equals \$2606.10.

84. The recurrence relation is $A(n) = (1.01)A(n - 1) - 80$ with a base case of $A(1) = 5000$. This is a linear, first-order recurrence relation with constant coefficients, so Equation (8) applies and gives the solution $A(n) = (1.01)^{n-1}(5000) - 80[1.01^{n-2} + 1.01^{n-3} + \ldots + 1.01 + 1] = (1.01)^{n-1}(5000) - 80[1 - (1.01)^{n-1}]/[1 - 1.01]$ from the formula for the sum of a geometric sequence. At the end of 18 months (the beginning of the 19th month), the loan balance remaining is $(1.01)^{18}(5000) - 80[1 - (1.01)^{18}]/(1 - 1.01)$ which equals \$4411.56.

85. The recurrence relation is $S(n) = 0.98S(n - 1) - 10{,}000$ with a base case of $S(1) = 1{,}000{,}000$. This is a linear, first-order recurrence relation with constant coefficients, so Equation (8) applies and gives the solution $S(n) = (0.98)^{n-1}(1{,}000{,}000) - 10{,}000[(0.98)^{n-2} + (0.98)^{n-3} + \ldots + (0.98) + 1] = (0.98)^{n-1}(1{,}000{,}000) - 10{,}000[1-(0.98)^{n-1}]/(1 - 0.98)$ from the formula for the sum of a geometric sequence. At the end of 9 years (the beginning of the 10th year), the population is $(0.98)^9(1{,}000{,}000) - 10000[1-(0.98)^9]/(1-0.98)$ which equals 750622.

86. The recurrence relation for the total number of infected machimes each day is $T(n) = T(n - 1) + 5T(n - 1) - 6^{n-2} = 6T(n - 1) - 6^{n-2}$ with a base case of $T(1) = 3$. This is a linear, first-order recurrence relation with constant coefficients, so Equation (8) applies; $c = 6$ and $g(n) = -6^{n-2}$. By Equation (8), the solution is

$$\begin{aligned} T(n) &= 6^{n-1}(3) + \sum_{i=2}^{n} 6^{n-i}\left(-6^{i-2}\right) \\ &= 6^{n-1}(3) - \sum_{i=2}^{n} 6^{n-2} \\ &= 6^{n-1}(3) - (n-1)6^{n-2} \\ &= 6^{n-2}[6 * 3 - (n-1)] \end{aligned}$$

Now setting $6*3 - (n - 1)$ equal to 0 and solving for n, the result is $n = 19$. The virus disappears after 19 days.

87. a. M(1) = 1 (one move for one disk)
M(n) = 2M(n – 1) + 1 (restack the top n – 1 disks on the third peg, requiring M(n – 1) disk moves, move the bottom disk to the target peg, restack the n – 1 disks on the target peg, which also requires M(n – 1) disk moves)

b. This is a linear first order recurrence relation with constant coefficients, so Equation (8) applies and gives the solution

$$M(n) = 2^{n-1}M(1) + \sum_{i=2}^{n} 2^{n-i}(1)$$

$$= 2^{n-1} + 2^{n-2} + 2^{n-3} + \cdots + 2^0$$
$$= \frac{1 - 1 \cdot 2^n}{1-2}$$
$$= 2^n - 1$$

c. $(2^{64} - 1)/(60*60*24*365)$ is over 500 billion years!

*88. The recurrence relation is P(n) = P(n - 1) + n, with P(1) = 1. Equation (8) applies and gives the solution $P(n) = 1 + \sum_{i=2}^{n} i = 1 + 2 + 3 + \ldots + n = n(n+1)/2$.

89. The recurrence relation is P(n) = P(n - 1) + 3n - 2 with P(1) = 1. Equation (8) applies and gives the solution

$$P(n) = 1 + \sum_{i=2}^{n}(3i-2) = 1 + 4 + 7 + 10 + \ldots + 3n - 2 = (n/2)(3n-1).$$

90. Base step: When n = 1, Equation (8) becomes $c^0S(1) = S(1)$

Assume that $S(k) = c^{k-1}S(1) + \sum_{i=2}^{k} c^{k-1}g(i)$

Then $S(k+1) = cS(k) + g(k+1)$

$$= c[c^{k-1}S(1) + \sum_{i=2}^{k} c^{k-i}g(i)] + g(k+1)$$

$$= c^kS(1) + \sum_{i=2}^{k} c^{(k+1)-i}g(i) + g(k+1)$$

$$= c^kS(1) + \sum_{i=2}^{k+1} c^{(k+1)-i}g(i)$$

EXERCISES 2.5

1. The outer loop is executed n times; for each execution of the outer loop, the inner loop is executed n times, so a total of n^2 additions is done.

2. On the first pass through the outer loop, the inner loop is executed n times. On the second pass through the outer loop, the inner loop is executed n – 1 times, etc. The number of additions is therefore

$$n + (n - 1) + (n - 2) + \ldots + 1 = \frac{n(n+1)}{2}$$

3. The algorithm of Example 47 is rewritten here. Assignments occur at the lines marked A and comparisons are done at the line marked C.

```
for i = 1 to n do
    low = roster[i].quiz[1]              //A1
    sum = roster[i].quiz[1]              //A2

    for j = 2 to m do
        sum = sum + roster[i].quiz[j]    //A3
        if roster[i].quiz[j] < low then  //C1
            low = roster[i].quiz[j]      //A4
        end if
    end for
    sum = sum - low;                     //A5
    write("Total for student ", i, " is ", sum)
end for
```

To count the number of comparisons, note that comparisons are done within the inner loop, which executes m - 1 times for each student, that is, for each of the n passes of the outer loop. The total number of comparisons is therefore always n(m - 1). Assignments A1, A2, and A5 are clearly done once for each pass through the outer loop, for a total of 3n. Assignment A3, like the comparison, is done m - 1 times for each pass through the outer loop for a total of n(m - 1).

Assignment A4, however, is done a variable number of times. The best case, requiring the least amount of work, occurs when the first quiz grade is the lowest quiz grade for each student. Then the condition of the **if** statement is never true and A4 executes 0 times. The worst case occurs when all of the quiz grades go downhill from beginning to end for each student. Then each new quiz grade is lower than the previous one, so A4 executes each time, or n(m - 1) times.

Total assignments and comparisons
- Best case: 3n + 2n(m - 1)
- Worst case: 3n + 3n(m - 1)

*4. If the list is empty, write "not found", else compare x to the first item in the current list segment; if that item equals x, write "found", otherwise remove that first item and search the remainder of the list segment. Initially, the list segment is the entire list. More formally:

```
SequentialSearch(list L; integer i, n; itemtype x)
//searches list L from L[i] to L[n] for item x

   if i > n then
   write('not found')
   else
   if L[i] = x then
      write('found')
   else
      SequentialSearch(L, i+1, n, x)
   end if
   end if
```

5. $C(1) = 1$ (1 comparison to search a 1-element list)
$C(n) = 1 + C(n - 1)$ for $n \geq 2$ (1 comparison against the first element, then however many comparisons are required for the rest of the list)

This is a first-order, linear recurrence relation with constant coefficients. By Equation (8) in Section 2.4, the solution is

$$C(n) = (1)^{n-1}(1) + \sum_{i=2}^{n}(1)^{n-i}(1) = 1 + (n-1) = n$$

6. This is in the form of Equation (1) with c = 1 and g(n) = n. By Equation (6), the solution is

$$3 + \sum_{i=1}^{\log n}(1)2^i = 3 + (2^1 + 2^2 + \ldots + 2^{\log n}) = 2 + 2^0 + 2^1 + 2^2 + \ldots + 2^{\log n} = 2 + 2^{(\log n)+1} - 1 = 2 + 2(n) - 1 = 2n + 1$$

*7. This is in the form of Equation (1) with c = 2 and g(n) = 3. By Equation (6), the solution is

$$2^{\log n}(1) + \sum_{i=1}^{\log n} 2^{(\log n)-i}(3) = n + 3[2^{(\log n)-1} + 2^{(\log n)-2} + \ldots + 2^0] = n + 3[2^{\log n} - 1] = n + 3(n-1) = 4n - 3$$

8. This is in the form of Equation (1) with c = 2 and g(n) = n. By Equation (6), the solution is

$$2^{\log n}(1) + \sum_{i=1}^{\log n} 2^{(\log n)-i}(2^i) = n + \sum_{i=1}^{\log n} 2^{\log n} = n + (\log n)2^{\log n} = n + (\log n)n = (1 + \log n)n$$

9. This is in the form of Equation (1) with c = 2 and g(n) = n^2. By Equation (6), the solution is

$$2^{\log n}(1)+\sum_{i=1}^{\log n}2^{(\log n)-i}(2^i)^2 = n+\sum_{i=1}^{\log n}2^{\log n}\cdot 2^{-i}\cdot 2^{2i} = n+\sum_{i=1}^{\log n}n\cdot 2^i = n+n\sum_{i=1}^{\log n}2^i$$

$$= n+n[2^1+2^2+\cdots+2^{\log n}] = n[1+2+2^2+\cdots+2^{\log n}] = n[2^{\log n+1}-1] = n[n\cdot 2-1]$$

*10. n - 1 compares are always needed - every element after the first must be considered a potential new maximum

11. $C(1) = 0$ (no comparisons are required; a 1-element list is already sorted)

$C(n) = (n - 1) + C(n - 1)$ for $n \geq 2$ (n - 1 compares to find the maximum element plus however many compares are required to sort the list minus the last element)

12. This is a first-order, linear recurrence relation with constant coefficients. By Equation (8) of Section 2.4, the solution is

$$C(n) = 1^{n-1}(0)+\sum_{i=2}^{n}(1)^{n-i}(i-1) = 0+\sum_{i=2}^{n}(i-1) = 1+2+\ldots+(n-1) = \frac{(n-1)n}{2}$$

*13. a. the merged list is 1, 4, 5, 6, 8, 9
3 comparisons - 6 vs. 1, 6 vs. 4, and 6 vs. 5

b. the merged list is 1, 2, 3, 4, 5, 8
4 comparisons - 1 vs. 2, 5 vs. 2, 5 vs. 3, 5 vs. 4

c. the merged list is 0, 1, 2, 3, 4, 7, 8, 9, 10
8 comparisons - 0 vs. 1, 2 vs. 1, 2 vs. 8, 3 vs. 8, 4 vs. 8, 7 vs. 8, 10 vs. 8, 10 vs. 9

14. The maximum number of comparisons takes place when each list runs out of elements at virtually the same time, as in (c) of Exercise 13. In this case, every element that is added to the merged list, except the last one, requires a comparison, so (r + s) - 1 comparisons are done.

*15. $C(1) = 0$ (no comparisons are required; a 1-element list is already sorted)

$C(n) = 2C(n/2) + (n - 1)$ for $n = 2^m$, $n \geq 2$ (C(n/2) comparisons are required for each half, and n - 1 comparisons are required to merge the two sorted halves.)

16. This is in the form of Equation (1), with c = 2 and g(n) = n - 1. By Equation (6), the solution is

$$C(n) = 2^{\log n}(0) + \sum_{i=1}^{\log n} 2^{(\log n)-i}(2^i - 1) = \sum_{i=1}^{\log n}(2^{\log n} - 2^{(\log n)-i}) = \sum_{i=1}^{\log n} 2^{\log n} - \sum_{i=1}^{\log n} 2^{(\log n)-i}$$

$$= (\log n)n - [2^{(\log n)-1} + 2^{(\log n)-2} + \ldots + 2^0] = (\log n)n - [2^{\log n} - 1] = (\log n)n - [n-1] = n(\log n) - n + 1$$

17. For selection sort, $C(n) = \frac{n(n-1)}{2}$. For mergesort, C(n) = n(log n) - n + 1

	selection sort	mergesort
n = 4	6	5
n = 8	28	17
n = 16	120	49
n = 32	496	129

18.

Position at which x occurs	Number of comparisons
1	1
2	2
3	3
...	...
n	n

Total number of comparisons = 1 + 2 + ··· + n = n(n + 1)/2. Each of the n cases is equally likely so divide by n to get an average of (n + 1)/2.

19. Total number of comparisons = (1 + 2 + ··· + n) [x in the list] + n [x not in the list] = n(n + 1)/2 + n = $(n^2 + 3n)/2$. Each of the n + 1 cases is equally likely so divide by n + 1to get an average of $(n^2 + 3n)/2(n + 1)$.

20. For m = 1

F(m + 2) = F(3) = 2
F(m + 1) = F(2) = 1

We need to show that if 1 division is required to find gcd(a, b), then a ≥ 2 and b ≥ 1. If b = 0, then by the Euclidean algorithm, gcd(a, 0) = a with 0 divisions required, which does not satisfy m = 1. Hence b ≥ 1 and, because a > b, a ≥ 2.

Assume that if k divisions are required, a ≥ F(k + 2), b ≥ F(k + 1). Show that if k + 1 divisions are required, then a ≥ F(k + 3), b ≥ F(k + 2). The first step of the algorithm in computing gcd(a, b) is to divide a by b, so

a = qb + r, 0 ≤ r < b

This is 1 division. The algorithm finishes the computation by finding gcd(b, r), which will therefore require k divisions. By the inductive hypothesis, $b \geq F(k + 2)$ and $r \geq F(k + 1)$. Then $a = qb + r \geq b + r$ ($q \geq 1$ since $a > b$) $\geq F(k + 2) + F(k + 1) = F(k + 3)$.

*21. From Exercise 20, $a = n \geq F(m + 2)$. By Exercise 23 of Section 2.4, $F(m+2) > \left(\frac{3}{2}\right)^{m+1}$. Therefore $\left(\frac{3}{2}\right)^{m+1} < F(m+2) \leq n$.

22. From Exercise 21,

$$\left(\frac{3}{2}\right)^{m+1} < n$$

so, taking the logarithm to the 3/2 of both sides,

$$m + 1 < \log_{1.5} n \quad \text{or} \quad m < \log_{1.5} n - 1$$

23. a. The divisions required are 89/55, 55/34, 34/21, 21/13, 13/8, 8/5, 5/3, 3/2, 2/1. The algorithm requires 9 divisions to find that gcd(89, 55) = 1. (Notice that these values are all Fibonacci numbers.)
b. $2 \log 89 \approx 13$
c. $(\log_{1.5} 89) - 1 \approx 10$

CHAPTER 3: Sets, Combinatorics, and Probability

Most students have seen sets and set operations before, and the material in Section 3.1 presents little difficulty, with the exception of the short section on countable and uncountable sets. Cantor's diagonalization method usually requires several passes before the light dawns. This material is not central, and can be omitted if desired, although it's nice for computer science students to know that there are different kinds of infinity.

Counting problems (Sections 3.2-3.4) are something that students usually think are fun. Occasional frustration occurs because there are sometimes very plausible-sounding wrong approaches. I've made a special effort to address this difficulty in the subsection "Eliminating Duplicates" in Section 3.4. Another stumbling block is the ability to distinguish whether the problem is a *combinations* problem or a *permutations* problem (hence determining what formula to use). I reiterate numerous times that if there is ordering involved, the problem is a permutations problem, not a combinations problem. Decision trees are used in Chapter 5 to establish worst-case lower bounds for searching and sorting. The use of decision trees in Section 3.2 for counting outcomes provides an easy introduction.

Section 3.5 introduces probability as a natural extension of counting. It covers the basics of finite probability, probability distributions, conditional probability, and expected value using, I am afraid, the usual coins and dice. An example of average case analysis of algorithms ties this material back to computer science.

The binomial theorem, Section 3.6, offers a nice occasion to see both an inductive proof and a combinatorial proof, but nonetheless I generally omit this section. The students all know how to square a binomial, and that's really about all they need.

EXERCISES 3.1

*1. a. T b. F c. F d. F

2. a. T b. F c. T d. F ($\sqrt{2} \notin Q$)

3. Four:
$\{2, 3, 4\} = \{x | x \in N \text{ and } 2 \leq x \leq 4\} = \{3, 4, 2\}$
{a, b, c} = {x|x is the first letter of cat, bat, or apple}
∅ = {x|x is the first letter of cat, bat, and apple}
{2, a, 3, b, 4, c}

*4. a. {0, 1, 2, 3, 4}
b. {4, 6, 8, 10}
c. {Washington, Adams, Jefferson}
d. ∅
e. {Maine, Vermont, New Hampshire, Massachusetts, Connecticut, Rhode Island}
f. {-3, -2, -1, 0 , 1, 2, 3}

5. a. {2, 3}
   b. $\{\sqrt{7}, -\sqrt{7}\}$
   c. {4}

6. a. $\{x | x \in N \text{ and } 1 \le x \le 5\}$
   b. $\{x | x \in N$ and x is odd}
   c. {x|x is one of the Three Wise Men}
   d. {x|x is a nonnegative integer written in binary form}

7. a. {4, 6}
   b. {1, 2, 3}
   c. $\{x | x \in N$ and x is an odd number}

*8. If $A = \{x | x = 2^n$ for n a positive integer}, then $16 \in A$. But if $A = \{x | x = 2 + n(n - 1)$ for n a positive integer}, then $16 \notin A$. In other words, there is not enough information to answer the question.

9. a. 2 b. 2 c. 1 d. 3 e. 3

10. a. T b. T c. F d. T e. T f. F g. F h. T

11. *a. F; $\{1\} \in S$ but $\{1\} \notin R$
   *b. T
   *c. F; $\{1\} \in S$, not $1 \in S$
   *d. F; 1 is not a set; the correct statement is $\{1\} \subseteq U$
   *e. T
   *f. F; $1 \notin S$
   g. T
   h. T
   i. T
   j. F; $3 \notin U$ and $\pi \notin U$
   k. T
   l. T
   m. T
   n. F

12. a. T
   b. T
   c. F; neither member of C is a member of A
   d. T
   e. T
   f. F; this 2-element set is not an element of A.
   g. T
   h. F; $a \notin C$
   i. T

13. Let $x \in A$. Then $x \in R$ and $x^2 - 4x + 3 < 0$ or $(x - 1)(x - 3) < 0$. The possible solutions are:

    $x - 1 < 0$ and $x - 3 > 0$
    $x < 1$ and $x > 3$, which cannot be satisfied
    or
    $x - 1 > 0$ and $x - 3 < 0$
    $x > 1$ and $x < 3$
    $1 < x < 3$

But if $x \in R$ and $1 < x < 3$, then $x \in R$ and $0 < x < 6$, so $x \in B$. Therefore $A \subseteq B$. The number $5 \in B$ but $5 \notin A$, so $A \subset B$.

*14. Let (x, y) $\in$ A. Then (x, y) lies within 3 units of the point (1, 4), so by the distance formula, $\sqrt{(x-1)^2+(y-4)^2} \le 3$, or $(x-1)^2+(y-4)^2 \le 9$, which means $(x-1)^2+(y-4)^2 \le 25$, so (x, y) $\in$ B. The point (6, 4) satisfies the inequality $(x-1)^2+(y-4)^2 \le 25$, so (6, 4) $\in$ B, but (6, 4) is not within 3 units of (1, 4), so (6, 4) does not belong to A.

15. a. For a = 1, b = -2, c = -24, the quadratic equation is $x^2 - 2x - 24 = 0$ or $(x + 4)(x - 6) = 0$, with solutions 6 and -4. Each of these is an even integer between -100 and 100, so each belongs to E.

b. Here Q = {6, -4}, but E = {-4, -2, 0, 2, 4}, and Q $\not\subseteq$ E.

16. Let x $\in$ A. Then cos (x/2) = 0. But cos (x/2) = 0 if and only if x/2 = $\pm\pi/2$, $\pm 3\pi/2$, $\pm 5\pi/2$, ..., or x = $\pm\pi$, $\pm 3\pi$, $\pm 5\pi$, ... and, for any multiple of π, the sine function is 0. Thus x $\in$ B.

17. *a. T *b. F *c. F *d. T *e. T *f. F g. F h. T i. F j. F

18. Let x $\in$ A. Then, because A $\subseteq$ B, x $\in$ B. Because B $\subseteq$ C, x $\in$ C. Thus A $\subseteq$ C.

19. Assume that A' $\subseteq$ B' but that B $\not\subseteq$ A. Then there is an element x such that x $\in$ B but x $\notin$ A. Thus x $\in$ A' and x $\notin$ B', which contradicts A' $\subseteq$ B'.

20. a. The proof uses mathematical induction.
n = 2: A set with 2 elements has exactly 1 subset with 2 elements, namely the set itself. Putting n = 2 into the formula n(n – 1)/2 gives the value 1. This proves the base case.
Assume that any set with k elements has k(k – 1)/2 subsets with exactly 2 elements. Show that any set with k + 1 elements has (k + 1)k/2 subsets with exactly 2 elements.

Let x be a member of a set with k + 1 elements. Temporarily removing x from the set gives a set of k elements that, by the inductive hypothesis, has k(k – 1)/2 subsets with exactly 2 elements. These are all of the 2-element subsets of the original set that do not include x. All 2-element subsets of the original set that do include x can be found by pairing x in turn with each of the remaining k elements, giving k subsets. The total number of 2-element subsets is therefore

$$\frac{k(k-1)}{2}+k=\frac{k(k-1)+2k}{2}=\frac{k^2-k+2k}{2}=\frac{k^2+k}{2}=\frac{(k+1)k}{2}$$

b. Again, the proof is by induction.
$n = 3$: A set with 3 elements has exactly 1 subset with 3 elements, namely the set itself. Putting $n = 3$ into the formula $n(n - 1)(n - 2)/6$ gives the value 1. This proves the base case.
Assume that any set with k elements has $k(k - 1)(k - 2)/6$ subsets with exactly 3 elements.
Show that any set with k + 1 elements has $(k + 1)k(k - 1)/6$ subsets with exactly 3 elements.

Let x be a member of a set with k + 1 elements. Temporarily removing x from the set gives a set of k element that, by the inductive hypothesis, has $k(k - 1)(k - 2)/6$ subsets with exactly 3 elements. These are all of the 3-element subsets of the original set that do not include x. All 3-element subsets of the original set that do include x can be found by pairing x in turn with each of the 2-element subsets of the k-element set. From part (a), there are $k(k - 1)/2$ such subsets. The total number of 3-element subsets is therefore

$$\frac{k(k-1)(k-2)}{6}+\frac{k(k-1)}{2}=\frac{k(k-1)(k-2)+3k(k-1)}{6}=\frac{k(k-1)(k-2+3)}{6}=\frac{(k+1)k(k-1)}{6}$$

21. $\wp(S) = \{\emptyset, \{a\}\}$

*22. $\wp(S) = \{\emptyset, \{1\}, \{2\}, \{3\}, \{4\}, \{1, 2\}, \{1, 3\}, \{1, 4\}, \{2, 3\}, \{2, 4\}, \{3, 4\}, \{1, 2, 3\}, \{1, 2, 4\}, \{1, 3, 4\}, \{2, 3, 4\}, \{1, 2, 3, 4\} \}$; $2^4 = 16$ elements

23. $\wp(S) = \{\emptyset,\{\emptyset\}\}$

24. $\wp(S) = \{\emptyset, \{\emptyset\}, \{\{\emptyset\}\}, \{\{\emptyset, \{\emptyset\}\}\}, \{\emptyset, \{\emptyset\}\}, \{\emptyset, \{\emptyset, \{\emptyset\}\}\}, \{\{\emptyset\}, \{\emptyset, \{\emptyset\}\}\}, \{\emptyset, \{\emptyset\}, \{\emptyset, \{\emptyset\}\}\} \}$

25. $\wp(\wp(S)) = \{\emptyset, \{\emptyset\}, \{\{a\}\}, \{\{b\}\}, \{\{a, b\}\}, \{\emptyset, \{a\}\}, \{\emptyset, \{b\}\}, \{\emptyset, \{a, b\}\}, \{\{a\}, \{b\}\}, \{\{a\}, \{a, b\}\}, \{\{b\}, \{a, b\}\}, \{\emptyset, \{a\}, \{b\}\}, \{\emptyset, \{a\}, \{a, b\}\}, \{\emptyset, \{b\}, \{a, b\}\}, \{\{a\}, \{b\}, \{a, b\}\}, \{\emptyset, \{a\}, \{b\}, \{a, b\}\} \}$

*26. $A = \{x, y\}$

27. $\wp(A)$ contains 3 elements, and $3 \neq 2^n$ for any n, so such a set A does not exist.

28. Let $X \in \wp(A)$. Then X is a subset of A, hence, since $A \subseteq B$, X is a subset of B and $X \in \wp(B)$.

*29. Let $x \in A$. Then $\{x\} \in \wp(A)$, so $\{x\} \in \wp(B)$ and $x \in B$. Thus $A \subseteq B$. A similar argument shows that $B \subseteq A$ so that $A = B$.

30. a. $x = 1, y = 5$
  b. $x = 8, y = 7$
  c. $x = 1, y = 4$

31. a. If $x = u$ and $y = v$, then clearly $\{\{x\},\{x, y\}\} = \{\{u\},\{u, v\}\}$.
Now assume $\{\{x\},\{x, y\}\} = \{\{u\},\{u, v\}\}$. Then $\{x\} = \{u\}$ or $\{x\} = \{u, v\}$. If $\{x\} = \{u, v\}$, then $u = v = x$; also $\{u\} = \{x, y\}$, and $x = y = u$. Thus $x = u$ and $y = v$.
If $\{x\} = \{u\}$, then $x = u$; also $\{x, y\} = \{u, v\}$ and, since $x = u$, $y = v$.

  b. For example, the ordered triples (1, 1, 2) and (1, 2, 1) expressed in set form would be
$\{\{1\},\{1, 1\},\{1, 1, 2\}\} = \{\{1\},\{1, 2\}\}$ and
$\{\{1\},\{1, 2\},\{1, 2, 1\}\} = \{\{1\},\{1, 2\}\}$, respectively,
and distinct ordered triples would have the same representation.

32. *a. binary operation
  *b. no; $0 \circ 0 \notin N$
  *c. binary operation
  d. no; ln x undefined for $x \leq 0$
  e. unary operation
  f. no; closure fails
  g. no; uniqueness fails (two different fractions could have the same denominator)
  h. binary operation

33. a. no; operation undefined for $x = 0$
  b. binary operation
  c. unary operation
  d. binary operation
  e. no; $x \circ y$ is undefined - there is no "greatest" common multiple because a larger common multiple can always be found
  f. no; closure fails because the sum of two random Fibonacci numbers is not always a Fibonacci number. For example, $1 + 3 = 4$ and 4 is not a Fibonacci number although 1 and 3 are.
  g. unary operation
  h. no; closure fails. The sum of two irrational numbers can be rational, for example, $\sqrt{2} + (-\sqrt{2}) = 0$

34. n^{n^2}; each of the n^2 entries in the n x n matrix has n possible answers. We multiply the number of possibilities. Thus there are $\underbrace{n \cdot n \cdot \ldots \cdot n}_{n^2} = n^{n^2}$ ways to complete the table.

	x_1	x_n
x_1		
x_n		

35. a. $(A + B)*(C - D) = ((A + B)*(C - D)) \rightarrow$ AB+CD-*
  b. $A**B - C*D = ((A**B) - (C*D)) \rightarrow$ AB**CD*-
  c. $A*C + B/(C + D*B) = ((A*C) + (B/(C + (D*B)))) \rightarrow$ AC*BCDB*+/+

*36. a. 13 b. 2 c. 28

37. a. {t} b. {p, q, r, s, t, u} c. {q, r, v, w) d. ∅ e. {r, v} f. {u, w}
g. {(p, r), (p, t), (p, v), (q, r), (q, t), (q, v), (r, r), (r, t), (r, v), (s, r), (s, t), (s, v)}
h. {q, r, v}

38.*a. {1, 2, 4, 5, 6, 8, 9} *b. {4, 5}
*c. {2, 4} d. {1, 2, 3, 4, 5, 9}
e. {2, 6, 8} f. {0, 1, 3, 7, 9}
g. ∅ *h. {0, 1, 2, 3, 6, 7, 8, 9}
i. {2, 3} j. {0, 1, 3, 4, 7, 9}
k. {2, 6, 8} 1. {2, 3}
m. {(1, 2), (1, 3), (1, 4), (4, 2), (4, 3), (4, 4), (5, 2), (5, 3), (5, 4), (9, 2), (9, 3), (9, 4)}

39. a. {a} b. {∅, {a}, {a, {a}} c. {∅, a, {a}, {{a}}, {a, {a}} } = S
d. ∅ e. {a, {a}} f. {∅, {a, {a}} } g. {∅}

40. a. {Adams, Jefferson, Grant} b. {Washington} c. ∅

41. a. F b. T c. T d. T

*42. a. B' b. $B \cap C$
c. $A \cap B$ d. $B' \cap C$
e. $B' \cap C'$ or $(B \cup C)'$ or $B' - C$

43. a. A' b. $A \cap B$ c. $A - B$

44. *a. C' *b. $B \cap D$ *c. $A \cap B$
d. $A \cap D'$ e. $B' \cap D'$ f. $C \cap A'$
g. $C \cup D$

*45. $D \cap R'$

46. $O \cap (G \cup R) \cap V'$

47. $(N \cup P) \cap A$

48.*a. T
b. T
*c. F (Let A = {1, 2, 3}, B = {1, 3, 5}, S = (1, 2, 3, 4, 5). Then
$(A \cap B)'$ = {2, 4, 5} but $A' \cap B'$ = {4, 5} ∩ {2, 4} = {4}.)
d. T
*e. F (Take A, B, and S as in (c), then A - B = {2}, (B - A)' = {1, 2, 3, 4}.)
f. T
g. F
h. F (Order matters in ordered pairs, so if A = {1, 2}, B = {3, 4}, then
B × A = {(3, 1), (3, 2), (4, 1), (4, 2)} and A × B = {(1, 3), (1, 4), (2, 3), (2, 4)}.)
i. T
j. T

k. F (Let $A = \{1, 2\}$, $B = \{2, 3, 4\}$, $C = \{1, 3\}$. Then $(A - B) \cup (B - C) = \{1\} \cup \{2, 4\} = \{1, 2, 4\}$ but $A - C = \{2\}$.)

l. T

49.*a. $B \subseteq A$ b. $A \subseteq B$

c. $A = \varnothing$ d. $B \subseteq A$

e. $A = B$ f. $A = B$

50. a. 12 b. 9 c. 16 d. 3 e. 4

51. Let $x \in A \cap B$. Then $x \in A$ and $x \in B$, so $x \in A$.

52. Let $x \in A$. Then $x \in A$ or $x \in B$ so $x \in A \cup B$.

*53. Let $C \in \wp(A) \cap \wp(B)$. Then $C \in \wp(A)$ and $C \in \wp(B)$, from which $C \subseteq A$ and $C \subseteq B$, so $C \subseteq A \cap B$ or $C \in \wp(A \cap B)$. Therefore $\wp(A) \cap \wp(B) \subset \wp(A \cap B)$. The same argument works in reverse.

54. Let $C \in \wp(A) \cup \wp(B)$. Then $C \in \wp(A)$ or $C \in \wp(B)$, which means $C \subseteq A$ or $C \subseteq B$. In either case, $C \subseteq A \cup B$, or $C \in \wp(A \cup B)$.

55. Suppose $B \neq \varnothing$. Let $x \in B$. Then $x \in A \cup B$ but $x \notin A - B$, which contradicts the equality of $A \cup B$ and $A - B$.

*56. Suppose $A \cap B \neq \varnothing$. Let $x \in A \cap B$. Then $x \in A$ and $x \in B$ so $x \in A \cup B$ but $x \notin A - B$ and $x \notin B - A$, so $x \notin (A - B) \cup (B - A)$, which contradicts the equality of $A \cup B$ and $(A - B) \cup (B - A)$.

57. i) Let $C \subseteq A$. Then $A \cup C = A$. Also,

$(A \cap B) \cup C = C \cup (A \cap B) = (C \cup A) \cap (C \cup B) = (A \cup C) \cap (B \cup C)$
$= A \cap (B \cup C)$

ii) Assume that $(A \cap B) \cup C = A \cap (B \cup C)$ and let $x \in C$. Suppose $x \notin A$. Then $x \in (A \cap B) \cup C$ but $x \notin A \cap (B \cup C)$, which is a contradiction. Therefore $x \in A$ and $C \subseteq A$.

58. a.

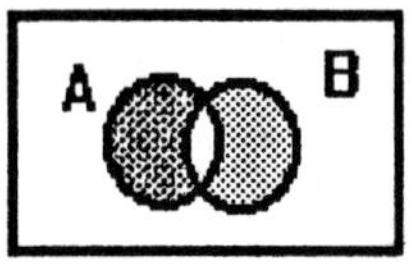

b. $\{2, 4, 6, 7, 9\}$

c. $x \in (A \cup B) - (A \cap B) \leftrightarrow x \in (A \cup B)$ and $x \in (A \cap B)'$

$\leftrightarrow$ ($x \in A$ or $x \in B$) and $x \notin A \cap B$

$\leftrightarrow$ ($x \in A$ and $x \notin A \cap B$) or ($x \in B$ and $x \notin A \cap B$)

$\leftrightarrow$ ($x \in A$ and $x \notin B$) or ($x \in B$ and $x \notin A$)

$\leftrightarrow x \in (A - B) \cup (B - A)$

d. $\varnothing$; A

e. $A \oplus B = (A \cup B) - (A \cap B) = (B \cup A) - (B \cap A) = B \oplus A$

f. First note that $(A \oplus B)' = (A \cap B) \cup (A' \cap B')$. Then

$$\begin{aligned}(A \oplus B) \oplus C &= [(A \oplus B) - C] \cup [C - (A \oplus B)]\\ &= [(A \oplus B) \cap C'] \cup [C \cap (A \oplus B)']\\ &= [((A - B) \cup (B - A)) \cap C'] \cup [C \cap ((A \cap B) \cup (A' \cap B'))]\\ &= (A \cap B' \cap C') \cup (B \cap A' \cap C') \cup (C \cap A \cap B) \cup (C \cap A' \cap B')\\ &= (A \cap B \cap C) \cup (A \cap B' \cap C') \cup (B \cap C' \cap A') \cup (C \cap B' \cap A')\\ &= A \cap [(B \cap C) \cup (B' \cap C')] \cup [(B \cap C') \cup (C \cap B')] \cap A'\\ &= [A \cap (B \oplus C)'] \cup [(B \oplus C) \cap A']\\ &= (A - (B \oplus C)) \cup ((B \oplus C) - A)\\ &= A \oplus (B \oplus C)\end{aligned}$$

59. a. F b. T c. T d. F e. F

60. (1a) $x \in A \cup B \leftrightarrow x \in A$ or $x \in B \leftrightarrow x \in B$ or $x \in A \leftrightarrow x \in B \cup A$

(1b) $x \in A \cap B \leftrightarrow x \in A$ and $x \in B \leftrightarrow x \in B$ and $x \in A \leftrightarrow x \in B \cap A$

(2a) $x \in (A \cup B) \cup C \leftrightarrow x \in (A \cup B)$ or $x \in C \leftrightarrow (x \in A$ or $x \in B)$ or $x \in C$
$\leftrightarrow x \in A$ or $x \in B$ or $x \in C$ $\leftrightarrow x \in A$ or $(x \in B$ or $x \in C)$ $\leftrightarrow x \in A$ or $x \in (B \cup C)$
$\leftrightarrow x \in A \cup (B \cup C)$

(2b) $x \in (A \cap B \cap C) \leftrightarrow x \in (A \cap B)$ and $x \in C \leftrightarrow (x \in A$ and $x \in B)$ and $x \in C$
$\leftrightarrow x \in A$ and $x \in B$ and $x \in C \leftrightarrow x \in A$ and $(x \in B$ and $x \in C)$
$\leftrightarrow x \in A$ and $x \in (B \cap C) \leftrightarrow x \in A \cap (B \cap C)$

(3b) $x \in A \cap (B \cup C) \leftrightarrow x \in A$ and $x \in (B \cup C) \leftrightarrow x \in A$ and $(x \in B$ or $x \in C)$
$\leftrightarrow (x \in A$ and $x \in B)$ or $(x \in A$ and $x \in C) \leftrightarrow x \in (A \cap B)$ or $x \in (A \cap C)$
$\leftrightarrow x \in (A \cap B) \cup (A \cap C)$

(4b) $x \in A \cap S \rightarrow x \in A$ and $x \in S \rightarrow x \in A$
$x \in A \rightarrow x \in A$ and $x \in S$ because $A \subseteq S \rightarrow x \in A \cap S$

(5a) $x \in A \cup A' \rightarrow x \in A$ or $x \in A' \rightarrow x \in S$ or $x \in S$ because $A \subseteq S$, $A' \subseteq S \rightarrow x \in S$
$x \in S \rightarrow (x \in S$ and $x \in A)$ or $(x \in S$ and $x \notin A) \rightarrow x \in A$ or $x \in A' \rightarrow x \in A \cup A'$

(5b) For any x such that $x \in A \cap A'$, it follows that $x \in A$ and $x \in A'$, or x belongs to A and x does not belong to A. This is a contradiction, so no x belongs to $A \cap A'$, and $A \cap A' = \varnothing$

61. a. $x \in (A \cup B)' \leftrightarrow x \notin (A \cup B) \leftrightarrow$ x does not belong to either A or B
$\leftrightarrow x \notin A$ and $x \notin B \leftrightarrow x \in A'$ and $x \in B' \leftrightarrow x \in A' \cap B'$

b. $x \in (A \cap B)' \leftrightarrow x \notin A \cap B \leftrightarrow$ x does not belong to both A and B
$\leftrightarrow x \notin A$ or $x \notin B \leftrightarrow x \in A'$ or $x \in B' \leftrightarrow x \in A' \cup B'$

*c. $x \in A \cup (B \cap A) \leftrightarrow x \in A$ or $x \in (B \cap A) \leftrightarrow x \in A$ or $(x \in B$ and $x \in A) \leftrightarrow x \in A$

d. $x \in (A \cap B')' \cup B \leftrightarrow x \in (A \cap B')'$ or $x \in B \leftrightarrow x \in (A' \cup B)$ or $x \in B$ by part (b) and $(B')' = B \leftrightarrow x \in A'$ or $x \in B$ or $x \in B \leftrightarrow x \in A'$ or $x \in B \leftrightarrow x \in A' \cup B$

e. $x \in (A \cap B) \cup (A \cap B') \leftrightarrow x \in A \cap B$ or $x \in A \cap B'$
$\leftrightarrow$ ($x \in A$ and $x \in B$) or ($x \in A$ and $x \in B'$) $\leftrightarrow x \in A$

f. $x \in (A \cap (B \cup C))' \leftrightarrow x \notin (A \cap (B \cup C)) \leftrightarrow x \notin A$ or $x \notin (B \cup C)$
$\leftrightarrow x \in A'$ or ($x \notin B$ and $x \notin C$) $\leftrightarrow x \in A'$ or ($x \in B'$ and $x \in C'$)
$\leftrightarrow x \in (A' \cup (B' \cap C')$

62.*a. $((A \cup B) \cap (A \cup B') = A \cup (B \cap B')$ (3a)
$= A \cup \varnothing$ (5b)
$= A$ (4a)
The dual is $(A \cap B) \cup (A \cap B') = A$

b. $[((A \cap C) \cap B) \cup ((A \cap C) \cap B')] \cup (A \cap C)'$
$= [(A \cap C) \cap (B \cup B')] \cup (A \cap C)'$ (3b)
$= [(A \cap C) \cap S] \cup (A \cap C)'$ (5a)
$= (A \cap C) \cup (A \cap C)'$ (4b)
$= S$ (5a)
The dual is $[((A \cup C) \cup B) \cap ((A \cup C) \cup B')] \cap (A \cup C)' = \varnothing$

c. $(A \cup C) \cap [(A \cap B) \cup (C' \cap B)]$
$= (A \cup C) \cap [(B \cap A) \cup (B \cap C')]$ (1b)
$= (A \cup C) \cap [B \cap (A \cup C')]$ (3b)
$= (A \cup C) \cap [(A \cup C') \cap B]$ (1b)
$= [(A \cup C) \cap (A \cup C')] \cap B$ (2b)
$= [A \cup (C \cap C')] \cap B$ (3a)
$= (A \cup \varnothing) \cap B$ (5b)
$= A \cap B$ (4a)
The dual is $(A \cap C) \cup [(A \cup B) \cap (C' \cup B)] = A \cup B$

63. a. $A \cup A = \{x \mid x \in A \text{ or } x \in A\} = \{x \mid x \in A\} = A$
b. $A \cap A = \{x \mid x \in A \text{ and } x \in A\} = \{x \mid x \in A\} = A$
c. $A \cap \varnothing = \{x \mid x \in A \text{ and } x \in \varnothing\}$ but since no x is a member of $\varnothing$, this is the empty set.

d. $A \cup S \subseteq S$ because $A \subseteq S$; $S \subseteq A \cup S$ by Exercise 52.
*e. $x \in (A')' \leftrightarrow x \notin A' \leftrightarrow x \notin \{y \mid y \notin A\} \leftrightarrow x \in A$

64. *a. $A \cap (B \cup A') = (A \cap B) \cup (A \cap A')$ (3b)
$= (A \cap B) \cup \varnothing$ (5b)
$= A \cap B$ (4a)
$= B \cap A$ (1b)

b. $(A \cup B) - C = (A \cup B) \cap C'$ (defn. set diff.)
$= C' \cap (A \cup B)$ (1b)
$= (C' \cap A) \cup (C' \cap B)$ (3b)
$= (A \cap C') \cup (B \cap C')$ (1b)
$= (A - C) \cup (B - C)$ (defn. set diff.)

c. $(A - B) - C = (A - B) \cap C'$ (defn. set diff.)
$= (A \cap B') \cap C'$ (defn. set diff.)
$= C' \cap (A \cap B')$ (1b)
$= (C' \cap A) \cap B'$ (2b)
$= (A \cap C') \cap B'$ (1b)
$= (A - C) \cap B'$ (defn. set diff.)
$= (A - C) - B$ (defn. set diff.)

d. $((A' \cup B') \cap A')' = ((A \cap B)' \cap A')'$ (DeMorgan's Laws)
$= ((A \cap B)')' \cup (A')'$ (DeMorgan's Laws)
$= (A \cap B) \cup A$ (part (a))
$= A \cup (A \cap B)$ (1a)
$= A \cup (B \cap A)$ (1b)
$= A$ (Exer. 56c)

e. $(A - B) - C = (A - B) \cap C'$ (defn. set diff.)
$= (A \cap B') \cap C'$ (defn. set diff.)
$= A \cap (B' \cap C')$ (2b)
$= A \cap (C' \cap B')$ (1b)
$= (A \cap C') \cap B'$ (2b)
$= ((A \cap C') \cap B') \cup \varnothing$ (4a)
$= ((A \cap C') \cap B') \cup (A \cap \varnothing)$ (Exercise 63c)
$= ((A \cap C') \cap B') \cup (A \cap (C \cap C'))$ (5b)
$= ((A \cap C') \cap B') \cup (A \cap (C' \cap C))$ (1b)
$= ((A \cap C') \cap B') \cup ((A \cap C') \cap C)$ (2b)
$= (A \cap C') \cap (B' \cup C)$ (3b)
$= (A \cap C') \cap (B' \cup (C')')$ (Exercise 63e)
$= (A \cap C') \cap (B \cap C')'$ (De Morgan's Laws)
$= (A \cap C') - (B \cap C')$ (defn. set diff.)
$= (A - C) - (B - C)$ (defn. set diff.)

f. $A - (A - B) = A \cap (A - B)'$ (defn. set diff.)
$= A \cap (A \cap B')'$ (defn. set diff.)
$= A \cap (A' \cup (B')')$ (De Morgan's Laws))
$= A \cap (A' \cup B)$ (Exercise 63e)
$= (A \cap A') \cup (A \cap B)$ (3b)
$= \varnothing \cup (A \cap B)$ (5b)
$= (A \cap B) \cup \varnothing$ (1a)
$= A \cap B$ (4a)

65. a. $A_1 \cup A_2 \cup \ldots \cup A_n = \{x | x \text{ belongs to some } A_i \text{ for } 1 \le i \le n\}$

b. $A_1 \cup A_2 = \{x | x \in A_1 \text{ or } x \in A_2\}$ $\quad n = 2$

$A_1 \cup A_2 \cup \ldots \cup A_n = (A_1 \cup \ldots \cup A_{n-1}) \cup A_n$ $\quad n > 2 \quad (1)$

*66. The proof is by induction on n. For n = 3,

$(A_1) \cup (A_2 \cup A_3) = (A_1 \cup A_2) \cup A_3$ by set identity 2a

$= A_1 \cup A_2 \cup A_3$ by Equation (1) of the answer to Exercise 65(b)

Assume that for n = k and $1 \le p \le k - 1$,

$(A_1 \cup \ldots \cup A_p) \cup (A_{p+1} \cup \ldots \cup A_k) = A_1 \cup \ldots \cup A_k$

Then for $1 \le p \le k$,

$(A_1 \cup \ldots \cup A_p) \cup (A_{p+1} \cup \ldots \cup A_{k+1})$

$= (A_1 \cup \ldots \cup A_p) \cup [(A_{p+1} \cup \ldots \cup A_k) \cup A_{k+1}]$ by Equation (1)

$= [(A_1 \cup \ldots \cup A_p) \cup (A_{p+1} \cup \ldots \cup A_k)] \cup A_{k+1}$ by set identity 2a

$= (A_1 \cup \ldots \cup A_k) \cup A_{k+1}$ by inductive hypothesis

$= A_1 \cup \ldots \cup A_{k+1}$ by Equation (1)

67. a. $A_1 \cap A_2 \cap \ldots \cap A_n = \{x | x \text{ belongs to every } A_i \text{ for } 1 \le i \le n\}$

b. $A_1 \cap A_2 = \{x | x \in A_1 \text{ and } x \in A_2\}$ $\quad n = 2$

$A_1 \cap A_2 \cap \ldots \cap A_n = (A_1 \cap \ldots \cap A_{n-1}) \cap A_n$ $\quad n > 2 \quad (2)$

68. The proof is by induction on n. For n = 3,

$(A_1) \cap (A_2 \cap A_3) = (A_1 \cap A_2) \cap A_3$ by set identity 2b

$= A_1 \cap A_2 \cap A_3$ by Equation (2) of the answer to Exercise 67(b)

Assume that for n = k and $1 \le p \le k - 1$,

$(A_1 \cap \ldots \cap A_p) \cap (A_{p+1} \cap \ldots \cap A_k) = A_1 \cap \ldots \cap A_k$

Then for $1 \le p \le k$,

$(A_1 \cap \ldots \cap A_p) \cap (A_{p+1} \cap \ldots \cap A_{k+1})$

$= (A_1 \cap \ldots \cap A_p) \cap [(A_{p+1} \cap \ldots \cap A_k) \cap A_{k+1}]$ by Equation (2)

$= [(A_1 \cap \ldots \cap A_p) \cap (A_{p+1} \cap \ldots \cap A_k)] \cap A_{k+1}$ by set identity 2b

$= (A_1 \cap \ldots \cap A_k) \cap A_{k+1}$ by inductive hypothesis

$= A_1 \cap \ldots \cap A_{k+1}$ by Equation (2)

69. *a. Proof is by induction on n.

For n = 2, $B \cup (A_1 \cap A_2) = (B \cup A_1) \cap (B \cup A_2)$ by identity 3a.

Assume that $B \cup (A_1 \cap \ldots \cap A_k) = (B \cup A_1) \cap \ldots \cap (B \cup A_k)$

Then $B \cup (A_1 \cap ... \cap A_{k+1})$

$= B \cup ((A_1 \cap ... \cap A_k) \cap A_{k+1})$ by Exercise 67b

$= (B \cup (A_1 \cap ... \cap A_k)) \cap (B \cup A_{k+1})$ by identity 3a

$= ((B \cup A_1) \cap ... \cap (B \cup A_k)) \cap (B \cup A_{k+1})$ by inductive hyp.

$= (B \cup A_1) \cap ... \cap (B \cup A_{k+1})$ by Exercise 67b

b. Proof is by induction on n.
For $n = 2$, $B \cap (A_1 \cup A_2) = (B \cap A_1) \cup (B \cap A_2)$ by identity 3b.

Assume that $B \cap (A_1 \cup ... \cup A_k) = (B \cap A_1) \cup ... \cup (B \cap A_k)$

Then $B \cap (A_1 \cup ... \cup A_{k+1})$

$= B \cap ((A_1 \cup ... \cup A_k) \cup A_{k+1})$ by Exercise 65b

$= (B \cap (A_1 \cup ... \cup A_k)) \cup (B \cap A_{k+1})$ by identity 3b

$= ((B \cap A_1) \cup ... \cup (B \cap A_k)) \cup (B \cap A_{k+1})$ by inductive hyp.

$= (B \cap A_1) \cup ... \cup (B \cap A_{k+1})$ by Exercise 65b

70. a. Proof is by induction on n
For $n = 2$, $(A_1 \cup A_2)' = A_1' \cap A_2'$ by Exercise 61a

Assume that $(A_1 \cup ... \cup A_k)' = A_1' \cap ... \cap A_k'$

Then $(A_1 \cup ... \cup A_k \cup A_{k+1})' = ((A_1 \cup ... \cup A_k) \cup A_{k+1})'$ by Exercise 65b

$= (A_1 \cup ... \cup A_k)' \cap A_{k+1}'$ by Exercise 61a

$= (A_1' \cap ... \cap A_k') \cap A_{k+1}'$ by inductive hyp.

$= A_1' \cap ... \cap A_k' \cap A_{k+1}'$ by Exercise 67b

b. Proof is by induction on n
For $n = 2$, $(A_1 \cap A_2)' = A_1' \cup A_2'$ by Exercise 61b

Assume that $(A_1 \cap ... \cap A_k)' = A_1' \cup ... \cup A_k'$

Then $(A_1 \cap ... \cap A_k \cap A_{k+1})' = ((A_1 \cap ... \cap A_k) \cap A_{k+1})'$ by Exercise 67b

$= (A_1 \cap ... \cap A_k)' \cup A_{k+1}'$ by Exercise 61b

$= (A_1' \cup ... \cup A_k') \cup A_{k+1}'$ by inductive hyp.

$= A_1' \cup ... \cup A_k' \cup A_{k+1}'$ by Exercise 65b

71. a. $\bigcup_{i \in I} A_i = \{x | x \in (-1,1)\}$; $\bigcap_{i \in I} A_i = \{0\}$

b. $\bigcup_{i \in I} A_i = \{x | x \in [-1,1]\}$; $\bigcap_{i \in I} A_i = \{0\}$

72. a. AIDS $\cup$ ALZHEIMERS = {genetics 0.7, virus 0.8, nutrition 0.3, bacteria 0.4, environment 0.4}

b) AIDS $\cap$ ALZHEIMERS = {genetics 0.2, virus 0.4, nutrition 0.1, bacteria 0.3, environment 0.3}

c) AIDS' = {genetics 0.8, virus 0.2, nutrition 0.9, bacteria 0.6, environment 0.7}

73. P(1) is true - every member of T is greater than 1, otherwise 1 would be the smallest member of T. Assume that P(k) is true, i.e., every member of T is greater than k. Consider P(k + 1), that every member of T is greater than k + 1. If P(k + 1) is not true, then there is some member of T $\leq$ k + 1. By the inductive hypothesis, every member of T is greater than k, therefore some member of T equals k + 1, and this is the smallest member of T. This is a contradiction, because we assumed T has no smallest member. Therefore P(k + 1) is true. By the first principle of induction, P(n) is true for all n, and T must be empty. This contradicts the fact that T is a non-empty set.

*74. If T is a non-empty set, then by the principle of well-ordering, T has a smallest member t_0. Then $P(t_0)$ is not true, so by statement 1', $t_0 \neq 1$. Also P(r) is true for all r, $1 \leq r \leq t_0 - 1$. This contradicts the implication in 2', so T is the empty set and therefore P(n) is true for all positive integers n.

*75. An enumeration of the set is 1, 3, 5, 7, 9, 11, ...

76. An enumeration of Z is 0, 1, -1, 2, -2, 3, -3, 4, -4, 5, -5, ...

*77. An enumeration of the set is a, aa, aaa, aaaa, ...

78. An enumeration of the set is shown by the arrow through the array

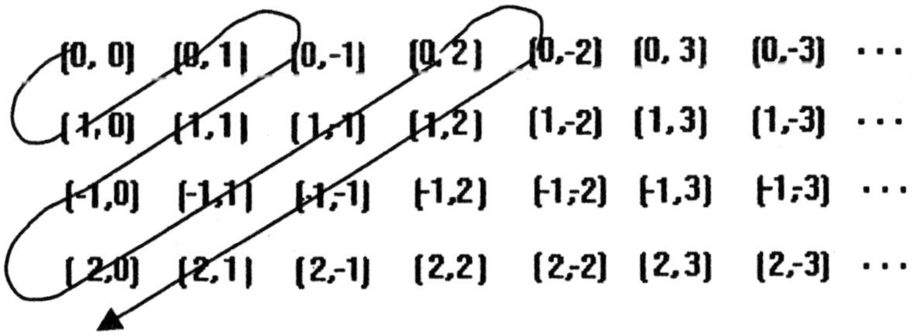

79. a. n = 0.249999....

100n = 24.9999....

99n = 100n − n = 24.9999.... − 0.2499999 = 24.75

n = 24.75/99 = 0.25

b. m = 0.250000...

100m = 25.0000...

99m = 100m − m = 25.0000... - 0.250000.... = 24.75

m = 24.75/99 = 0.25

c. n = m

80. Assume that the set has an enumeration

$z_{11}, z_{12}, z_{13}, z_{14}, \ldots$
$z_{21}, z_{22}, z_{23}, z_{24}, \ldots$
$z_{31}, z_{32}, z_{33}, z_{34}, \ldots$
$\vdots$

Now construct an infinite sequence Z of positive integers with $Z = z_1, z_2, z_3, \ldots$ such that $z_i \neq z_{ii}$ for all i. Then Z differs from every sequence in the enumeration, yet is a member of the set. This is a contradiction, so the set is uncountable.

81. Assume that the set has an enumeration

$s_{11}s_{12}s_{13}s_{14} \ldots$
$s_{21}s_{22}s_{23}s_{24} \ldots$
$s_{31}s_{32}s_{33}s_{34} \ldots$
$\vdots$

where each s_{ii} is either a or b. Now construct an infinite string $s = s_1s_2s_3s_4 \ldots$ such that s_i = a if s_{ii} = b, and s_i = b if s_{ii} = a. Then s differs from every string in the enumeration, yet is a member of the set. This is a contradiction, so the set is uncountable.

82. Let A be a countable set. Then A is finite or countably infinite. If A is finite and $B \subseteq A$, then B is finite, hence countable. If A is countably infinite, let a_1, a_2, a_3, be an enumeration of A. Using this same list but eliminating elements in A - B gives an enumeration of B.

*83. Let A and B be denunmerable sets with enumerations
$A = a_1, a_2, a_3, \ldots$ and $B = b_1, b_2, b_3, \ldots$
Then use the list $a_1, b_1, a_2, b_2, a_3, b_3, \ldots$ and eliminate any duplicates. This will be an enumeration of $A \cup B$, which is therefore denumerable.

84. $B = \{S | S \text{ is a set and } S \notin S\}$. (Perhaps you think that no set S can be an element of itself, in which case B is empty. But we can still talk about set B.) Then either $B \in B$ or $B \notin B$. If $B \in B$, then B has the property of all members of B, namely $B \notin B$. Hence both $B \in B$ and $B \notin B$ are true. If $B \notin B$, then B has the property characterizing members of B, hence $B \in B$. Therefore both $B \notin B$ and $B \in B$ are true.

EXERCISES 3.2

*1. $5 \cdot 3 \cdot 2 = 30$

*2. $4 \cdot 2 \cdot 2 = 16$

3. $4 \cdot 8 \cdot 6 = 92$

4. $4^{20} \cdot 5^{10}$

5. $26^3 \cdot 10^2$

6. $52^3 \cdot 10^2$

*7. $45 \cdot 13 = 585$

8. $3 \cdot 2\,(A - B - D) + 2 \cdot 4\,(A - C - D) = 14$

*9. 10^9

10. No - the number of different codes is $10 \cdot 10 = 100$, so not every apartment has its own code.

*11. $26 \cdot 26 \cdot 26 \cdot 1 \cdot 1 = 17{,}576$

12. $2 \cdot 4 \cdot 4 = 32$

13. $2 \cdot 2 \cdot 2 \cdot 2$ (fill in the 4 rows of the truth table with T or F)

*14. (3, 1)

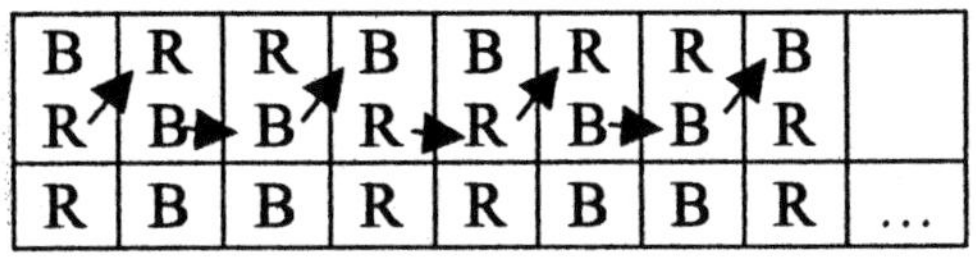

The cycles affect the (bottom) 2 elements of the stack, then the bottom 1 element of the stack. The stack number is 21.

(1, 3)

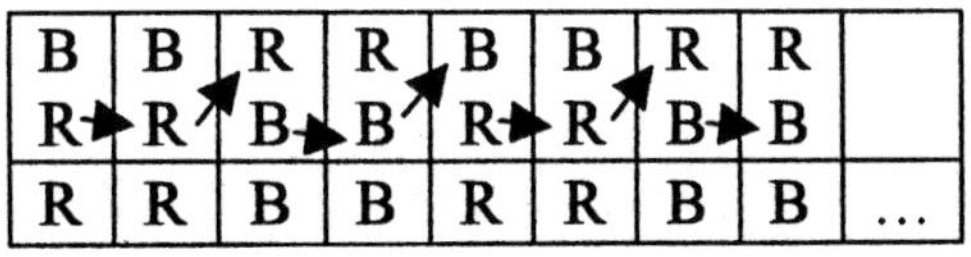

The cycles affect the bottom 1 element of the stack, then the (bottom) 2 elementS of the stack. The stack number is 12.

(2, 2)

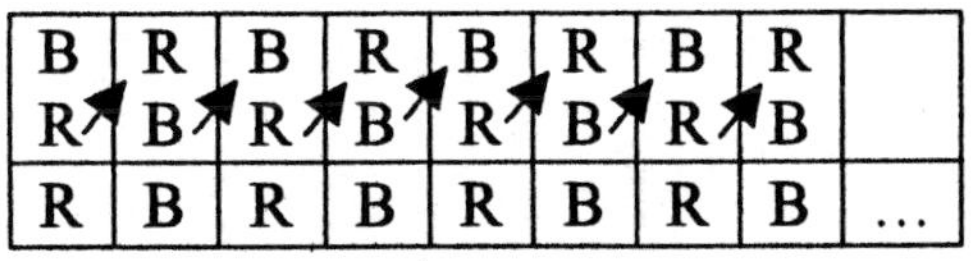

The cycles always affect the (bottom) 2 elements of the stack. The stack number is 22.

(1, 1)

R	R	R	R	R	R	R	R	
R	R	R	R	R	R	R	R	...

The cycles always affect the (bottom) 1 element of the stack. The stack number is 11.

15. a. (1, 5)
    b.

R	R	G	G	B	B	R	R	G	G	B	B	
B	B	R	R	G	G	B	B	R	R	R	R	
G	G	B	B	R	R	G	G	B	B	G	G	
G	G	B	B	R	R	G	G	B	B	R	R	...

The cycles affect the bottom 1 element of the stack, then the (bottom) 3 elements of the stack. The stack number is 13.

c. Using the stack number 221, we can recreate the stacks

B	R	B	B	R	B	B	R	B	B	...
R	B	R	R	B	R	R	B	R	R	

and then the juggling pattern, which is

B	R	B	B	R	B	B	R	B	B	...
R	B	R	R	B	R	R	B	R	R	
R	B	R	R	B	R	R	B	R	R	...
2	3	1	2	3	1	2	3	1	2	...

or (2, 3, 1).

16. a. $3^2 = 9$
    b. 11, 12, 13, 21, 22, 23, 31, 32, 33
    c. From Exercise 14, the stack numbers 11, 12, 21, and 22 represent all the juggling patterns of length 2 using 2 balls. The remaining 5 stack numbers must involve 3 balls. Using them, the stacks and then the juggling patterns can be recreated:

stack number 13:

R	R	G	G	B	B	R	R	G	G	B	B	
B	B	R	R	G	G	B	B	R	R	R	R	
G	G	B	B	R	R	G	G	B	B	G	G	
G	G	B	B	R	R	G	G	B	B	R	R	...

pattern (1, 5) [see Exercise 15a and 15b)

stack number 23:

R	R	B	B	R	R	B	B	R	R	B	B	
B	G	R	G	B	G	R	G	B	G	R	G	
G	B	G	R	G	B	G	R	G	B	G	R	
G	B	G	R	G	B	G	R	G	B	G	R	...

pattern (2, 4)

stack number 31:

R	G	G	B	B	R	R	G	G	B	B	R	
B	R	R	G	G	B	B	R	R	G	G	B	
G	B	B	R	R	G	G	B	B	R	R	G	
G	B	B	R	R	G	G	B	B	R	R	G	...

pattern (5, 1)

stack number 32:

R	G	G	R	R	G	G	R	R	G	G	R	
B	R	B	G	B	R	B	G	B	R	B	G	
G	B	R	B	G	B	R	B	G	B	R	B	
G	B	R	B	G	B	R	B	G	B	R	B	...

pattern (4, 2)

stack number 33

R	G	B	R	G	B	R	G	B	R	G	B	
B	R	G	B	R	G	B	R	G	B	R	G	
G	B	R	G	B	R	G	B	R	G	B	R	
G	B	R	G	B	R	G	B	R	G	B	R	...

pattern (3, 3)

17. $4^3 = 64$

*18. $26 + 26 \cdot 10 = 286$

19. $4 \cdot 3 \cdot 2 = 24$

20. $17 \cdot 16 + 24 \cdot 23 = 824$

*21. $5 \cdot 3 \cdot 4 \cdot 3 = 180$

22. $5 \cdot 4 \cdot 3 + 3 \cdot 4 \cdot 3 = 96$

23. $10 \cdot 7 \cdot 3 \cdot 2 \cdot 2 \cdot 2 = 1680$

24. $10 \cdot 7 \cdot 1 \cdot 2 \cdot 2 \cdot 2 + 10 \cdot 7 \cdot 2 \cdot 2 \cdot 2 \cdot 1 = 1120$

*25. $9 \cdot 10 \cdot 26 \cdot 10 \cdot 10 + 9 \cdot 10 \cdot 26 \cdot 10 \cdot 10 \cdot 10 + 9 \cdot 10 \cdot 26 \cdot 10 \cdot 10 \cdot 10 \cdot 10 = 25{,}974{,}000$

26. $2 \cdot 2 \cdot 2 \cdot 2$ (hamburger alone) + $2 \cdot 2 \cdot 2$ (fish sandwich alone) + 5 (beverage alone) + $2 \cdot 2 \cdot 2 \cdot 2 \cdot 2 \cdot 2 \cdot 2$ (hamburger and fish) + $2 \cdot 2 \cdot 2 \cdot 2 \cdot 5$ (hamburger and beverage) + $2 \cdot 2 \cdot 2 \cdot 5$ (fish and beverage) + $2 \cdot 2 \cdot 2 \cdot 2 \cdot 2 \cdot 2 \cdot 2 \cdot 5$ (all three) = 917

*27. $5 \cdot 3 = 15$

28. $2 \cdot 11 \cdot 6 = 132$

29. $9 \cdot 10 \cdot 2 = 180$ (2 choices, 0 or 5, for the third digit)

*30. 900 - 180 = 720

31. $9 \cdot 5 \cdot 3$ (middle digit even, third digit 0, 4, or 8) + $9 \cdot 5 \cdot 2$ (middle digit odd, third digit 2 or 6) = 135 + 90 = 225

32. 225 (numbers divisible by 4) + $9 \cdot 5 \cdot 1$ (middle digit odd, third digit 0) + $9 \cdot 10 \cdot 1$ (third digit 5) = 225 + 45 + 90 = 360

33. $9 \cdot 5 \cdot 1$ (middle digit even, third digit 0) = 45

34. 900 - 360 = 540

*35. $2^8 = 256$

36. $2^6 = 64$

*37. $1 \cdot 2^7$ (begin with 0) + $1 \cdot 2^6 \cdot 1$ (begin with 1, end with 0) = $2^7 + 2^6 = 192$

38. $2 \cdot 1 \cdot 2^6 = 2^7 = 128$

39. $1 \cdot 1 \cdot 1 \cdot 2^5 = 32$

40. 8 (one for each digit at which the 0 occurs)

41. $1 \cdot 1 \cdot 2^6$ (begin with 10) + $1 \cdot 1 \cdot 1 \cdot 2^5$ (begin with 110) + $1 \cdot 1 \cdot 1 \cdot 2^5$ (begin with 010) + $1 \cdot 1 \cdot 1 \cdot 2^5$ (begin with 000) = $2^6 + 3 \cdot 2^5 = 160$

42. $2 \cdot 2 \cdot 2 \cdot 2 \cdot 1 \cdot 1 \cdot 1 \cdot 1 = 2^4 = 16$

*43. 8 (This is the same problem as Exercise 40.)

44. 256 - 8 (contain exactly one 0) - 1 (contain no 0s) = 247

45. $6 \cdot 6 = 36$

*46. $6 \cdot 1 = 6$

47. $1 \cdot 1 = 1$

48. 2 + 2 + 2 (two ways to get each of 6 & 1, 5 & 2, 4 & 3) + 2 (two ways to get 6 & 5) = 8

49. $5 \cdot 5 = 25$

50. $4 \cdot 5 \cdot 3 \cdot 3 \cdot 2 \cdot 2 = 720$

*51. $4 \cdot 1 \cdot 3 \cdot 3 \cdot 2 \cdot 2 = 144$

52. $1 \cdot 5 \cdot 3 \cdot 3 \cdot 1 \cdot 1 = 45$

53. $5 \cdot 3 \cdot 3 \cdot 2 \cdot 2 = 180$

54. $4 \cdot 3 \cdot 3 \cdot 3 \cdot 2 = 216$

*55. $52 \cdot 52 = 2704$

56. $4 \cdot 4 = 16$

57. $12 \cdot 12 = 144$

*58. $4 \cdot 4 = 16$ ways to get 2 of one kind; there are 13 distinct "kinds", so by the Addition Principle, the answer is $16 + 16 + \ldots + 16 = 13 \cdot 16 = 208$

59. $4 \cdot 48$ (flower king, bird nonking) + $4 \cdot 48$ (bird king, flower nonking) = 384

60. Face value of 5 can occur in 4 disjoint ways:

flower face value	bird face value
1	4
2	3
3	2
4	1

Each has $4 \cdot 4$ ways of occurring, so the total is $4 \cdot 16 = 64$

61. Face value of less than 5 can occur in the disjoint ways shown below:

flower face value	bird face value
1	1, 2, or 3
2	1 or 2
3	1

Each has $4 \cdot 4$ ways of occurring, so the total is $6 \cdot 16 = 96$

62. $40 \cdot 40 = 1600$

*63. $12 \cdot 52$ (flower face card, any bird card) + $40 \cdot 12$ (flower non-face card, bird face card) = 1104

or

$52 \cdot 52$ (total number of hands - Exercise 55) - $40 \cdot 40$ (hands with no face cards - Exercise 62) = 1104

64. 2704 - $48 \cdot 48$ (total number of hands - hands with no kings) = 400

65. 00111000 11001001 00100001 00001010

66. 0111000

*67. Using the formula

Total IP addresses = (# of A netids)(# of A hostids) + (# of B netids)(# of B hostids) + (# of C netids)(# of C hostids)

and eliminating the special cases gives $(2^7 - 1)(2^{24} - 2) + (2^{14})(2^{16} - 2) + (2^{21})(2^8 - 2)$ = (127)(16,777,214)+ (16,384)(65,534) + (2,097,152)(254) = 2,130,706,178 + 1,073,709,056 + 532,676,608 = 3,737,091,842

68. A(1) = 1,300,000

A(n) = 2A(n – 1) for $n \geq 2$

This is a linear first-order recurrence relation with constant coefficients; the solution is

$A(n) = 2^{n-1}A(1) = 2^{n-1}(1,300,000)$

$3,737,091,842 = 2^{n-1}(1,300,000)$

$2^{n-1} = 3,737,091,842/1,300,000$

$n \cong \log_2(3737/1.3) + 1 \cong 12.49$ years, so in about 2006.

69.

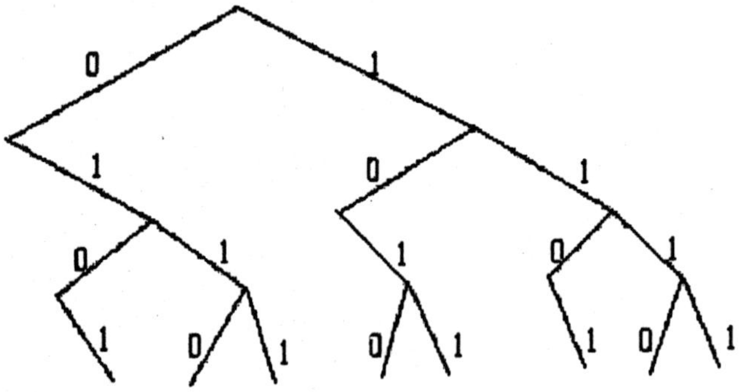

8 outcomes

*70.

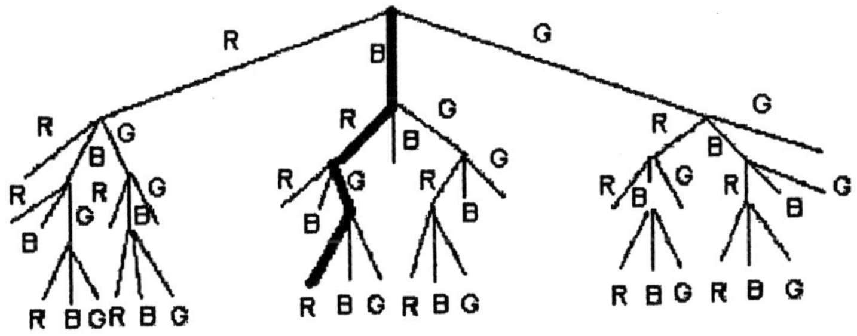

33 ways; the outcome that is highlighted is BRGR.

71. A win = left branch, B win = right branch. There are 70 ways; the outcome labeled here would be BBABAB

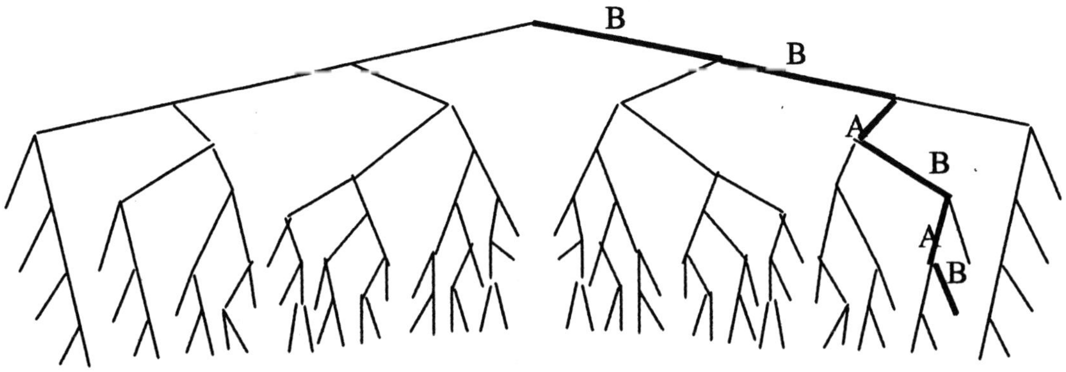

72. For m = 2, the result follows from the Multiplication Principle.
Assume that for m = k, there are $n_1 \cdots n_k$ possible outcomes for the sequence of events 1 to k.
Let m = k + 1. Then the sequence of events 1 to k + 1 consists of the sequence of events 1 to k followed by event k + 1. The sequence of events 1 to k has $n_1 \cdots n_k$ possible outcomes by the inductive hypothesis. The sequence 1 to k followed by event k + 1then has $(n_1 \cdots n_k)n_{k+1}$ outcomes by the Multiplication Principle, which equals $n_1 \cdots n_{k+1}$.

73. For m = 2, the result follows from the Addition Principle.
Assume that for m = k there are $n_1 + ... + n_k$ possible outcomes from choosing one of these k events.
Let m = k + 1. Then choosing one of the k + 1 disjoint events can be thought of as choosing one of the k disjoint events from 1 to k or choosing event k + 1. The k disjoint events have $n_1 + ... + n_k$ possible outcomes by the inductive hypothesis, so by the Addition Principle, there are $(n_1 + ... + n_k) + n_{k+1} = n_1 + ... + n_{k+1}$ possible outcomes.

74. a.

P(1) = 1 (trivial case)
P(2) = 1 (only one way to multiply 2 terms)

For n > 2, let the last multiplication occur at position k, $1 \leq k \leq n - 1$. The product is then split into two products of k and (n – k) factors, respectively, which can be parenthesized in P(k) and P(n – k) ways, respectively. By the Multiplication Principle, there are P(k)P(n – k) ways to parenthesize for a fixed k. Each value for k gives a different set of parentheses, so by the Addition Principle,

$$P(n) = P(1)P(n-1) + P(2)P(n-2) + \cdots + P(n-1)P(1)$$
$$= \sum_{k=1}^{n-1} P(k)P(n-k)$$

b. The proof will use the Second Principle of Induction.

P(1) = 1 = C(0)
P(2) = 1 = C(1)

Assume that P(r) = C(r – 1) for $1 \leq r \leq m$ and consider P(m + 1):

$$P(m+1) = \sum_{k=1}^{m} P(k)P(m+1-k) \text{ by the definition of P from part (a)}$$
$$= \sum_{k=1}^{m} C(k-1)C(m-k) \text{ inductive hypothesis}$$
$$= C(m) \text{ definition of C(m) from Exercise 33 of Section 2.4.}$$

75. *a. A pair of straight lines that shaves off two corners reduces the triangulation problem to a polygon with 2 less sides (see figures).

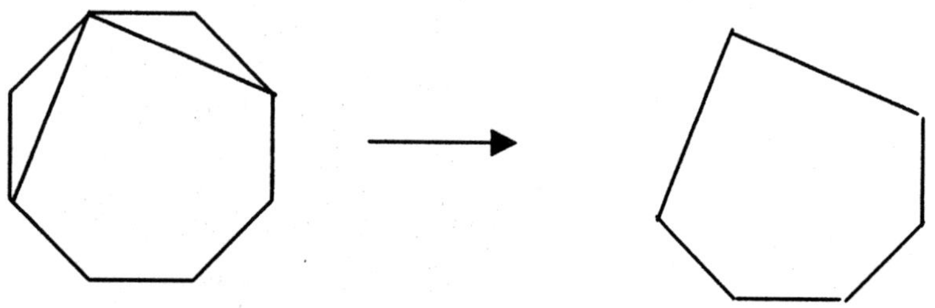

If n is odd, we want to do this process k times to reduce the polygon to a single triangle, so that

(n + 2) – 2k = 3
2k = n – 1

If n is even, we want to do this process k times to reduce the polygon to a 4-sided figure, so that

(n + 2) – 2k = 4
2k = n – 2

One final line triangulates the 4-sided figure, making the total number of lines needed to be

$$n - 2 + 1 = n - 1$$

b. $T(0) = 1$ (trivial case)
$T(1) = 1$ (number of ways to triangulate a 3-sided polygon)

For a fixed k, the (k + 1)-sided polygon can be triangulated in $T(k - 1)$ ways and the (n - k + 2)-sided polygon can be triangulated $T(n - k)$ ways, giving, by the Multiplication Principle, $T(k - 1)T(n - k)$ triangulations. Each value for k gives a different set of triangulations, so by the Addition Principle,

$$T(n) = T(0)T(n-1) + T(1)T(n-2) + \ldots T(n-1)T(0)$$
$$= \sum_{k=1}^{n} T(k-1)T(n-k)$$

c. The recurrence relations for T(n) and C(n) are identical.

EXERCISES 3.3

1. Let A = those who speak English
   B = those who speak French
   Then $|A \cup B| = 42$, $|A| = 35$, $|B| = 18$.
   $|A \cap B| = |A| + |B| - |A \cup B| = 35 + 18 - 42 = 11$

*2. Let A = guests who drink coffee
   B = guests who drink tea
   Then $|A| = 13$, $|B| = 10$, and $|A \cap B| = 4$.
   $|A \cup B| = |A| + |B| - |A \cap B| = 13 + 10 - 4 = 19$

3. Yes - 5

4. 18

*5. Let A = breath set, B = gingivitis set, C = plaque set
   a. $|A \cap B \cap C| = 2$
   b. $|A - C| = |A| - |A \cap C| = 12 - 6 = 6$

6. Let A = CS 120 set, B = CS 180 set, C = CS 215 set. Then $|A \cup B \cup C| = 32 + 27 + 35 - 7 - 16 - 3 + 2 = 70$. Therefore 83 - 70 = 13 students are not eligible to enroll.

7. Let A = checking account set, B = regular savings set, C = money market savings set.
   a. $|A \cap C| = 93$.
   b. $|A - (B \cup C)| = |A| - |A \cap (B \cup C)|$ by Example 29
      $= |A| - |(A \cap B) \cup (A \cap C)|$
      $= |A| - (|A \cap B| + |A \cap C|)$ by Example 28 because $(A \cap B)$ and $(A \cap C)$ are disjoint
      $= 189 - (69 + 93) = 27$

*8. Let A = auto set, B = bike set, C = motorcycle set
   a. $|B - (A \cup C)| = |B| - |B \cap (A \cup C)|$ by Example 32
      $= |B| - |(B \cap A) \cup (B \cap C)|$
      $= |B| - (|B \cap A| + |B \cap C| - |B \cap A \cap C|)$
      $= 97 - (53 + 7 - 2) = 39$

   b. $|A \cup B \cup C| = 97 + 83 + 28 - 53 - 14 - 7 + 2 = 136$, so $150 - 136 = 14$ do not own any of the three.

9. From the Principle of Inclusion and Exclusion,
   $87 = 68 + 34 + 30 - 19 - 11 - 23 + |A \cap B \cap C|$, so $|A \cap B \cap C| = 8$.

10. No. Letting A, B, and C be the odor, lather, and ingredients sets, the Principle of Inclusion and Exclusion says that the union of the three sets would contain 491 people, yet only 450 were surveyed.

11. $|A| + |B| + |C| + |D| - |A \cap B| - |A \cap C| - |A \cap D| - |B \cap C| - |B \cap D| - |C \cap D|$
    $+ |A \cap B \cap C| + |A \cap B \cap D| + |A \cap C \cap D| + |B \cap C \cap D| - |A \cap B \cap C \cap D|$

12. The terms in the expansion are equivalent to all of the subsets of $\{A_1, \ldots, A_n\}$ except for the empty set. Therefore there are $2^n - 1$ terms.

13. 5 (Use the Pigeonhole Principle, where suits are bins, cards are items)

*14. No - there are 13 different denominations (bins), so 12 cards could all be different.

15. 51; there are 50 potential pairs (bins)

16. 3; there are two genders (bins).

17. 367

18. Yes - there are 12 bins, so more than 24 items means that at least one bin has more than 2 items.

*19. There are 3 pairs - 1 and 6, 2 and 5, 3 and 4 - that add up to 7. Each element in the set belongs to one of these pairs. Apply the Pigeonhole Principle, where the pairs are the bins, and the numbers are the items.

20. 6

21. This follows from the Pigeonhole Principle, where the n possible remainders (the numbers 0 through n - 1) are the bins.

EXERCISES 3.4

1. *a. 42 b. 6720 c. 360 d. $\frac{n!}{[n-(n-1)]!} = \frac{n!}{1!} = n!$

2. $9! = 362,880$

3. $14! \sim 87,178,291,000$

4. $8! = 40,320$; $7!\cdot 3 = 15,120$ (there are only 3 choices for the last character)

*5. 5! (total permutations)
3! (arrangement of the 3 R's for each distinguished permutation)
$= 5\cdot 4 = 20$

6. Seat one person in one chair anywhere in the circle – the position doesn't matter. Choose seats for the remaining 5 people from the remaining 5 positions relative to the first person, which gives 5! = 120 arrangements.

*7. $P(15, 3) = \frac{15!}{12!} = 15\cdot 14\cdot 13 = 2730$

8. a. $(26)^3$
b. $P(26, 3) = 26\cdot 25\cdot 24 = 15600$

9. 19!

*10. $(2!)(11!)(8!) = 2(39,916,800)(40,320)$

(2!): men first or women first

11. $11!\cdot C(12, 8)\cdot 8!$

8!: arrange women

C(12, 8): select 8 of 12 slots between the men to fill with women

11!: arrange men

12. Seat one person in one chair anywhere in the circle – the position doesn't matter. Choose seats for the remaining 18 people from the remaining 18 positions relative to the first person, which gives 18!

13. Seat one man in one chair anywhere in the circle – the position doesn't matter. Choose seats for the remaining 10 men in positions relative to the first man, which gives 10!. Each woman must be seated to a man's right, giving 11 locations for women of which 8 must be chosen – C(11,8). Then arrange the 8 women in these 8 chosen locations, giving 8! arrangements of women. The answer is thus 10!·C(11,8)·8!

14. *a. 120 b. 36 c.28 d. $\frac{n!}{(n-1)!1!} = n$

15. $C(n, n-1) = \frac{n!}{(n-1)!1!} = n$. The number of ways to select n - 1 objects from n objects is the number of ways to exclude 1 object, C(n, 1).

*16. $C(300, 25) = \frac{300!}{25!275!}$

17. C(18, 11)

*18. C(17, 5)·C(23, 7) = (6188)(245,157)

19. C(21, 4)·C(11, 3)

*20. C(7, 1) · C(14, 1) · C(4, 1) · C(5, 1) · C(2, 1) · C(3, 1) = 7·14·4·5·2·3 = 11,760

21. C(14, 2)·C(21, 4) (2 from manufacturing, 4 from the others)

22. C(3, 1)·C(30, 5) (1 from marketing, 5 from non-accounting and non-marketing)

*23. all committees - (none or 1 from manufacturing) = C(35, 6) - [C(21, 6) + C(14, 1)·C(21, 5)]

24. 1·1·1·1·C(48, 1) = 48 (one way to choose the four queens, a pool of 48 left for the 5th card)

*25. C(13, 3)·C(13, 2)

26. C(13, 5)

27. 4·C(13, 2)·C(13, 1)·C(13, 1)·C(13, 1) (pick the suit that has 2 cards, select the 2 cards, then select the 1 from each remaining suit)

28. C(12, 5) (select all 5 cards from the 12 face cards)

29. 13·C(4,2)·C(12,3)·C(4,1)·C(4,1)·C(4,1) = 1,098,240 (select the kind for the pair, select the pair of that kind, select the three remaining kinds, select 1 card from each of those kinds)

*30. C(13,2)·C(4,2)·C(4,2)·C(44,1) = 123,552 (select the two kinds for the pairs, select the pair of one kind, select the pair of the other kind, select the fifth card from the unused kinds)

31. 13·C(4,3)·C(12,2)·C(4,1)·C(4,1) = 54,912 (select the kind, select the three of that kind, select two remaining kinds, select the fourth and fifth card from the unused kinds)

32. 10·4·4·4·4·4 = $10\cdot(4)^5$ = 10,240 (there are 10 possible starting points for the sequence, then a choice of 1 card from each of the four suits for each of the five cards)

*33. 4·C(13, 5) = 5,148 (select the suit, then the 5 cards of that suit)

34. 13·C(4,3)·12·C(4,2) =3,744 (select a kind, select three of that kind, select the second kind, select a pair of that kind)

35. 13·C(4,4)·C(48,1) = 624 (select the kind, select the 4 cards of that kind, select the remaining card)

36. 4·10·1·1·1·1·1 = 40 (select the suit, select the starting point, select the 5 cards)

37. 4·1·1·1·1·1 = 4 (select the suit, select the 5 cards)

*38. C(48, 14)

39. C(32, 14)

40. C(16, 8)·C(32, 6)

41. C(43, 14)

*42. C(32, 2)·C(16, 12) (Choose the two processors from B, then the remaining 12 modules are assigned to cluster A.)

*43. C(12, 4) = 495

44. C(5, 2)·C(7, 2) = (10)(21) = 210

45. C(5, 4) + C(7, 4) = 5 + 35 = 40

46. C(7, 3)·C(5, 1) + C(7, 4) = 210

47. C(60, 2)

*48. C(60, 1) + C(60, 2)

49. C(59, 7)

50. C(2, 1)·C(58, 6)

*51. C(12, 3) = 220

52. all committees - no independent = C(12, 3) - C(8, 3) = 164

*53. no Democrats + no Republicans - all independents (so as not to count twice)
= C(7, 3) + C(9, 3) - C(4, 3) = 115

54. all committees - (those without both) = 220 - 115 (from Exercises 47 and 49) = 105

*55. C(14, 6) = 3003

56. all groups - (those without both types) = 3003 - (no bores + no interesting)
= 3003 - [C(8, 6) + C(8, 6)] = 3003 - 56 = 2947

57. all - both = C(14, 6) - C(12, 4) = 2508
or
neither + exactly one = C(12, 6) + C(2, 1)·C(12, 5)

58. both + neither = C(12, 4) + C(12, 6) = 1419

*59. C(25, 5) - C(23, 5) (all committees - those with neither)
(Not 1·C(24, 4) + 1·C(24, 4) - this number is too big, it counts some combinations more than once)

60. C(12, 5) - C(10, 5) (all combinations - those without American history or English literature) or C(10,4) + C(10,4) +C(10,3) (history, no English + English, no history + both history and English)
(Not 1·C(11, 4) + 1·C(11, 4) - this number is too big, it counts some combinations more than once)

61. C(4, 4)·(36) = 36

C(4, 4): choose the 4 aces

(36): 39 non-clubs - 3 aces

(Not C(4, 4)·C(12, 1) - this number is too small; one of the aces is already the single club allowed, so the last card can be any non-club, non-ace)

62. $C(3, 3) \cdot C(12, 2) + 1 \cdot C(3, 2) \cdot [C(12, 1) \cdot 36]$

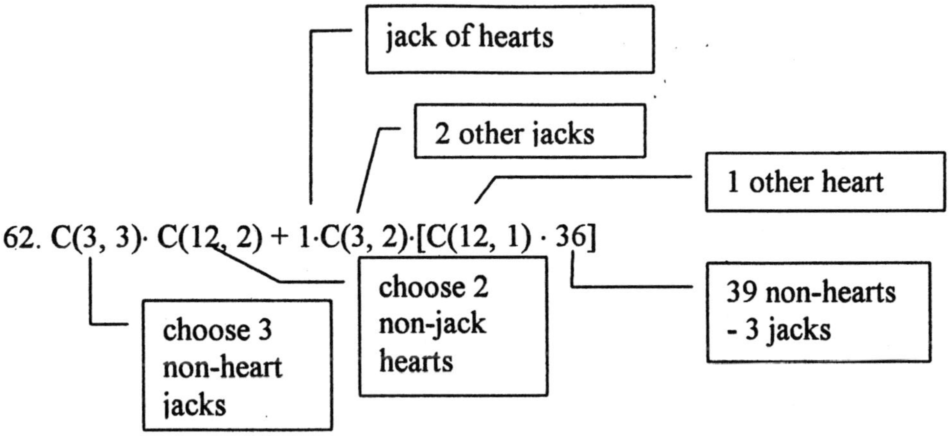

(Not $C(4,3) \cdot C(13,2)$ - this number is too small, one of the jacks could be a heart)

*63. a. $\frac{8!}{3!2!}$

b. $\frac{7!}{3!2!}$

64. a. $\frac{12!}{4!2!2!}$

b. $\frac{11!}{4!2!}$

65. $\frac{12!}{5!3!4!}$

66. 5 distinct characters would result in $5! = 5 \cdot 4 \cdot 3 \cdot 2$ code words. Since there are only 10 code words, this number has been divided by $4 \cdot 3$, which can only be written as $2! \cdot 3!$. Therefore there are two copies of one character and three copies of another.

*67. C(7, 5)

68. C(15, 12)

69. C(81, 48)

70. C(26, 12)

*71. a. C(8, 6) (choose 6 from 3 with repetitions)

b. C(7, 6) (choose 6 from 2 with repetitions)

c. C(5, 3) (3 of the 6 items are fixed, choose the remaining 3 from among 3, with repetitions)

72. a. C(10, 8)
   b. C(7, 5) (3 of the 8 objects are fixed, choose the remaining 5 from among 3, with repetitions)
   c. C(9, 8) (choose 8 from 2 with repetitions)
   d. C(8, 6) (2 of the 8 objects are fixed, choose the remaining 6 from among 3, with repetitions)
   e. C(9, 8) + C(8, 7) + C(7, 6) (zero chocolate chip cookies used - choose 8 from 2 with repetitions) + (one chocolate chip cookie used - choose remaining 7 from among 2 with repetitions) + (two chocolate chip cookies used - choose remaining 6 from among 2 with repetitions)

*73. a. C(16, 10)
   b. C(9, 3) (7 of the 10 assignments are fixed, choose the remaining 3 from among 7, with repetitions)

74. a. C(10, 8)
   b. C(8, 7) (1 assignment is fixed, choose the remaining 7 from among 2, with repetitions)

*75. C(13,10)

76. C(6, 4) (3 assignments are fixed, choose the remaining 4 from among 3, with repetitions)

77.

$$\frac{n!}{(n-1)!}+\frac{n!}{(n-2)!}=\frac{n!(n-2)!+n!(n-1)!}{(n-1)!(n-2)!}=\frac{n![(n-2)!+(n-1)!]}{(n-1)!(n-2)!}$$

$$=\frac{n[(n-2)!(1+(n-1))]}{(n-2)!}=n[1+(n-1)]=n\cdot n=n^2$$

78. $C(n,r)=\dfrac{n!}{r!(n-r)!}=\dfrac{n!}{(n-r)!(n-(n-r))!}=C(n,n-r)$

Whenever r objects are chosen from n, n - r objects are not chosen. Therefore the number of ways to choose r objects out of n is the same as the number of ways to choose n - r objects out of n.

79. C(r, 2) + C(n - r, 2) + r(n - r) =

$$\frac{r!}{(r-2)!2!}+\frac{(n-r)!}{(n-r-2)!2!}+r(n-r)$$

$$=\frac{r(r-1)}{2}+\frac{(n-r)(n-r-1)}{2}+r(n-r)$$

$$=\frac{r(r-1)+(n-r)(n-r-1)+2r(n-r)}{2}$$

$$=\frac{n^2-n}{2}=\frac{n!}{(n-2)!2!}$$

= C(n, 2)

*80. Consider selecting r elements from a set of n and putting those in bucket A, then selecting k of those r to put in bucket B. The left side multiplies the number of outcomes from those two sequential tasks.

Alternatively, we can select k elements from n and place them in bucket B, then r – k elements from the remaining n – k and place them in bucket A. The right side multiplies the number of outcomes from these two sequential tasks.

81. The union of the two sets is a set of size n + m, and there are C(n + m, r) ways to select r objects from it. Keeping the two sets separate, another way to choose r elements from the union is to choose k objects from one and r – k from the other, where $0 \le k \le r$. For a fixed k, using the Multiplication Principle gives C(n, k)C(m, r - k) outcomes. By the Addition Principle, all the ways to choose r objects is

$$\sum_{k=0}^{r} C(n,k)C(m,r-k)$$

Therefore both C(n + m, r) and $\sum_{k=0}^{r} C(n,k)C(m,r-k)$ represent the number of ways to choose r objects from the union of the two sets, and are thus equal in value.

82. $C(2) = \frac{1}{3}C(4,2) = \frac{1}{3}\frac{4!}{2!\cdot 2!} = \frac{4\cdot 3}{3\cdot 2} = 2$

$C(3) = \frac{1}{4}C(6,3) = \frac{1}{4}\frac{6!}{3!\cdot 3!} = \frac{6\cdot 5\cdot 4}{4\cdot 2\cdot 3} = 5$ $\quad C(4) = \frac{1}{5}C(8,4) = \frac{1}{5}\frac{8!}{4!\cdot 4!} = \frac{8\cdot 7\cdot 6\cdot 5}{5\cdot 2\cdot 3\cdot 4} = 14$

83. a. A path sequence has length 2n and is composed of n R's and n D's in any order. There are C(2n, n) ways to select the positions for the n R's.
b. C(2n, n) = (n + 1)C(n)

EXERCISES 3.5

*1. |{HHH, HHT, HTH, HTT, THH, THT, TTH, TTT}| = 8
2. 3/8
3. 1/8
*4. 2/8 = 1/4
5. 0
*6. 6·6 = 36
7. 1/36
8. 6/36 = 1/6

*9.

A	B
1	6
2	5
3	4
4	3
5	2
6	1

Probability is 6/36 = 1/6

10. (6 + 6 – 1)/36 = 11/36 (using Principle of Inclusion and Exclusion)

11.

A	B
5	6
6	5
6	6

Probability is 3/36 = 1/12.

12. An odd sum can happen by (odd, even) or (even, odd). Probability = $(3\cdot3 + 3\cdot3)/36$ = 18/36 = 1/2. (Intuitively, half the outcomes would be an odd number.)

*13. C(52, 2) = 1326.

14. C(13, 2) / C(52, 2) = 78/1326 = 1/17

*15. C(39, 2) / C(52, 2) = 741/1326 = 19/34

16. $C(13, 1)\cdot C(39, 1)/C(52, 2) = 13\cdot 39 / 1326 = 13/34$ (one spade, one non-spade)

17. $(C(13, 2) + 13\cdot 39) / C(52, 2) = 585/1326 = 15/34$ (both spades or one spade, one non-spade)
or
(C(52, 2) - C(39, 2)) / C(52, 2) = 585/1326 =15/34 (all hands minus both non-spades)

18. C(12, 2) / C(52, 2) = 66/1326 = 11/221

19. C(3, 2) / C(52, 2) = 3/1326 = 1/442

*20. (C(12, 2) + C(13, 2) – C(3, 2)) / C(52, 2) = 141/1326 = 47/442 (both face cards + both spades – both spade face cards)

*21 The size of the sample space is $10\cdot10\cdot10 = 1000$
a. 1/1000 (only one winning 3-digit number)
b. $1\cdot 3!/1000$ (all permutations of the 3-digit number drawn) = 6/1000 $\cong$ 1/167
c. (3!/2!)/1000 (all distinct permutations of the 3-digit number drawn) = 3/1000 $\cong$ 1/333

22. The size of the sample space is $10 \cdot 10 \cdot 10 \cdot 10 = 10{,}000$
   a. 1/10,000 (only one winning 4-digit number)
   b. $1 \cdot 4!/10{,}000$ (all permutations of the 4-digit number drawn) $= 24/10{,}000 \cong 1/417$
   c. $(4!/2!)/10{,}000$ (all distinct permutations of the 4-digit number drawn) $= 12/10{,}000 \cong 1/833$
   d. $(4!/2!\cdot 2!)/10{,}000$ (all distinct permutations of the 4-digit number drawn) $= 6/10{,}000 \cong 1/1667$
   e. $(4!/3!)/10{,}000$ (all distinct permutations of the 4-digit number drawn) $= 4/10{,}000 \cong 1/2500$

23. The size of the sample space is $36 \cdot 35 \cdot 34 \cdot 33 \cdot 32 = 45{,}239{,}040$
   a. $5!/45{,}239{,}040 = 1/376992$. $5!$ = all permutations of the one winning set of 5 numbers.
   b. $31 \cdot 5 \cdot 5!/45{,}239{,}040 \cong 1/2432$. If 4 of the numbers must equal those actually drawn, there are 31 choices for the 5th number (it must be different from the other 4, and it cannot equal the 5th number actually drawn). There are 5 "wild card" positions for the 5th number. And for each set of numbers, there are 5! permutations.
   c. $C(31, 2) \cdot C(5, 2) \cdot 5! /45{,}239{,}040 \cong 1/81$. If 3 of the numbers must equal those actually drawn and the remaining two cannot equal the remaining 2 actually drawn, there is a pool of 31 allowable values for the remaining two numbers. There are 2 positions out of 5 that can be "wild card" positions for the two non-matching numbers. And for each set of numbers, there are 5! permutations.

24. The size of the sample space is $49 \cdot 48 \cdot 47 \cdot 46 \cdot 45 \cdot 42 = 9{,}610{,}695{,}360$
   a. $5!/ 9{,}655{,}934{,}400 = 1/80{,}089{,}128$. $5!$ = exactly one winning set of 6 numbers but all permutations of the group of 5.
   b. $C(44, 5) \cdot 5! \cdot 1/9{,}655{,}934{,}400 \cong 1/74$. The 5 numbers must be chosen from the 44 numbers not drawn in the group of 5; each set of numbers has 5! permutations; the 1 is the single choice for the Powerball number.

For Exercises 25-33, the size of the sample space is $C(52, 5) = 2{,}598{,}960$

25. $13 \cdot C(4, 2) \cdot C(12, 3) \cdot C(4, 1) \cdot C(4, 1) \cdot C(4, 1) / C(52, 5) = 1{,}098{,}240/2{,}598{,}960 \cong 0.423$

*26. $C(13, 2) \cdot C(4, 2) \cdot C(4, 2) \cdot C(44, 1) / C(52, 5) = 123{,}552/2{,}598{,}960 \cong 0.048$

27. $13 \cdot C(4, 3) \cdot C(12, 2) \cdot C(4, 1) \cdot C(4, 1) / C(52, 5) = 54{,}912/2{,}598{,}960 \cong 0.021$

28. $10 \cdot 4 \cdot 4 \cdot 4 \cdot 4 \cdot 4 / C(52, 5) = 10{,}240/2{,}598{,}960 \cong 0.004$

*29. $4 \cdot C(13, 5) / C(52, 5) = 5{,}148/2{,}598{,}960 \cong 0.002$

30. $13 \cdot C(4, 3) \cdot 12 \cdot C(4, 2) / C(52, 5) = 3{,}744/2{,}598{,}960 \cong 0.0014$

31. $13 \cdot C(4, 4) \cdot C(48, 1) / C(52, 5) = 624/2{,}598{,}960 \cong 0.0002$

32. $4 \cdot 10 \cdot 1 \cdot 1 \cdot 1 \cdot 1 \cdot 1 / C(52, 5) = 40/2{,}598{,}960 \cong 0.000015$
33. $4 \cdot 1 \cdot 1 \cdot 1 \cdot 1 \cdot 1 / C(52, 5) = 4/2{,}598{,}960 \cong 0.0000015$

34. a. The probability of a straight flush, 0.0000015, is much less than the probability of a full house, 0.0014, and so is a better hand.
    b. The probability of four of a kind, 0.0002, is much less than the probability of a straight, 0.004, and so is a better hand.

*35. 365^n (each of the n persons has one of 365 possible birthdays)

36. $|E| = (365)(364)(363)\ldots(365 - n + 1)$ (each of the n persons has a different birthday), so $P(E) = (365)(364)(363)\ldots(365 - n + 1)/(365)^n$

37. $B = E'$, so from Exercise 36, $P(B) = 1 - P(E) = 1 - (365)(364)(363)\ldots(365 - n + 1)/(365)^n$

38. When $n = 22$, using the expression for P(B) from Exercise 37 gives $P(B) \cong 0.475695$, When $n = 23$, $P(B) \cong 0.507297$,

39. a. $P(E_2') = 1 - P(E_2) = 1 - 0.45 = 0.55$
    b. $P(E_1 \cup E_2) = P(E_1) + P(E_2) - P(E_1 \cap E_2) = 0.37 + 0.45 - 0.14 = 0.68$
    c. $P((E_1 \cup E_2)') = 1 - P(E_1 \cup E_2) = 1 - 0.68 = 0.32$

*40. a. $P(E_1) = p(1) + p(3) + p(5) = 0.6$
    b. $P(E_2) = p(3) + p(6) = 0.25$
    c. $P(E_3) = p(4) + p(5) + p(6) = 0.65$
    d. $P(E_2 \cap E_3) = p(6) = 0.15$
    e. $P(E_1 \cup E_3) = P(E_1) + P(E_3) - P(E_1 \cap E_3) = 0.6 + 0.65 - p(5) = 0.95$
    or
    $P(E_1 \cup E_3) = p(1) + p(3) + p(4) + p(5) + p(6) = 0.95$

41. a.

D	R	I
8/13	4/13	1/13

    b. 8/13
    c. 9/13

42. a. $P(A \mid B) = \dfrac{P(A \cap B)}{P(B)} = \dfrac{0.08}{0.34} \cong 0.24$
    b. $P(A \cup B) = P(A) + P(B) - P(A \cap B) = 0.17 + 0.34 - 0.08 = 0.43$
    c. $P(A \cup B)' = 1 - P(A \cup B) = 1 - 0.43 = 0.57$

For Exercises 43-49, the sample space is {GGG, GGB, GBG, GBB, BGG, BGB, BBG, BBB}

43. $4/8 = 1/2$

*44. 3/8

45. 1/8

46. P(at least one girl) = P(no girls)' = 1 – 1/8 = 7/8

47. 1/8

*48. P(three girls | first two girls) = P(three girls)/P(first two girls) = (1/8) / (2/8) = 1/2

49. P(at least 1 boy and at least 1 girl | at least 1 boy) = (6/8) / (7/8) = 6/7

50. a. $P(E_i \mid F) = \dfrac{P(E_i \cap F)}{P(F)}$ (1)

$P(F \mid E_i) = \dfrac{P(F \cap E_i)}{P(E_i)}$ or $P(F \cap E_i) = P(F \mid E_i)\, P(E_i)$ (2)

$P(F \cap E_i) = P(E_i \cap F)$ so substitute from Equation (2) into Equation (1), giving

$$P(E_i \mid F) = \frac{P(F \mid E_i)\, P(E_i)}{P(F)}$$

b. The events $E_i, 1 \le i \le n$, are all disjoint, therefore the events $F \cap E_i, 1 \le i \le n$, are all disjoint.

$$F = F \cap S = F \cap (E_1 \cup E_2 \cup \ldots \cup E_n) = (F \cap E_1) \cup (F \cap E_2) \cup \ldots \cup (F \cap E_n)$$

Because the probability of the union of disjoint events is the sum of the probabilities of each event, $P(F) = \sum_{k=1}^{n} P(F \cap E_k)$

c. From Equation (2) of part (a),

$$P(F \cap E_i) = P(F \mid E_i)\, P(E_i)$$

Substituting into the result of part (b) gives

$$P(F) = \sum_{k=1}^{n} P(F \mid E_k) P(E_k)$$

d. Substituting the result of part (c) into the result of part (a) gives

$$P(E_i \mid F) = \frac{P(F \mid E_i) P(E_i)}{\sum_{k=1}^{n} P(F \mid E_k) P(E_k)}$$

51. From the given information, $P(E_1) = 0.62$ and $P(E_2) = 0.38$; also $P(F \mid E_1) = 0.014$ and $P(F \mid E_2) = 0.029$. Using Bayes' Theorem,

$$P(E_1 \mid F) = \frac{P(F \mid E_1)P(E_1)}{\sum_{k=1}^{2} P(F \mid E_k)P(E_k)} = \frac{(0.014)(0.62)}{(0.014)(0.62) + (0.029)(0.38)} \cong 0.44$$

$$P(E_2 \mid F) = \frac{P(F \mid E_2)P(E_2)}{\sum_{k=1}^{2} P(F \mid E_k)P(E_k)} = \frac{(0.029)(0.38)}{(0.014)(0.62) + (0.029)(0.38)} \cong 0.56$$

The patient is more likely to be on medication Y.

*52. a.

X	2	3	4	5	6	7	8	9	10	11	12
p	1/36	2/36	3/36	4/36	5/36	6/36	5/36	4/36	3/36	2/36	1/36

b. $E(X) = 2(1/36) + 3(2/36) + 4(3/36) + 5(4/36) + 6(5/36) + 7(6/36) + 8(5/36) + 9(4/36) + 10(3/36) + 11(2/36) + 12(1/36) = 252/36 = 7$.

53. a. The sample space is {red, green, blue}. The random variable is the prize amount, and the probability is computed from the number of each color in the box.

x_i	red	green	blue
$X(x_i)$	3	6	10
$p(x_i)$	43/78	27/78	8/78

$E(X) = 3(43/78) + 6(27/78) + 10(8/78) = 371/78 = 4.75$

b. The expected value of the prize money is \$0.25 less than the cost of the game card, so after 100 games, the casino could expect a profit of 100(0.25) = \$25.00.

54. Letting G = good (virus-free) file and B = bad file,

x_i	G	BG	BBG	BBBG
$X(x_i)$	1	2	3	4
$p(x_i)$	$\left(\frac{9}{12}\right)$	$\left(\frac{3}{12}\right)\left(\frac{9}{11}\right)$	$\left(\frac{3}{12}\right)\left(\frac{2}{11}\right)\left(\frac{9}{10}\right)$	$\left(\frac{3}{12}\right)\left(\frac{2}{11}\right)\left(\frac{1}{10}\right)\left(\frac{9}{9}\right)$

$E(X) = 1(9/12) + 2(27/132) + 3(54/1320) + 4(6/1320)$
$= 1(990/1320) + 2(270/1320) + 3(54/1320) + 4(6/1320) = 1716/1320 = 1.3$

*55. The sample space of input values has n + 1 elements (n elements from the list, or not in the list), all equally likely.

x_i	L_1	L_2	...	L_n	L_{n+1} (not in list)
$X(x_i)$	1	2	...	n	n
$p(x_i)$	1/(n + 1)	1/(n + 1)	...	1/(n + 1)	1/(n + 1)

Then $E(X) = \sum_{i=1}^{n+1} X(x_i)p(x_i) =$

$$\sum_{i=1}^{n} i\left(\frac{1}{n+1}\right) + n\left(\frac{1}{n+1}\right) = \frac{1}{n+1}\sum_{i=1}^{n} i + \frac{n}{n+1} = \frac{1}{n+1}(1+2+\ldots+n) + \frac{n}{n+1} = \frac{1}{n+1}\frac{n(n+1)}{2} + \frac{n}{n+1}$$

$$= \frac{n^2+3n}{2(n+1)}$$

56. The sample space of input values has n + 1 elements (n elements from the list, or not in the list). The probability that the target is not in the list is 0.8. The probability that the target is in the list is 0.2, divided equally among the n positions in the list.

x_i	L_1	L_2	...	L_n	L_{n+1} (not in list)
$X(x_i)$	1	2	...	n	n
$p(x_i)$	0.2/n	0.2/n	...	0.2/n	0.8

Then $E(X) = \sum_{i=1}^{n+1} X(x_i)p(x_i) =$

$$\sum_{i-1}^{n} i\left(\frac{0.2}{n}\right) + n(0.8) = \frac{0.2}{n}\sum_{i-1}^{n} i + 0.8n = \frac{0.2}{n}(1+2+\ldots+n) + 0.8n = \frac{0.2}{n}\frac{n(n+1)}{2} + 0.8n$$

$$= 0.1(n+1) + 0.8n = 0.9n + 0.1$$

EXERCISES 3.6

1. *a. $(a + b)^5 = C(5, 0)a^5 + C(5, 1)a^4b + C(5, 2)a^3b^2 + C(5, 3)a^2b^3 + C(5, 4)ab^4 + C(5, 5)b^5 = a^5 + 5a^4b + 10a^3b^2 + 10a^2b^3 + 5ab^4 + b^5$

b. $(x + y)^6 = C(6, 0)x^6 + C(6, 1)x^5y + C(6, 2)x^4y^2 + C(6, 3)x^3y^3 + C(6, 4)x^2y^4 + C(6, 5)xy^5 + C(6,6)y^6 = x^6 + 6x^5y + 15x^4y^2 + 20x^3y^3 + 15x^2y^4 + 6xy^5 + y^6$

*c. $(a + 2)^5 = C(5, 0)a^5 + C(5, 1)a^4(2) + C(5, 2)a^3(2)^2 + C(5, 3)a^2(2)^3 + C(5, 4)a(2)^4 + C(5, 5)(2)^5 = a^5 + 10a^4 + 40a^3 + 80a^2 + 80a + 32$

d. $(a - 4)^4 = C(4, 0)a^4 + C(4, 1)a^3(-4)^2 + C(4, 2)a^2(-4) + C(4, 3)a(-4)^3 + C(4, 4)(-4)^4$
$= a^4 - 16a^3 + 96a^2 - 256a + 256$

*e. $(2x + 3y)^3 = C(3, 0)(2x)^3 + C(3, 1)(2x)^2(3y) + C(3, 2)(2x)(3y)^2 + C(3, 3)(3y)^3$
$= 8x^3 + 36x^2y + 54xy^2 + 27y^3$

f. $(3x - 1)^5 = C(5, 0)(3x)^5 + C(5, 1)(3x)^4(-1) + C(5, 2)(3x)^3(-1)^2 + C(5, 3)(3x)^2(-1)^3$
$+ C(5, 4)(3x)(-1)^4 + C(5, 5)(-1)^5 = 243x^5 - 405x^4 + 270x^3 - 90x^2 + 15x - 1$

g. $(2p - 3q)^4 = C(4, 0)(2p)^4 + C(4, 1)(2p)^3(-3q) + C(4, 2)(2p)^2(-3q)^2 + C(4, 3)(2p)(-3q)^3$
$+ C(4, 4)(-3q)^4 = 16p^4 - 96p^3q + 216p^2q^2 - 216pq^3 + 81q^4$

h. $(3x + \frac{1}{2})^5 = C(5, 0)(3x)^5 + C(5, 1)(3x)^4(\frac{1}{2}) + C(5, 2)(3x)^3(\frac{1}{2})^2 + C(5, 3)(3x)^2(\frac{1}{2})^3$
$+ C(5, 4)(3x)(\frac{1}{2})^4 + C(5, 5)(\frac{1}{2})^5 = 243x^5 + \frac{405}{2}x^4 + \frac{132}{2}x^3 + \frac{45}{4}x^2 + \frac{15}{16}x + \frac{1}{32}$

2. $120a^7b^3$

3. $924x^6y^6$

*4. $C(9, 5)(2x)^4(-3)^5 = -489{,}888x^4$

5. $15{,}120a^3b^4$

*6. $C(8, 8)(-3y)^8 = 6561y^8$

7. $729x^6$

*8. $C(5, 2)(4x)^3(-2y)^2 = 2560x^3y^2$

9. $-1701x^5$

10. $(a + b + c)^3 = ((a + b) + c)^3 = C(3, 0)(a + b)^3 + C(3, 1)(a + b)^2c + C(3, 2)(a + b)c^2 +$
$C(3, 3)c^3 = [C(3, 0)a^3 + C(3, 1)a^2b + C(3, 2)ab^2 + C(3, 3)b^3]$
$+ 3(a^2 + 2ab + b^2)c + 3(a + b)c^2 + c^3$
$= a^3 + 3a^2b + 3ab^2 + b^3 + 3a^2c + 6abc + 3b^2c + 3ac^2 + 3bc^2 + c^3$

11. $(1 + 0.1)^5 = C(5, 0)(1) + C(5, 1)(0.1) + C(5, 2)(0.1)^2 + C(5, 3)(0.1)^3 + C(5, 4)(0.1)^4 +$
$C(5, 5)(0.1)^5 = 1 + 0.5 + 10(0.01) + 10(0.001) + 5(0.0001) + 0.00001$
$= 1 + 0.5 + 0.1 + 0.01 + 0.0005 + 0.00001 = 1.61051$

*12. $C(8, 1)(2x - y)^7 5^1 = C(8, 1)(5)[C(7, 4)(2x)^3(-y)^4] = C(8, 1)C(7,4)(2)^3(5)x^3y^4 =$
$11{,}200x^3y^4$

13. $C(9, 2)(x + y)^7(2z)^2 = C(9, 2)[C(7, 2)x^5y^2](2z)^2 = 3024x^5y^2z^2$

14. $C(n + 2, r) = C(n + 1, r - 1) + C(n + 1, r)$ (Pascal's formula)
$= C(n, r - 2) + C(n, r - 1) + C(n, r - 1) + C(n, r)$ (Pascal's formula again)
$= C(n, r) + 2C(n, r - 1) + C(n, r - 2)$

15. Basis: $n = k$: $C(k, k) = C(k+1, k+1)$ true because both equal 1
Assume $C(k, k) + C(k+1, k) + \ldots + C(n-1, k) = C(n, k+1)$
Then $C(k, k) + \ldots + C(n-1, k) + C(n, k) = C(n, k+1) + C(n, k)$
$= C(n, k+1) + [C(n+1, k+1) - C(n, k+1)]$ by Pascal's formula
$= C(n+1, k+1)$

16. From the Binomial Theorem with $a = 1$, $b = (-1)$:
$C(n, 0) - C(n,1) + C(n,2) - \ldots + (-1)^nC(n,n) = (1 + (-1))^n = 0^n = 0$

*17. From the Binomial Theorem with $a = 1$, $b = 2$:
$C(n, 0) + C(n, 1)2 + C(n, 2)2^2 + \ldots + C(n, n)2^n = (1 + 2)^n = 3^n$

18. a. From the Binomial Theorem with $a = 1$, $b = 2^{-1}$,

$$C(n, 0) + C(n, 1)2^{-1} + C(n, 2)2^{-2} + \ldots + C(n, n)2^{-n} = (1 + 2^{-1})^n$$

so, multiplying by 2^n,

$$C(n, 0)2^n + C(n, 1)2^{n-1} + C(n, 2)2^{n-2} + \ldots + C(n, n)2^{n-n} = 2^n(1 + 2^{-1})^n$$

or

$$C(n, 0)2^n + C(n, 1)2^{n-1} + C(n, 2)2^{n-2} + \ldots + C(n, n)$$
$$= 2^n\left(1 + \frac{1}{2}\right)^n = 2^n\left(\frac{2+1}{2}\right)^n = 2^n\left(\frac{3^n}{2^n}\right) = 3^n$$

b. From the symmetry of Pascal's triangle (or from Exercise 74 of Section 3.4), $C(n, r) = C(n, n - r)$. Thus the coefficients in the sum of Exercise 17 can be "reversed" - $C(n, 0)$ gets replaced with $C(n, n)$ and so forth, giving the desired result.

19. a. $C(n, 0) + C(n, 1)x + C(n, 2)x^2 + C(n, 3)x^3 + \ldots + C(n, n)x^n$

b. Differentiating both sides of the equation
$(1 + x)^n = C(n, 0) + C(n, 1)x + C(n, 2)x^2 + \ldots + C(n, n)x^n$
gives
$n(1 + x)^{n-1} = C(n, 1) + 2C(n, 2)x + 3C(n, 3)x^2 + \ldots + nC(n, n)x^{n-1}$

c. follows from (b) with x = 1

d. follows from (b) with x = -1

20. $(1 + x)^n = C(n, 0) + C(n, 1)x + C(n, 2)x^2 + C(n, 3)x^3 + ... + C(n, n)x^n$

Integrating both sides with respect to x,

$$\frac{(1+x)^{n+1}}{n+1} = C(n,0)x + \frac{C(n,1)x^2}{2} + \frac{C(n,2)x^3}{3} + \ldots + \frac{C(n,n)x^{n+1}}{n+1} + C$$

To evaluate the constant of integration C, let x = 0:

$$\frac{1}{n+1} = 0 + 0 + \ldots + 0 + C$$

so

$$\frac{(1+x)^{n+1}}{n+1} = C(n,0)x + \frac{C(n,1)x^2}{2} + \frac{C(n,2)x^3}{3} + \ldots + \frac{C(n,n)x^{n+1}}{n+1} + \frac{1}{n+1}$$

or

$$\frac{(1+x)^{n+1} - 1}{n+1} = C(n,0)x + \frac{C(n,1)x^2}{2} + \frac{C(n,2)x^3}{3} + \ldots + \frac{C(n,n)x^{n+1}}{n+1}$$

Let x = 1. Then

$$\frac{2^{n+1} - 1}{n+1} = C(n,0) + \frac{1}{2}C(n,1) + \frac{1}{3}C(n,2) + \ldots + \frac{1}{n+1}C(n,n)$$

Let x = -1. Then

$$\frac{-1}{n+1} = -C(n,0) + \frac{1}{2}C(n,1) - \frac{1}{3}C(n,2) + \ldots + \frac{(-1)^{n+1}}{n+1}C(n,n)$$

or

$$\frac{1}{n+1} = C(n,0) - \frac{1}{2}C(n,1) + \frac{1}{3}C(n,2) + \ldots + \frac{(-1)^n}{n+1}C(n,n)$$

21. a. Out of all the intersections of m sets, $1 \le m \le k$, we want the ones that pick all m elements from the k elements in B. There are C(k,m) ways to do this.
b. $C(k, 1) - C(k, 2) + C(k, 3) - \ldots + (-1)^{k+1}C(k, k)$
c. From Exercise 16,

$$C(k, 0) - C(k, 1) + C(k, 2) - \ldots + (-1)^k C(k, k) = 0$$

or

$$C(k, 1) - C(k,2) + \ldots (-1)^{k+1}C(k, k) = C(k,0)$$

but C(k,0) = 1

CHAPTER 4: Relations, Functions, and Matrices

Many of the definitions we give in mathematics are in symbolic form, and students often don't understand that a symbolic expression represents a *pattern,* which they need to be able to recognize in various guises. I begin to stress this back in Chapter 1; tautologies and derivation rules are patterns that must be recognized even when the letters change, or when the letters themselves become expressions. The same notion occurs in this chapter. When a binary relation is defined by $x \rho y \leftrightarrow x$ is even, I encourage students to translate that to "an ordered pair is related if the first component (whatever its name) is even." A function whose action upon the domain elements is described by the equation $f(x) = x^2$ is giving a pattern for what to do with its argument (whatever its name), not what to do with "x."

There are always some misunderstandings about the properties of binary relations. For example, having *some* ordered pairs of the form (x,x) does not make the relation reflexive. The definition of transitivity does not imply that x, y, and z must be distinct elements. A relation that is not symmetric is not necessarily antisymmetric. And relations can satisfy properties in somewhat trivial ways because of the truth table for the logical connective implication. Indeed, exploring the various properties of binary relations in detail provides a good review of both the implication connective, and the meaning of universal and existential quantifiers. Looking into what it means to fail to be reflexive, for example, introduces the negation of a universally quantified expression, which turns into an existentially quantified negated expression to which DeMorgan's laws can be applied, etc.

"What are the properties of a binary relation [or an equivalence relation]?" becomes my opening question for about a week! I also like to assure students, when we first talk about equivalence classes, that they began using equivalence classes in fourth grade, and then I talk about rational numbers and equivalent fractions (Example 14).

The material in Sections 4.2 and 4.3 is not central and can be omitted without difficulty. Topological sorting (Section 4.2) is explained as the process of turning a partial ordering into a total ordering. Another approach to this task is mentioned in Section 6.4 as an application of depth-first search. Section 4.3, relational databases, provides a nice application of the ideas of an n-ary relation, practical examples of set operations of union, intersection, and difference and of predicate logic notation (in relational calculus).

Students have worked with functions for years, but they still have a surprising amount of trouble with the concepts of one-to-one and onto. I really emphasize how to begin when asked to prove that a function is one-to-one or onto. I make up lots of non-mathematical examples in class and ask whether they are one-to-one and/or onto, such as "S is the set of people in this room, T is the set of shoes in this room, f associates with each person his or her left shoe."

Another technique that causes confusion is composition of cycles. Students have some difficulty at first in understanding that this is really composition, i.e., they apply the first cycle to some element and then don't know what to do with the rest of the cycles.

Order of magnitude of functions is important in analysis of algorithms. The text describes the order of magnitude of sequential search and binary search, which were analyzed in Chapter 2; order-of-magnitude analyses will be done on the graph algorithms of Chapter 6.

The final section in Chapter 4, on matrices, I leave as a review. Most students have seen this material (although the idea of a Boolean matrix is probably new) and need only look it over to recall the definition of matrix multiplication.

EXERCISES 4.1

*1. a. (1, 3), (3, 3) b. (4, 2), (5, 3)
c. (5, 0), (2, 2) d. (1, 1), (3, 9)

2. a. (1, -1), (-3, 3)
b. (19, 7), (41, 16)
c. (-3, -5), (-4, 1/2), (1/2, 1/3)
d. ((1, 3), (3, 2))

3. *a.

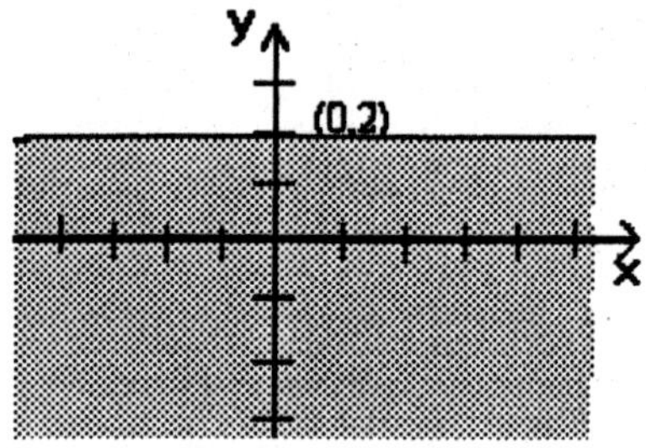

b.

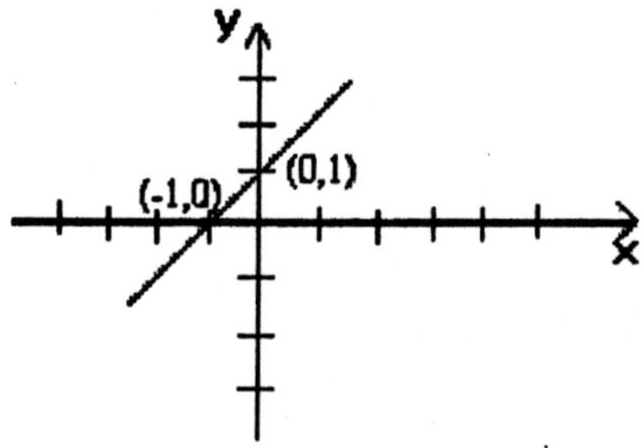

c.

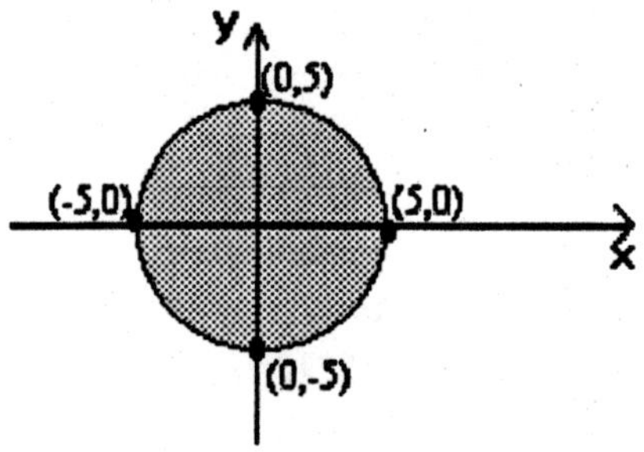

d.

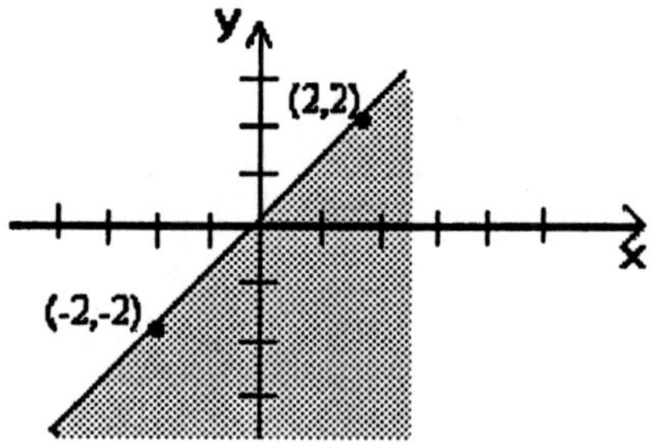

4. a. $x \rho y \leftrightarrow x > -1$
b. $x \rho y \leftrightarrow -2 \leq y \leq 2$
c. $x \rho y \leftrightarrow x \leq 2 - y$
d. $x \rho y \leftrightarrow x^2 + 4y^2 \leq 4$

*5. a. many-to-many
b. many-to-one
c. one-to-one
d. one-to-many

6. a. one-to-one
b. many-to-one
c. many-to-many
d. one-to-many

*7. a. (2, 6), (3, 17), (0, 0)
b. (2, 12)
c. none
d. (1, 1), (4, 8)

8. a. reflexive, antisymmetric
b. symmetric
c. symmetric, transitive
d. reflexive, symmetric, transitive
e. symmetric, antisymmetric, transitive

9. a. reflexive, transitive
b. antisymmetric (because x taller than y and y taller than x is always false, the implication is true), transitive
c. reflexive, symmetric, transitive
d. antisymmetric (false antecedent)
e. antisymmetric, transitive (false antecedent in each case)
f. symmetric
g. reflexive, symmetric, transitive
h. none (not transitive - $x \rho y$ and $y \rho x$ does not imply $x \rho x$)

10. *a. reflexive, transitive
*b. reflexive, symmetric, transitivc
*c. symmetric
d. transitive
e. reflexive, symmetric, transitive
f. reflexive, symmetric, transitive
g. symmetric
h. reflexive, symmetric, transitive
i. symmetric
j. reflexive, antisymmetric, transitive

11. (b); the equivalence classes are
    $[0] = \{\ldots, -9, -6, -3, 0, 3, 6, 9, \ldots\}$
    $[1] = \{\ldots, -8, -5, -2, 1, 4, 7, 10, \ldots\}$
    $[2] = \{\ldots, -7, -4, -1, 2, 5, 8, 11, \ldots\}$
    (e); the equivalence classes are sets consisting of squares with equal length sides
    (f); the equivalence classes are sets consisting of strings with the same number of characters
    (h); the equivalence classes are sets consisting of sets with the same number of elements

12. For example:
    a. S = set of all lines in the plane, $x \rho y \leftrightarrow$ x coincides with y or x is perpendicular to y. Then ρ is reflexive (x coincides with x) and symmetric (x coincides with y $\rightarrow$ y coincides with x or $x \perp y \rightarrow y \perp x$) but not transitive ($x \perp y$ and $y \perp z$ only implies x and z parallel).

    b. S = set of integers, $x \rho y \leftrightarrow x^2 \leq y^2$. Then ρ is reflexive ($x^2 \leq x^2$) and transitive ($x^2 \leq y^2$ and $y^2 \leq z^2 \rightarrow x^2 \leq z^2$) but not symmetric ($2 \rho 3$ but not $3 \rho 2$).

    c. S = set of nonnegative integers, $x \rho y \leftrightarrow x < y$. Then ρ is not reflexive (x is not less than x), not symmetric ($x < y \not\rightarrow y < x$), but is transitive ($x < y$ and $y < z \rightarrow x < z$).

    d. S = set of integers, $x \rho y \leftrightarrow x \leq |y|$. Then ρ is reflexive ($x \leq |x|$), but not symmetric ($-2 \rho 3$ but not $3 \rho -2$) and not transitive ($3 \rho -4$ and $-4 \rho 2$ but not $3 \rho 2$).

13. a. yes, yes b. yes, yes
    c. no, yes d. no, yes

14. a. reflexive closure = ρ itself
    symmetric closure - add (1, 0), (2, 1), (4, 2), (6, 4)
    transitive closure - add (0, 2), (1, 4), (2, 6), (0, 4), (0, 6), (1, 6)
    b. reflexive closure - add (0, 0), (1, 1), (2, 2), (4, 4), (6, 6)
    symmetric closure = ρ itself
    transitive closure - add (0, 0), (1, 1), (2, 2), (4, 4), (6, 6)
    c. reflexive closure - add (4, 4), (6, 6)
    symmetric closure = transitive closure = ρ itself
    d. ρ is its own closure with respect to all three properties
    e. reflexive closure - add (0, 0), (1, 1), (2, 2), (4, 4), (6, 6)
    symmetric closure = transitive closure = ρ itself

*15. $x \rho^* y \leftrightarrow$ one can fly from x to y (perhaps by multiple hops) on Take-Your-Chance Airlines

16.*a. $\rho = \{(1, 1)\}$

b. $\rho = \{(1, 2), (2, 1), (1, 3)\}$

c. Assume ρ is asymmetric and not irreflexive. Then for some $x \in S$, $(x, x) \in \rho$. But then $(x, x) \in \rho$ and $(x, x) \in \rho$, which contradicts asymmetry.

d. Assume ρ is irreflexive and transitive but not asymmetric. Then for some $x, y \in S$, $(x, y) \in \rho$ and $(y, x) \in \rho$. By transitivity, $(x, x) \in \rho$, which contradicts irreflexivity.

e. Assume ρ is nonempty, symmetric, and transitive. Let $(x, y) \in \rho$. Then $(y, x) \in \rho$ by symmetry, and $(x, x) \in \rho$ by transitivity. Therefore $(x, x) \in \rho$ for some $x \in S$ and ρ is not irreflexive.

17. a. No - if the relation is irreflexive, it is its own irreflexive closure. If the relation is not irreflexive, there must be some $x \in S$ with (x, x) in the relation; extending the relation will not remove this pair, so no extension can be irreflexive.

b. No - if the relation is asymmetric, it is its own asymmetric closure. If the relation is not asymmetric, there must be two pairs (x, y) and (y, x) in the relation; extending the relation will not remove these pairs, so no extension can be asymmetric.

18. A binary relation is a subset of $S \times S$. The number of different binary relations is the size of the power set of $S \times S$. There are n^2 ordered pairs in $S \times S$, so the size of the power set is 2^{n^2}.

19. a. Let $x \in \#A$. Then $x \rho y$ for all $y \in A$, so by symmetry, $y \rho x$ for all $y \in A$, and $x \in A\#$. Therefore $\#A \subseteq A\#$. By a similar argument, $A\# \subseteq \#A$ and $\#A = A\#$.

b. Assume $A \subseteq B$ and let $x \in \#B$. Then $x \rho y$ for all $y \in B$ and since $A \subseteq B$, $x \rho y$ for all $y \in A$. Thus $x \in \#A$, and $\#B \subseteq \#A$. Similarly $B\# \subseteq A\#$.

c. Let $x \in A$ and let $z \in \#A$. Then $z \rho y$ for all $y \in A$, so in particular, $z \rho x$. Since z was arbitrary, $z \rho x$ holds for all z in #A, and therefore $x \in (\#A)\#$ and $A \subseteq (\#A)\#$.

d. Let $x \in A$ and let $z \in A\#$. Then $y \rho z$ for all $y \in A$, so in particular, $x \rho z$. Since z was arbitrary, $x \rho z$ holds for all z in A#, and therefore $x \in \#(A\#)$ and $A \subseteq \#(A\#)$.

20. a.

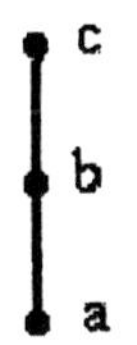

*b.

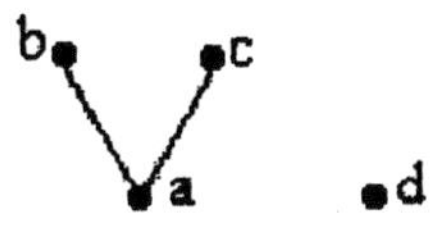

c.

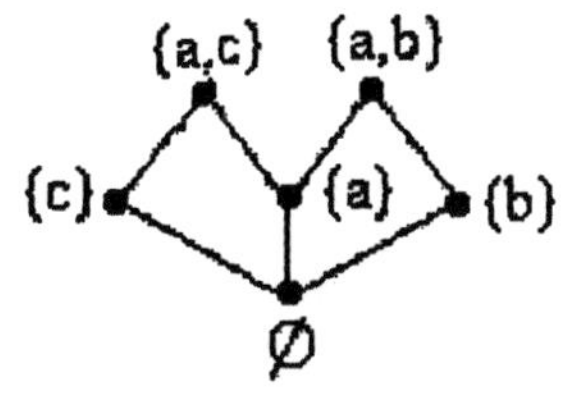

21. a. a is minimal and least
c is maximal and greatest
b. a and d are minimal
b, c, and d are maximal
c. ∅ is minimal and least
{a, c} and {a, b} are maximal

22. Reflexivity: If x ∈ A, then x ∈ S, so (x ≼ x) because ≼ is a reflexive relation on S.
Symmetry: if x, y ∈ A and x ≼y, then x, y ∈ S and x ≼ y, so y ≼ x because ≼ is symmetric on S.
Transitivity: if x, y, z ∈A and x ≼ y and y ≼ z, then x, y, z ∈ S, x ≼ y, and y ≼ z, so x ≼z because ≼ is transitive on S.

23. a.

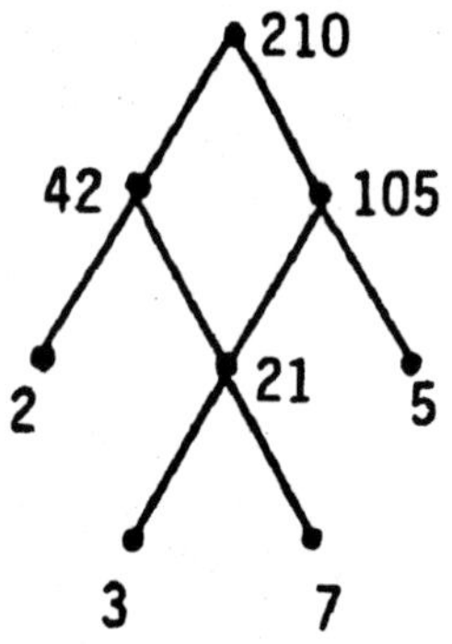

No least element; minimal elements of 2, 3, 5, 7; greatest element = maximal element = 210. Totally ordered subsets:
{3, 21, 105, 210}, {3, 21, 42, 210}, {7, 21, 105, 210}, {7, 21, 42, 210}.

b.

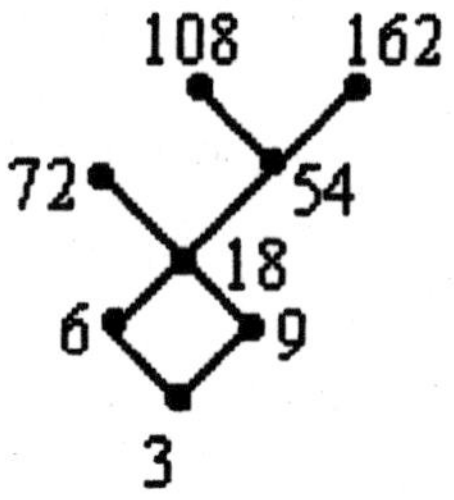

3 is both least and minimal; there is no greatest element; 72, 108 and 162 are maximal; 6 and 9 are unrelated, as are 72 and 54, 72 and 108, 72 and 162, and 108 and 162.

*24.

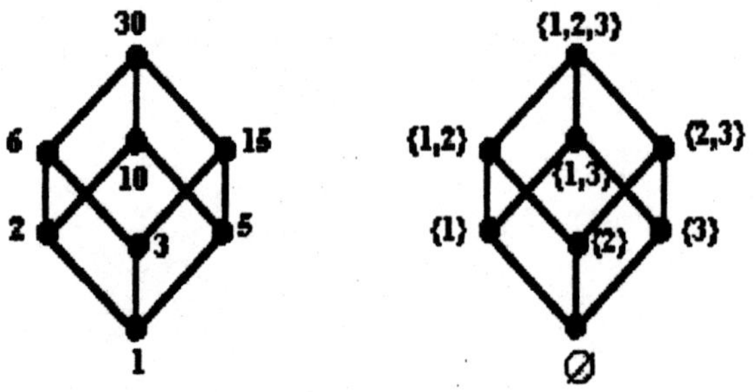

The two graphs are identical in structure.

25.*a. $\rho = \{(1, 1), (2, 2), (3, 3), (4, 4), (5, 5), (1, 3), (3, 5), (1, 5), (2, 4), (4, 5), (2, 5)\}$
b. $\rho = \{(a, a), (b, b), (c, c), (d, d), (e, e), (f, f), (a, d), (b, e), (c, f)\}$
c. $\rho = \{(1, 1), (2, 2), (3, 3), (4, 4), (5, 5), (1, 2), (2, 4), (4, 5), (1, 4), (1, 5), (2, 5), (1, 3), (3, 4), (3, 5)\}$

26. Reflexive: $(s_1, t_1)\ \mu\ (s_1, t_1)$ because both $s_1\ \rho\ s_1$ and $t_1\ \sigma\ t_1$ due to reflexivity of ρ and σ.
Antisymmetric: $(s_1, t_1)\ \mu\ (s_2, t_2)$ and $(s_2, t_2)\ \mu\ (s_1, t_1) \rightarrow s_1\ \rho\ s_2$ and $s_2\ \rho\ s_1$, $t_1\ \sigma\ t_2$ and $t_2\ \sigma\ t_1 \rightarrow s_1 = s_2$ and $t_1 = t_2$ due to antisymmetry of ρ and $\sigma \rightarrow (s_1, t_1) = (s_2, t_2)$.
Transitive: $(s_1, t_1)\ \mu\ (s_2, t_2)$ and $(s_2, t_2)\ \mu\ (s_3, t_3) \rightarrow s_1\ \rho\ s_2$ and $s_2\ \rho\ s_3$, $t_1\ \sigma\ t_2$ and $t_2\ \sigma\ t_3 \rightarrow s_1\ \rho\ s_3$ and $t_1\ \sigma\ t_3$ due to transitivity of ρ and $\sigma \rightarrow (s_1,t_1)\ \mu\ (s_3,t_3)$.

27. a. $\rho^{-1} = \{(2, 1), (3, 2), (3, 5), (5, 4)\}$
b. If ρ is reflexive, then $x\ \rho\ x$ for all $x \in S$, so $x\ \rho^{-1}\ x$ for all $x \in S$.
c. Let $x\ \rho^{-1}\ y$. Then $y\ \rho\ x$ and, because ρ is symmetric, $x\ \rho\ y$. Therefore $y\ \rho^{-1}\ x$.
d. Let $x\ \rho^{-1}\ y$ and $y\ \rho^{-1}\ x$. Then $y\ \rho\ x$ and $x\ \rho\ y$ so, because ρ is antisymmetric, $x = y$.
e. Let $x\ \rho^{-1}\ y$ and $y\ \rho^{-1}\ z$. Then $y\ \rho\ x$ and $z\ \rho\ y$. By the transitivity of ρ, $z\ \rho\ x$ and therefore $x\ \rho^{-1}\ z$.
f. Let ρ be irreflexive. Then for all $x \in S$, $(x, x) \notin \rho$, so for all $x \in S$, $(x, x) \notin \rho^{-1}$.
g. Let $x\ \rho^{-1}\ y$. Then $y\ \rho\ x$ and, by the asymmetric property of ρ, $(x, y) \notin \rho$ so $(y, x) \notin \rho^{-1}$.

*28. Assume that ρ is reflexive and transitive on S. Then for all $x \in S$, $(x, x) \in \rho$, which means $(x, x) \in \rho^{-1}$, so $(x, x) \in \rho \cap \rho^{-1}$ and $\rho \cap \rho^{-1}$ is reflexive.

Let $(x, y) \in \rho \cap \rho^{-1}$. Then $(x, y) \in \rho$ and $(x, y) \in \rho^{-1}$, which means $(x, y) \in \rho$ and $(y, x) \in \rho$. This implies $(y, x) \in \rho^{-1}$ and $(y, x) \in \rho$, so $(y, x) \in \rho \cap \rho^{-1}$ and $\rho \cap \rho^{-1}$ is symmetric.

Let $(x, y) \in \rho \cap \rho^{-1}$ and $(y, z) \in \rho \cap \rho^{-1}$. Then $(x, y) \in \rho$ and $(x, y) \in \rho^{-1}$ and $(y, z) \in \rho$ and $(y, z) \in \rho^{-1}$, so that $(x, y) \in \rho$ and $(y, x) \in \rho$ and $(y, z) \in \rho$ and $(z, y) \in \rho$. Because ρ is transitive, this means $(x, z) \in \rho$ and $(z, x) \in \rho$ or $(x, z) \in \rho$ and $(x, z) \in \rho^{-1}$, so $(x, z) \in \rho \cap \rho^{-1}$ and $\rho \cap \rho^{-1}$ is transitive.

29. a. If (S, ρ) is a partially ordered set, then ρ is reflexive, antisymmetric, and transitive. By parts (b), (d), and (e) of Exercise 27, ρ^{-1} is also reflexive, antisymmetric, and transitive on S, so (S, ρ^{-1}) is a partially ordered set.

b.

c. Let $(x, y) \in \rho^{-1} - X$. Then $(x, y) \in \rho^{-1}$ and $x \neq y$. Therefore $(y, x) \in \rho$. If $(x, y) \in \rho$, then, because ρ is antisymmetric, $x = y$, a contradiction. So $(x, y) \in \rho'$. This proves that $\rho^{-1} - X \subseteq \rho'$. Now let $(x, y) \in \rho'$. Then not $x \rho y$, so by total ordering, $y \rho x$ and $(x, y) \in \rho^{-1}$. If $x = y$, then $x \rho y$ by reflexivity of ρ, a contradiction. Thus $(x, y) \in \rho^{-1} - X$, and $\rho' \subseteq \rho^{-1} - X$.

30. a. Reflexive: $X \preccurlyeq X$ because $x_i = x_i$, $1 \leq i \leq k$.

Antisymmetric: Let $X \preccurlyeq Y$ and $Y \preccurlyeq X$. If $X \neq Y$, let $m + 1$ be the first index where $x_{m+1} \neq y_{m+1}$. Then $x_{m+1} \preccurlyeq y_{m+1}$ and $y_{m+1} \preccurlyeq x_{m+1} \rightarrow x_{m+1} = y_{m+1}$, a contradiction.

Transitive: Let $X \preccurlyeq Y$ and $Y \preccurlyeq Z$. Then $x_p \preccurlyeq y_p$ for some $p \leq k$ and $y_q \preccurlyeq z_q$ for some $q \leq k$. Let $m = \min(p, q)$. Then $x_m \preccurlyeq z_m$ and $X \preccurlyeq Z$.

Total: by "otherwise"

b. bah $\prec$ be $\prec$ boo $\prec$ bug $\prec$ bugg

*31. a. when; no; all but the last

b.

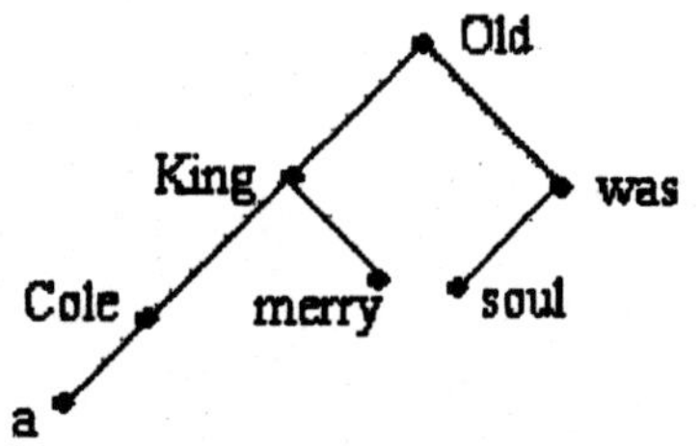

Maximal elements: a, merry, soul

32. The enumeration of A* is a, b, ..., z, aa, ab, ac, ... , az, ba, bb, ..., bz, ca, ... cz, ... , za, ..., zz, aaa, aab, aac, ... (each interval is finite)

*33. a. [a] = {a, c} = [c]

b. [3] = {1, 2, 3}

[4] = {4, 5}

c. [1] = {..., -5, -3, -1, 1, 3, 5, ...}

d. [-3] = {..., -13, -8, -3, 2, 7, 12, ...}

34. If $x \equiv y \pmod{n}$ then $x - y = k_1 n$ for some integer k_1, or $x = k_1 n + y$. If $z \equiv w \pmod{n}$ then $z - w = k_2 n$ for some integer k_2, or $z = k_2 n + w$.

*a. $x + z = (k_1 n + y) + (k_2 n + w) = y + w + (k_1 + k_2)n$, so $x + z - (y + w) = (k_1 + k_2)n$ where $k_1 + k_2$ is an integer, and $x + z \equiv y + w \pmod{n}$

b. $x - z = (k_1 n + y) - (k_2 n + w) = y - w + (k_1 - k_2)n$, so $x - z - (y - w) = (k_1 - k_2)n$ where $k_1 - k_2$ is an integer, and $x - z \equiv y - w \pmod{n}$

c. $x^s - y^s = (k_1n + y)^s - y^s =$

$$\left[\sum_{k=0}^{s} C(s,k)(k_1n)^{s-k}y^k\right] - y^s = \left[\sum_{k=0}^{s-1} C(s,k)(k_1n)^{s-k}y^k\right] + y^s - y^s = n\sum_{k=0}^{s-1} C(s,k)k_1^{s-k}n^{s-k-1}y^k = nk_2$$

where k_2 is an integer.

35. If $x \equiv y \pmod p$ then $x - y = kp$ for some integer k and $x^2 - y^2 = (x + y)(x - y) = (x + y)kp$ where $(x + y)k$ is an integer so $x^2 \equiv y^2 \pmod p$. Similarly, if $x \equiv -y \pmod p$ then $x + y = kp$ for some integer k and $x^2 - y^2 = (x + y)(x - y) = (x - y)kp$ where $(x - y)k$ is an integer so $x^2 \equiv y^2 \pmod p$.
Conversely, if $x^2 \equiv y^2 \pmod p$, then $x^2 - y^2 = (x + y)(x - y) = kp$ for some integer k. Therefore p divides $(x + y)(x - y)$, but because p is a prime, either p divides $(x + y)$ or p divides $(x - y)$. If p divides $(x + y)$ then $x + y$ is an integral multiple of p and $x \equiv -y \pmod p$. If p divides $(x - y)$ then $x - y$ is an integral multiple of p and $x \equiv y \pmod p$.

36. a. {(1, 1), (2, 2), (3, 3), (4, 4), (1, 2), (2, 1), (3, 4), (4, 3)}
b. {(a, a), (b, b), (c, c), (d, d), (e, e), (a, b), (b, a), (a, c), (c, a), (b, c), (c, b), (d, e), (e, d)}

37. reflexive - the color of x's cover is the same as the color of x's cover
symmetric - if x's cover is the same color as y's, then y's cover is the same color as x's
transitive - if x's cover is the same color as y's and y's cover is the same color as z's, then x's cover is the same color as z's
The equivalence classes are sets consisting of books with the same colored covers.

*38. reflexive - $(x, y)\ \rho\ (x, y)$ because $y = y$
symmetric - if $(x, y)\ \rho\ (z, w)$ then $y = w$ so $w = y$ and $(z, w)\ \rho\ (x, y)$
transitive - if $(x, y)\ \rho\ (z, w)$ and $(z, w)\ \rho\ (s, t)$ then $y = w$ and $w = t$ so $y = t$ and $(x, y)\ \rho\ (s, t)$
The equivalence classes are sets of ordered pairs with the same second components.

39. reflexive - $(x, y)\ \rho\ (x, y)$ because $x + y = x + y$
symmetric - if $(x, y)\ \rho\ (z, w)$ then $x + y = z + w$ so $z + w = x + y$ and $(z, w)\ \rho\ (x, y)$
transitive - if $(x, y)\ \rho\ (z, w)$ and $(z, w)\ \rho\ (s, t)$ then $x + y = z + w$ and $z + w = s + t$ so $x + y = s + t$ and $(x,y)\ \rho\ (s,t)$
The equivalence classes are sets of ordered pairs whose components add to the same value.

40. reflexive - $x^2 - x^2 = 0$, which is even
symmetric - if $x^2 - y^2 = 2n$ then $y^2 - x^2 = -2n$, which is even.
transitive - if $x^2 - y^2 = 2n$ and $y^2 - z^2 = 2m$, then $x^2 - z^2 = x^2 - y^2 + y^2 - z^2 = 2n + 2m = 2(n + m)$, which is even
The equivalence classes are the set of even integers and the set of odd integers.

*41. Clearly $P \leftrightarrow P$ is a tautology. If $P \leftrightarrow Q$ is a tautology, then P and Q have the same truth values everywhere, so $Q \leftrightarrow P$ is a tautology. If $P \leftrightarrow Q$ and $Q \leftrightarrow R$ are tautologies, then P, Q, and R have the same truth values everywhere, and $P \leftrightarrow R$ is a tautology. The equivalence classes are sets consisting of wffs with the same truth values everywhere.

42. Reflexive: $\pi_1 \preccurlyeq \pi_1$ because each block of π_1 is a subset of itself due to reflexivity of set inclusion. Antisymmetry and transitivity also follow from the corresponding set inclusion properties.

43. a. 1 The only way to partition a 1-element set is to use the whole set.
b. 5 Using a combinatorial argument, the partitions are

all in one block = 1
2 in one block, 1 in another = C(3, 2) = 3
each element separate = 1
Total = 5

or by directly counting
{a, b, c}
{a, b} {c} + {a, c} {b} + {b, c} {a}
{a} {b} {c}

c. 15 Using a combinatorial argument, the partitions are

all in one block = 1
3 in one block, 1 in another = C(4, 3) = 4
2 in one block, 2 in another = C(4, 2)/2 = 3
2 in one block, 2 in separate blocks = C(4, 2) = 6
each element separate = 1
Total = 15

or by directly counting
{a, b, c, d}
{a, b, c} {d} + {a, b, d} {c} + {a, c, d} {b} + {b, c, d} {a}
{a,,b} {c,,d} + {a,,c} {b,,d} + {a,,d} {b,,c}
{a, b} {c} {d} + {a, c} {b} {d} + {a, d} {b} {c} +
{c, d} {a} {b} + {b, d} {a} {c} + {b, c} {a} {d}
{a} {b} {c} {d}

*44. a. Partitions of 3 elements into 2 blocks can only be done with 2 elements in one block and 1 in the other, so the answer is the number of ways to select the 2 elements, or C(3, 2) = 3.

b. Partitions of 4 elements into 2 blocks can be done with 3 elements in one block and 1 in the other (pick the 3 elements out of 4), or two elements in each block (pick 2 elements out of 4 but this determines the other two elements). The answer is C(4, 3) + C(4, 2)/2 = 7.

45. The number of blocks in a partition can range from 1 (the whole set) to n (a single element in each block). The result follows by the definition of S(n,k) and the Addition Principle.

46. S(n, 1) = 1 because the only partition with 1 block is when the block is the entire set
S(n, n) = 1 because the only partition with n blocks is where each block consists of a single element
S(n + 1, k + 1) = S(n, k) + (k + 1)S(n, k + 1) because if the set without x is partitioned into k blocks, which can be done in S(n, k) ways, these blocks together with {x} form a partition of the original set with k + 1 blocks. If the set without x is partitioned into k + 1 blocks, which can be done S(n, k + 1) ways, then x can be added to any of these blocks (which can be done k + 1 ways) to give a partition of the original set with k + 1 blocks; by the Multiplication Principle, this can be done (k + 1)S(n, k + 1) ways. These are the only two possibilities, so the result follows from the Addition Principle.

*47. a. $S(3, 2) = S(2, 1) + 2S(2, 2) = 1 + 2\cdot 1 = 3$
b. $S(4, 2) = S(3, 1) + 2S(3, 2) = 1 + 2\cdot 3 = 7$

48.

				1				
			1		1			
		1		3		1		
	1		7		6		1	
1		15		25		10		1

*49. S(4, 3) = 6

50. S(5, 3) = 25

51. A general term of the sum has the form

$$C(n - 1, k)P_k$$

where $0 \le k \le n - 1$. This number counts all the partitions in which x appears in a block of size n - k, as follows: For each such block, the complement is a set of size k that does not include x. The number of sets of size k that do not include x is C(n - 1,k). Each such set can be partitioned in P_k ways. By the Multiplication Principle, there are $C(n - 1, k)P_k$ partitions where x appears in a block of size n - k, and by the Addition Principle, P_n is the sum over all block sizes n - k where $1 \le n - k \le n$ or $0 \le k \le n - 1$.

52. a. $P_1 = C(0, 0)P_0 = 1\cdot 1 = 1$
$P_2 = C(1, 0)P_0 + C(1, 1)P_1 = 1\cdot 1 + 1\cdot 1 = 2$
$P_3 = C(2, 0)P_0 + C(2, 1)P_1 + C(2, 2)P_2 = 1\cdot 1 + 2\cdot 1 + 1\cdot 2 = 5$
$P_4 = C(3, 0)P_0 + C(3, 1)P_1 + C(3, 2)P_2 + C(3, 3)P_3 = 1\cdot 1 + 3\cdot 1 + 3\cdot 2 + 1\cdot 5 = 15$

b. $P_1 = S(1, 1) = 1$
$P_2 = S(2, 1) + S(2, 2) = 1 + 1 = 2$
$P_3 = S(3, 1) + S(3, 2) + S(3, 3) = 1 + 3 + 1 = 5$
$P_4 = S(4, 1) + S(4, 2) + S(4, 3) + S(4, 4) = 1 + 7 + 6 + 1 = 15$

53. a. 25, 49
b. (3, 4, 5), (0, 5, 5), (8, 6, 10)
c. (-4, 4, 2, 0), (-6, 6, 0, -2)

EXERCISES 4.2

1. Yes; for example: 1, 2, 3, 8, 4, 5, 6, 7, 9

*2.

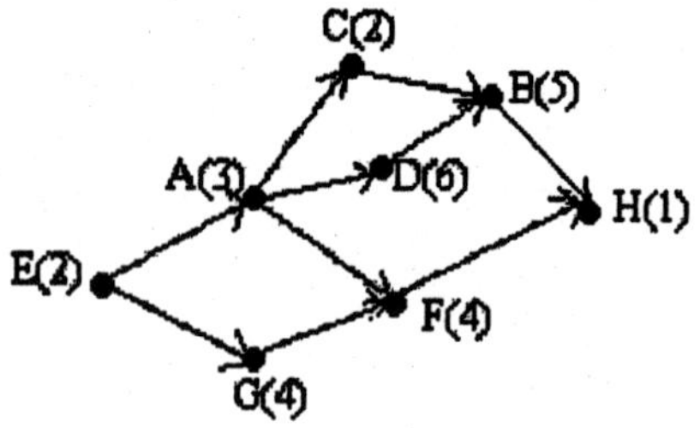

3.

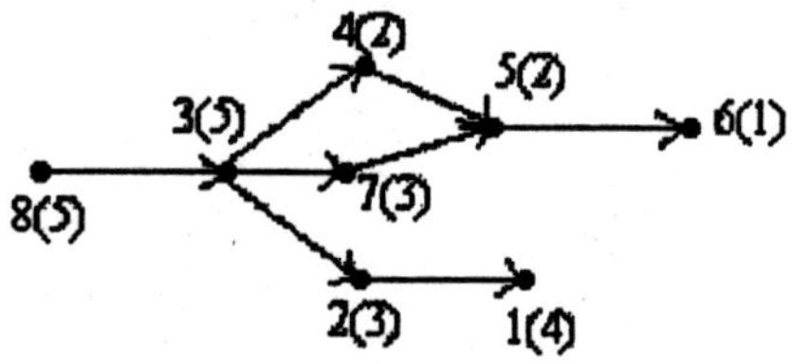

*4. Minimum time-to-completion is 17 time units. Critical path: E, A, D, B, H

5. Minimum time-to-completion is 16 time units. Critical path: 8, 3, 2, 1 or 8, 3, 7, 5, 6

6. For example: G, H, F, D, E, C, A, B

*7. For example: E, A, C, D, G, F, B, H

8. For example: 8, 3, 4, 7, 5, 6, 2, 1

9. The PERT chart is

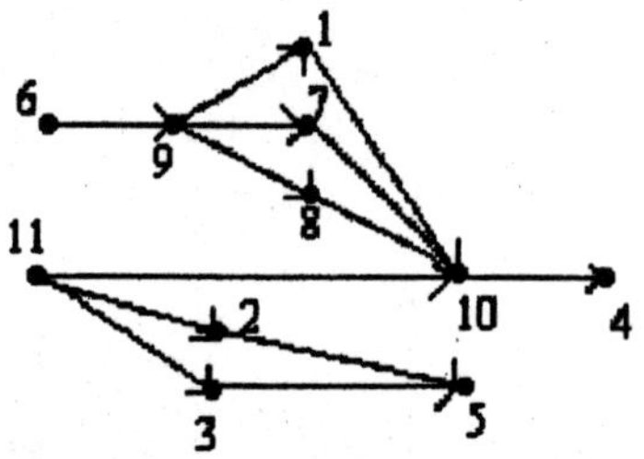

One topological ordering is 6, 9, 1, 7, 8, 11, 2, 3, 5, 10, 4

10. The PERT chart is

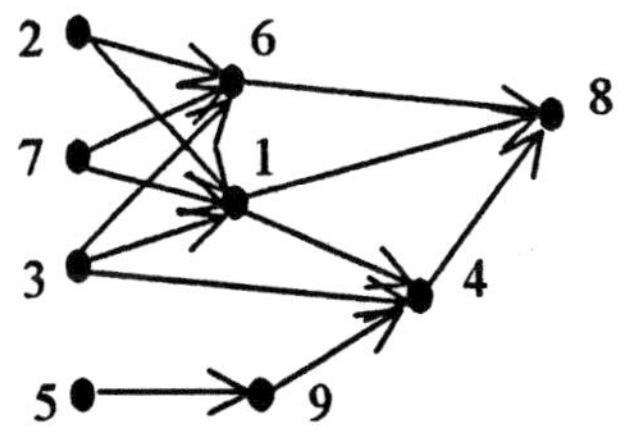

One topological ordering is 2, 3, 5, 9, 7, 1, 4, 6, 8

11. The PERT chart is

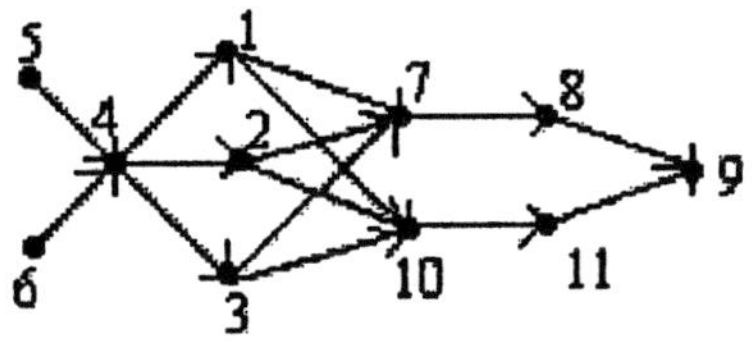

One topological ordering is 5, 6, 4, 1, 2, 3, 7, 10, 8, 11, 9

EXERCISES 4.3

*1.

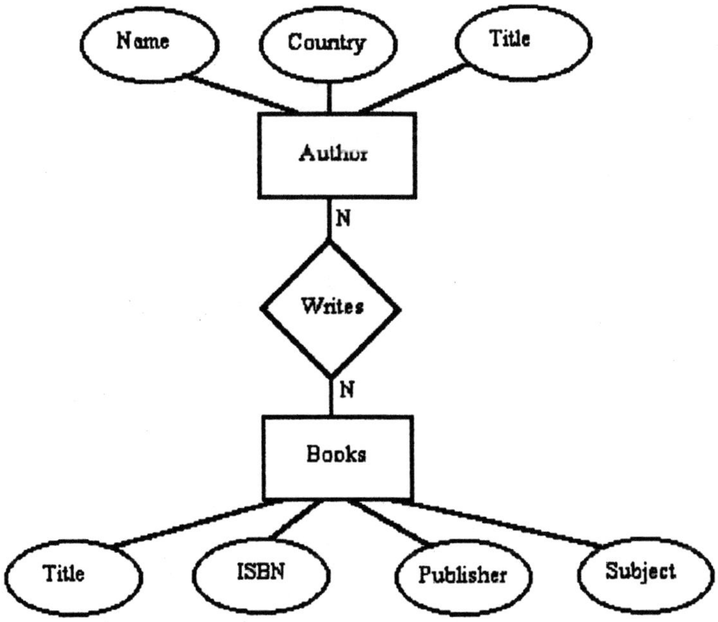

The "writes" relation is many-to-many, that is, one author can write many books, and one book can have more than one author.

2. No

3.

Author		
Name	Country	Title

Book			
Title	ISBN	Publisher	Subject

Writes		
Name	Title	ISBN

The Author relation requires a composite primary key because one author can write more than one book and one book can have more than one author. The implicit business rule is that one author does not write multiple books with the same title (otherwise the *ISBN* or some other unique book attribute would need to be an attribute in the author relation.) Tuples in the Book relation are uniquely identified by *ISBN*. For the Writes relation, both *Name* and *Title* are required as a foreign key into the Author relation, and *ISBN* is required as a foreign key into the Book relation. While a composite primary key is required, it could be either *Name* and *ISBN*, or *Name* and *Title* because of the implicit business rule.

*4.

Results1		
Name	*Country*	*Title*
Bert Kovalsco	U.S.	Baskets for Today
Jane East	U.S.	Springtime Gardening

5.

Results2		
Name	*Title*	*ISBN*
Dorothy King	Autumn Annuals	0-816-88506-0
Dorothy King	Springtime Gardening	0-816-35421-9

6.

Results3			
Title	*ISBN*	*Publisher*	*Subject*
Early Tang Paintings	0-364-87547-X	Bellman	Art
Springtime Gardening	0-56-000142-8	Swift-Key	Nature

7.

Results4			
Title	*ISBN*	*Publisher*	*Subject*
Baskets for Today	0-816-53705-4	Harding	Art

*8.

Results5	
Name	*Title*
Dorothy King	Springtime Gardening
Jon Nkoma	Birds of Africa
Won Lau	Early Tang Paintings
Bert Kovalsco	Baskets for Today
Tom Quercos	Mayan Art
Jimmy Chan	Early Tang Paintings
Dorothy King	Autumn Annuals
Jane East	Springtime Gardening

9.

Results6	
Name	*Country*
Dorothy King	England
Jon Nkoma	Kenya
Won Lau	China
Bert Kovalsco	U.S.
Tom Quercos	Mexico
Jimmy Chan	China
Jane East	U.S.

10.

Results7	
Publisher	*Subject*
Harding	Nature
Bellman	Art
Loraine	Nature
Swift-Key	Nature
Harding	Art

11.

Results8		
Title	*ISBN*	*Subject*
Springtime Gardening	0-816-35421-9	Nature
Early Tang Paintings	0-364-87547-X	Art
Birds of Africa	0-115-01214-1	Nature
Springtime Gardening	0-56-000142-8	Nature
Baskets for Today	0-816-53705-4	Art
Autumn Annuals	0-816-88506-0	Nature

*12.

Results9

Title	*ISBN*	*Publisher*	*Subject*	*Name*
Springtime Gardening	0-816-35421-9	Harding	Nature	Dorothy King
Early Tang Paintings	0-364-87547-X	Bellman	Art	Jimmy Chan
Early Tang Paintings	0-364-87547-X	Bellman	Art	Won Lau
Birds of Africa	0-115-01214-1	Loraine	Nature	Jon Nkoma
Springtime Gardening	0-56-000142-8	Swift-Key	Nature	Jane East
Baskets for Today	0-816-53705-4	Harding	Art	Bert Kovalsco
Autumn Annuals	0-816-88506-0	Harding	Nature	Dorothy King

13.

Results10

Name	*Country*	*Title*	*ISBN*
Dorothy King	England	Springtime Gardening	0-816-35421-9
Jon Nkoma	Kenya	Birds of Africa	0-115-01214-1
Won Lau	China	Early Tang Paintings	0-364-87547-X
Bert Kovalsco	U.S.	Baskets for Today	0-816-53705-4
Jimmy Chan	China	Early Tang Paintings	0-364-87547-X
Dorothy King	England	Autumn Annuals	0-816-88506-0
Jane East	U.S.	Springtime Gardening	0-56-000142-8

14. A tuple would result in which Dorothy King's name was associated with the ISBN for Springtime Gardening written by Jane East.

*15. a. **project(restrict** Author **where** Country = "U.S.")**over** Title **giving** Results11.
b. **SELECT** Title **FROM** Author
WHERE Author.Country = "U.S."
c. Range of x is Author
{x.title|x.Country = "U.S."}
d. Baskets for Today
Springtime Gardening

16. a. **project(join(restrict** Book **where** Publisher = "Harding") **and** Writes **over** ISBN)
over Name **giving** Results12.
b. **SELECT** Name **FROM** Writes, Book
WHERE Book.ISBN = Writes.ISBN
AND Book.Publisher = "Harding"
c. Range of x is Writes
Range of y is Book
{x.Name|**exists** y(y.Publisher = "Harding" **and** y.ISBN = x.ISBN)}
d. Dorothy King
Bert Kovalsco

17. a. **project(join(restrict** Book **where** Subject = "Nature") **and** Author **over** Title) **over** Name **giving** Results13.
    b. **SELECT** Name **FROM** Author, Book
       WHERE Author.Title = Book.Title
       AND Book.Subject = "Nature"
    c. Range of x is Author
       Range of y is Book
       {x.Name|**exists** y(y.Subject = "Nature" **and** y.Title = x.Title}
    d. Dorothy King
       Jon Nkoma
       Jane East

18. a. **project(join(join(restrict** Author **where** Country = "U.S.") **and** Writes **over** Name **and** Title) **and (restrict** Book **where** Subject = "Art") **over** ISBN) **over** Publisher **giving** Results14
    b. **SELECT** Publisher **FROM** Book, Author, Writes
       WHERE Author.Country = "U.S."
       AND Book.Subject = "Art"
       AND Author.Name = Writes.Name
       AND Author.Title = Writes.Title
       AND Writes.ISBN = Book.ISBN
    c. Range of x is Book
       Range of y is Author
       Range of z is Writes
       {x.Publisher|x.Subject = "Art" and **exists** y(y.Country = "U.S."
       and exists z(y.Name = z.Name **and** y.Title = z.Title **and** z.ISBN = x.ISBN))}
    d. Harding

19. Bert Kovalsco
    Jane East
    (Note that Suzanne Fleur does not satisfy the criteria for the query because Author.Country = "U.S." is False, Author.Country = NULL is NULL, and False or NULL is NULL, which then gets set to False.)

*20. a. The Cartesian product has cardinality p*q.
     b. If the common attribute is sorted in each table, then the join can be performed by doing something similar to a merge sort (see Exercise 13 in Section 2.5) on the common attribute, which means at most (p + q) rows would need to be examined.

EXERCISES 4.4

*1. a. Domain = {4, 5, 6, 7, 8} codomain = {8, 9, 10, 11} range = {8, 9, 10}
    b. 8, 10
    c. 6, 7
    d. no, no

2. (a) is a function, one-to-one but not onto
   (b) is not a function
   (c) is a one-to-one, onto function
   (d) is an onto function but not one-to-one

3. a. {(0, -1), (1, 1), (2, 3)}
   b. {(1, 1), (2, 3), (4, 7), (5, 9)}
   c. $\{(\sqrt{7}, 2\sqrt{7}-1), (1.5, 2)\}$

*4. a. f(A) = {3, 9, 15}
   b. $f(A) = \{x \mid x \in Z \text{ and } (\exists y)(y \in Z \text{ and } x = 6y)\}$

5. a. $f(\mathbb{N}) = \{0, 1, 4, 9, 16, \ldots\}$
   b. $f(\mathbb{Z}) = f(\mathbb{N})$
   c. $f(\mathbb{R}) = \{x \mid x \in \mathbb{R}, x \geq 0\}$

6. a. not a function from S to T (not a subset of S × T)
   b. function
   c. function; one-to-one and onto
   d. not a function from S to T (0 has no associated value)
   e. not a function (two values associated with 0)

7. For part (c), $f^{-1}: T \to S$, $f^{-1} = \{(3, 2), (7, 4), (1, 0), (5, 6)\}$

8. *a. function
   *b. not a function; undefined at x = 0
   *c. function; onto
   *d. bijection; $f^{-1}: \{p, q, r\} \to \{1, 2, 3\}$ where $f^{-1} = \{(q, 1), (r, 2), (p, 3)\}$
   *e. function; one-to- one
   *f. bijection; $h^{-1}: R^2 \to R^2$ where $h^{-1}(x, y) = (y-1, x-1)$
   g. function
   h. function; onto
   i. not a function (undefined for x = -1, no associated value in R for x < -1)
   j. bijection; $f^{-1}: N \to N$ where $f^{-1} = \begin{cases} x + 1 & \text{if } x \text{ is even} \\ x - 1 & \text{if } x \text{ is odd} \end{cases}$
   (note that the function f is its own inverse)
   k. not a function (no associated value in N for x = y = 0, z = 1)
   l. function; one-to-one

9. n odd

10. f performs the following mapping:
    $\varnothing \rightarrow 00$
    $\{a\} \rightarrow 10$
    $\{b\} \rightarrow 01$
    $\{a, b\} \rightarrow 11$
    Thus f maps each of the four elements in the power set of {a, b} to the four binary strings of length two, and is both one-to-one and onto.

11. f is neither one-to-one nor onto. f({a, b}) = f({b, c}) = 2, so f is not one-to-one. There will be no negative numbers in the range of f, so f is not onto.

12. f is neither one-to-one nor onto. f(xxy) = f(yyy) = 3, so f is not one-to-one. For any string s, $f(s) \geq 0$; there are no strings in A that map to negative values, so f is not onto.

13. f is both one-to-one and onto. If $f(s_1) = f(s_2)$ then $s_1 = s_2$ (just reverse the strings again and you get back where you started), so f is one-to-one. Given any string s in A*, let y be its reverse. Then f(y) = s, so f is onto.

14. f is one-to-one but not onto. If $f(s_1) = f(s_2)$, then $xs_1 = xs_2$, which means that $s_1 = s_2$, so f is one-to-one. Nothing maps to the empty string, so f is not onto.

15. For example, f(x) = 1/x

16. For example:
    a. f = {(a, x), (b, x), (c, y), (d, y)}
    b. f = {(a, x), (b, x), (c, y), (d, z)}
    c. no

17. x is an integer

*18. Let $k \leq x < k + 1$ where k is an integer. Then $\lfloor x \rfloor = k$. Also, $-k \geq -x > -k - 1$ so $\lceil -x \rceil = -k$ and $-\lceil -x \rceil = k$.

19. Let $n = \lceil x \rceil$. Then n is the smallest integer that is greater than or equal to x, so $n - 1 < x \leq n$. Therefore, adding 1 throughout the inequality, $n < x + 1 \leq n + 1$, and n + 1 is the smallest integer that is greater than or equal to x + 1. Therefore $\lceil x \rceil + 1 = n + 1 = \lceil x + 1 \rceil$.

20. a. Let x = 3.6. Then $\lceil \lfloor x \rfloor \rceil = \lceil 3 \rceil = 3 \neq x$.
    b. Let x = 4.8. Then $\lfloor 2x \rfloor = \lfloor 9.6 \rfloor = 9$ but $2\lfloor x \rfloor = 2(4) = 8$.
    c. Let x = 3.6, y = 4.8. Then $\lfloor x \rfloor + \lfloor y \rfloor = \lfloor 3.6 \rfloor + \lfloor 4.8 \rfloor = 3 + 4 = 7$ but $\lfloor x + y \rfloor = \lfloor 8.4 \rfloor = 8$.
    d. Case 1: $n \leq x < n + 1/2$. Then $2n \leq 2x < 2n + 1$ so $\lfloor 2x \rfloor = 2n$, and $n + 1/2 \leq x + 1/2 < n + 1$ so $\lfloor x + 1/2 \rfloor = n$. Therefore $\lfloor x \rfloor + \lfloor x + 1/2 \rfloor = n + n = 2n = \lfloor 2x \rfloor$.

Case 2: $n + 1/2 \le x < n + 1$. Then $2n + 1 \le 2x < 2n + 2$ so $\lfloor 2x \rfloor = 2n + 1$, and $n + 1 \le x + 1/2 < n + 1 + 1/2$ so $\lfloor x + 1/2 \rfloor = n + 1$. Therefore $\lfloor x \rfloor + \lfloor x + 1/2 \rfloor = n + (n + 1) = 2n + 1 = \lfloor 2x \rfloor$.

*21. If $2^k < n < 2^{k+1}$ then $\log(2^k) < \log n < \log(2^{k+1})$ or $k < \log n < k + 1$ and $\lfloor \log n \rfloor = k$, $\lceil \log n \rceil = k + 1$.

22. If $2^k \le n < 2^{k+1}$ then $\log(2^k) \le \log n < \log(2^{k+1})$ or $k \le \log n < k + 1$ so $k = \lfloor \log n \rfloor$, and $\lfloor \log n \rfloor + 1 = k + 1$. Also, $2^k < 2^k + 1 \le n + 1 \le 2^{k+1}$ so $\log(2^k) < \log(n + 1) \le \log(2^{k+1})$ or $k < \log(n + 1) \le k + 1$ and $\lceil \log(n + 1) \rceil = k + 1$.

23. *a. 9 *b. 0 c. 4 d. 2 $[-7 = (-3) \cdot 3 + 2]$

24. Let $x = q_1 n + r_1$, $0 \le r_1 < n$ and $y = q_2 n + r_2$, $0 \le r_2 < n$, so $x \bmod n = r_1$ and $y \bmod n = r_2$. Also,

$$x - y = (q_1 n + r_1) - (q_2 n + r_2) = (q_1 - q_2)n + (r_1 - r_2) \text{ with } -n < r_1 - r_2 < n.$$

Then $x \equiv y \pmod n$ if and only if $x - y = (q_1 - q_2)n$ and $r_1 - r_2 = 0$, which is true if and only if $r_1 = r_2$ or $x \bmod n = y \bmod n$.

25. a. (1, 1), (2, 0), (3, 1), (4, 0), (5, 1)

b. $c_{A \cap B}(x) = 1 \leftrightarrow x \in A$ and $x \in B \leftrightarrow c_A(x) = 1$ and $c_B(x) = 1 \leftrightarrow c_A(x) \cdot c_B(x) = 1$

c. If $c_{A'}(x) = 1$, then $x \in A'$ and $x \notin A$, so $c_A(x) = 0 = 1 - c_{A'}(x)$
If $c_{A'}(x) = 0$. then $x \notin A'$ and $x \in A$ so $c_A(x) = 1 = 1 - c_{A'}(x)$

d. No. Let $S = \{1, 2, 3\}$, $A = \{1, 2\}$, $B = \{2, 3\}$. Then $c_{A \cup B}(2) = 1$ but $c_A(2) + c_B(2) = 1 + 1$.

26. a. $2x$ b. 2^x c. 2^{16}

27. $g \circ f = \{(1,6), (2,7), (3,9), (4,9)\}$

*28. a. $(g \circ f)(5) = g(f(5)) = g(6) = 18$
b. $(f \circ g)(5) = f(g(5)) = f(15) = 16$
c. $(g \circ f)(x) = g(f(x)) = g(x + 1) = 3(x + 1) = 3x + 3$
d. $(f \circ g)(x) = f(g(x)) = f(3x) = 3x + 1$
e. $(f \circ f)(x) = f(f(x)) = f(x + 1) = (x + 1) + 1 = x + 2$
f. $(g \circ g)(x) = g(g(x)) = g(3x) = 3(3x) = 9x$

29. a. 25 b. 1470 (the index doesn't count) c. 9

30. a. $g \circ f = 12x^3$ $f \circ g = 48x^3$
b. $g \circ f = x^2 - 2x + 1$ $f \circ g = (4x^2 - 1)/2$
c. $g \circ f = \lceil x \rceil$ $f \circ g = \lfloor x \rfloor$

31. a. If $f(s_1) = f(s_2)$ then $g(f(s_1)) = g(f(s_2))$ so $(g \circ f)(s_1) = (g \circ f)(s_2)$. Because $g \circ f$ is one-to-one, $s_1 = s_2$ and therefore f is one-to-one.

b. For $u \in U$, there exists $s \in S$ such that $(g \circ f)(s) = u$, because $g \circ f$ is onto. Thus $g(f(s)) = u$ and $f(s)$ is a member of T that is a preimage of u under g, and g is onto.

c. Let $S = \{1, 2, 3\}$, $T = \{1, 2, 3, 4\}$, $U = \{1, 2, 3\}$, $f = \{(1, 1), (2, 2), (3, 3)\}$, $g = \{(1, 1), (2, 2), (3, 3), (4, 3)\}$. Then $f: S \rightarrow T$, $g: T \rightarrow U$, g is not one-to-one but $g \circ f = \{(1, 1), (2, 2), (3, 3)\}$ is one-to-one.

d. same example as for (c)

*32. a. $f^{-1}(x) = x/2$
b. $f^{-1}(x) = \sqrt[3]{x}$
c. $f^{-1}(x) = 3x - 4$

33. a. Assume f has a left inverse g, and that $f(s_1) = f(s_2)$. Then $g(f(s_1)) = g(f(s_2))$ or $(g \circ f)(s_1) = (g \circ f)(s_2)$ and $i_S(s_1) = i_S(s_2)$, thus $s_1 = s_2$ and f is one-to-one. Now let $f: S \rightarrow T$ with f one-to-one. We want to define $g: T \rightarrow S$. For $t \in T$ with $t \in f(S)$, define g(t) to be the unique preimage of t under f. For $t \in T$ with $t \notin f(S)$, let g(t) be any fixed element of S. Then $g: T \rightarrow S$ and for $s \in S$, $(g \circ f)(s) = g(f(s)) = g(t) = s$, so $g \circ f = i_S$.

b. Assume f has a right inverse g, and let $t \in T$. Then $t = i_T(t) = (f \circ g)(t) = f(g(t))$; $g(t) \in S$, so $t \in f(S)$ and f is onto. Now let $f: S \rightarrow T$ with f onto. Then every $t \in T$ has at least one preimage in S under f. Define $g: T \rightarrow S$ by g(t) = a fixed preimage s of t. Then $(f \circ g)(t) = f(g(t)) = f(s) = t$, so $f \circ g = i_T$.

c. For example:

$$g_1(x) = \begin{cases} x/3 & \text{for } x = 3k \quad k \text{ an integer} \\ 0 & \text{for } x \neq 3k \end{cases}$$

$$g_2(x) = \begin{cases} x/3 & \text{for } x = 3k \quad k \text{ an integer} \\ 1 & \text{for } x \neq 3k \end{cases}$$

d. $g_1(x) = 2x$
$g_2(x) = 2x - 1$

34. $f^{-1}: T \to S$, $g^{-1}: U \to T$, so $f^{-1} \circ g^{-1}: U \to S$. For $s \in S$, let $f(s) = t$ and $g(t) = u$. Then $(f^{-1} \circ g^{-1}) \circ (g \circ f)(s) = f^{-1}(g^{-1}(u)) = f^{-1}(t) = s$. Also for $u \in U$, $(g \circ f) \circ (f^{-1} \circ g^{-1})(u) = g(f(s)) = g(t) = u$. Then $(f^{-1} \circ g^{-1}) \circ (g \circ f) = i_S$ and $(g \circ f) \circ (f^{-1} \circ g^{-1}) = i_U$, so $f^{-1} \circ g^{-1} = (g \circ f)^{-1}$.

*35. a. (1, 3, 5, 2)
b. (1, 4, 3, 2, 5)

36. a. $\begin{pmatrix} a & b & c & d \\ c & b & d & a \end{pmatrix}$
b. $\begin{pmatrix} a & b & c & d \\ b & d & a & c \end{pmatrix}$
c. $\begin{pmatrix} a & b & c & d \\ d & a & c & b \end{pmatrix}$
d. $\begin{pmatrix} a & b & c & d \\ c & d & b & a \end{pmatrix}$

37. Both $h \circ (g \circ f)$ and $(h \circ g) \circ f$ have domain and codomain A. For $x \in A$, $(h \circ (g \circ f))(x) = h((g \circ f)(x)) = h(g(f(x))) = (h \circ g)(f(x)) = ((h \circ g) \circ f)(x)$.

*38. a. (1, 2, 5, 3, 4) b. $(1, 7, 8) \circ (2, 4, 6)$
c. $(1, 5, 2, 4) \circ (3, 6)$ d. $(2, 3) \circ (4, 8) \circ (5, 7)$

39. a. (1, 6, 4, 8, 3, 5, 2)
b. $(1, 3) \circ (2, 4) \circ (5, 13, 6)$
c. (1, 5, 4, 3, 2)

40. For example, $f: \mathbb{Z} \to \mathbb{Z}$ defined by $f(x) = x + 1$.

41. a. $\begin{pmatrix} 1 & 2 & 3 & 4 \\ 2 & 1 & 4 & 3 \end{pmatrix}$ b. For example, $\begin{pmatrix} 1 & 2 & 3 & 4 \\ 3 & 1 & 2 & 4 \end{pmatrix}$

*42. a. 3^4 b. 36

43. a. 4^3 b. 4!

44. a. for $|S| = 2$, $2! = 2$ and $2^2 - C(2, 1)(1)^2 = 4 - 2 = 2$
for $|S| = 3$, $3! = 6$ and $3^3 - C(3, 1)(2)^3 + C(3, 2)(1)^3 = 27 - 3 \cdot 8 + 3 = 6$
for $|S| = 4$, $4! = 24$ and $4^4 - C(4, 1)(3)^4 + C(4, 2)(2)^4 - C(4, 3)(1)^4 = 256 - 4 \cdot 81 + 6 \cdot 16 - 4 = 24$

b. Assume f is onto. If two distinct elements of S map to one element of S, then n - 2 elements are left to map onto n - 1 elements, which cannot be done. Therefore f is one-to-one. Now assume f is one-to-one. Then the n elements of S map to n distinct elements of S; thus every element of S is in the range of f, and f is onto.

c. For example, S = N, f: N → N given by f(x) = 2x.

d. For example, S = N, f: N → N given by $\begin{cases} f(0) = 0 \\ f(x) = x - 1, x \geq 1 \end{cases}$

*45. a. n^n
b. n!
c. n!
d. n!
e. $n!\left[1-\frac{1}{1!}+\frac{1}{2!}-\frac{1}{3!}+\cdots+(-1)^n\frac{1}{n!}\right] = n!\left[\frac{1}{2!}-\frac{1}{3!}+\cdots+(-1)^n\frac{1}{n!}\right] < n!\left[\frac{1}{2!}\right] = n!\cdot\frac{1}{2} < n!$
f. Number of derangements < n! < n^n. The total number of functions, with no restrictions, is the maximum. Only some of these functions are one-to-one and onto, but this is the definition of a permutation as well. Not all permutations are derangements, so the number of derangements is smaller still.

46. a. This is the number of onto functions from a set of 5 elements to a set of 3, which is 150.
b. If Maria does additional tasks, then the mapping from the test plan development to Maria is already determined, leaving the remaining 4 tasks to be assigned to 3 workers with each worker getting at least one task. This number is 36. If Maria does no additional task, then the mapping from the test plan development to Maria is already determined, leaving the remaining 4 tasks to be assigned to 2 workers with each worker getting at least one task. This number is 14. By the Addition Principle, the total number of outcomes is 36 + 14 = 50.

47. For a given onto function from a set with m elements to a set with n elements, any permutation of the n images would give a different onto function but would determine the same partition of m objects into n blocks. Hence dividing the number of onto functions by n!, the number of image permutations, will give S(m, n).

48. 24, 9

$$\begin{pmatrix} a & b & c & d \\ b & c & d & a \end{pmatrix} \begin{pmatrix} a & b & c & d \\ b & d & a & c \end{pmatrix} \begin{pmatrix} a & b & c & d \\ b & a & d & c \end{pmatrix} \begin{pmatrix} a & b & c & d \\ c & a & d & b \end{pmatrix} \begin{pmatrix} a & b & c & d \\ c & d & a & b \end{pmatrix}$$

$$\begin{pmatrix} a & b & c & d \\ c & d & b & a \end{pmatrix} \begin{pmatrix} a & b & c & d \\ d & a & b & c \end{pmatrix} \begin{pmatrix} a & b & c & d \\ d & c & b & a \end{pmatrix} \begin{pmatrix} a & b & c & d \\ d & c & a & b \end{pmatrix}$$

*49. This is the number of derangements of 7 items, which is 1854.

50. a. $e^x = 1+\frac{x}{1!}+\frac{x^2}{2!}+\frac{x^3}{3!}+\cdots$

b. $e^{-1} = 1 - \frac{1}{1!} + \frac{1}{2!} - \frac{1}{3!} + L$

c. $e^{-1} \sim 0.36788$

d. The expression in brackets in Equation (4) is the sum of the first n + 1 terms of the series representation for e^{-1}. Because the terms of the series approach zero as n gets larger, e^{-1} is close to the value of this expression for large n. Therefore the expression $n!e^{-1}$ is a good approximation to (4).

e. $7!e^{-1} \sim 1854.1$

f. $10!e^{-1} \sim 1334961$

51. a. The values are stored in locations 6, 14, 1, 7, 8, 2, 16, 9, 0.

b. 58 hashes to location 7, which contains another element (40), so, following the collision resolution scheme under which 58 would have stored, search the next table position, 8, which contains 24, then search the next table position, 9, which contains 58. 41 also hashes to location 7 in the table; proceeding as before, locations 8 and 9 are also checked, and do not contain 41. The next location to check is 10, which is empty. Therefore 41 is not in the table.

c. If, say, item 24, stored at location 8, is deleted, then in searching for 58, we would check location 7 and then location 8. Finding location 8 to be empty, we would conclude incorrectly that 58 is not in the table.

*52. a. 3

b. X

53. a. We must have $11 \cdot d \mod 8 = 1$, with $0 < d < 8$, so $d = 3$ ($11 \cdot 3 = 33 \mod 8 = 1$).

b. The code for 3 is $3^{11} \mod 15$. Doing successive reductions mod 15, $3^{11} \rightarrow 3^4 \cdot 3^4 \cdot 3^3 \rightarrow 81 \cdot 81 \cdot 27 \rightarrow 6 \cdot 6 \cdot 12 \rightarrow 6 \cdot 72 \rightarrow 6 \cdot 12 \rightarrow 72 \rightarrow 12$.

c. To decode, compute $(12)^3 \mod 15$. Doing successive reductions mod 15, $12^3 \rightarrow (12^2) \cdot 12 \rightarrow 144 \cdot 12 \rightarrow 9 \cdot 12 \rightarrow 108 \rightarrow 3$.

54. Reflexive: $S \rho S$ by the identify function.

Symmetric: If $S \rho T$ and f is a bijection from S to T, then $f^{-1}: T \rightarrow S$ and f^{-1} is a bijection, so $T \rho S$.

Transitive: if $S \rho T$ and $T \rho U$, $f: S \rightarrow T$, $g: T \rightarrow U$, f and g bijections, then $g \circ f : S \rightarrow U$ and $g \circ f$ is a bijection, so $S \rho U$.

55. $[A] = \{A\}$, $[B] = \{B, C, F\}$ $[D] = \{D, E\}$

56. a. Let $t \in f(A \cap B)$. Then $t = f(s)$ for some $s \in A \cap B$. Thus $t \in f(A)$ and $t \in f(B)$, so $t \in f(A) \cap f(B)$.

b. Assume f is one-to-one. By (a), $f(A \cap B) \subseteq f(A) \cap f(B)$. Let $t \in f(A) \cap f(B)$. Then $t = f(s_1)$ for some $s_1 \in A$ and $t = f(s_2)$ for some $s_2 \in B$. Because f is

one-to-one, $s_1 = s_2$ and $s_1 \in A \cap B$, so $t \in f(A \cap B)$. Thus $f(A) \cap f(B) \subseteq f(A \cap B)$ and the two sets are equal.
Now assume $f(A \cap B) = f(A) \cap f(B)$ for all subsets A and B of S, and let $s_1, s_2 \in S$ such that $f(s_1) = f(s_2)$. Let $t = f(s_1)$ and let $A = \{s_1\}$, $B = \{s_2\}$. Then $t \in f(A) \cap f(B)$, or $t \in f(A \cap B)$. Therefore $A \cap B \neq \varnothing$, and $s_1 = s_2$, so f is one-to-one.

*57. a. {m, n, o, p}
b. {n, o, p}; {o}

58. a. For $x \in S$, $f(x) = f(x)$, so $x \rho x$ and ρ is reflexive. For $x, y \in S$, if $x \rho y$ then $f(x) = f(y)$ and $f(y) = f(x)$ so $y \rho x$ and ρ is symmetric. For $x, y, z \in S$, if $x \rho y$ and $y \rho z$ then $f(x) = f(y)$ and $f(y) = f(z)$ so $f(x) = f(z)$, and $x \rho z$, so ρ is transitive.
b. $[4] = \{4, -4\}$

59. This algorithm does the maximum amount of work when the wff is a tautology, because it must examine every row of the truth table to see that each gives a true result for the wff. If the wff has n statement letters, there are 2^n rows to the truth table.

*60. For example, $n_0 = c_2 = 1$, $c_1 = 1/34$

61. For example, $n_0 = 2$, $c_1 = 1$, $c_2 = 6$

62. $\log(x^2 + 3) = \Theta(\log(x^2))$ (ignoring low-order terms) $= \Theta(2 \log x)$ (property of logarithms) $= \Theta(\log x)$ (ignoring constant coefficient)

63. Yes. For example, in Exercise 60, we could use the constants $n_0 = 1$, $c_1 = 1/34$, $c_2 = 1/10$. Then the envelope would be entirely below g(x), but it still follows the general "shape" of g(x).

64. Let $f_1 = \Theta(g_1)$ and $f_2 = \Theta(g_2)$. Then there exist positive constants $n_0, n_1, c_1, c_2, d_1, d_2$ with

$$c_1 g_1(x) \le f_1(x) \le c_2 g_1(x) \quad \text{for } x \ge n_0$$

and

$$d_1 g_2(x) \le f_2(x) \le d_2 g_2(x) \quad \text{for } x \ge n_1$$

Then for all $x \ge \max(n_0, n_1)$,
$\min(c_1, d_1)\max(g_1(x), g_2(x)) \le c_1 g_1(x) + d_1 g_2(x) \le f_1(x) + f_2(x)$
$\le c_2 g_1(x) + d_2 g_2(x) \le c_2 \max(g_1(x), g_2(x)) + d_2 \max(g_1(x), g_2(x))$
$= (c_2 + d_2)\max(g_1(x), g_2(x))$

and $f_1 + f_2 = \Theta(\max(g_1, g_2)$.

*65. $\lim_{x\to\infty} \frac{x}{17x+1} = \lim_{x\to\infty} \frac{1}{17} = \frac{1}{17}$

66. $\lim_{x\to\infty} \frac{3x^3-7x}{\frac{1}{2}x^3} = \lim_{x\to\infty} \frac{9x^2-7}{\frac{3}{2}x^2} = \frac{18x}{3x} = 6$

*67. $\lim_{x\to\infty} \frac{x}{x^2} = \lim_{x\to\infty} \frac{1}{2x} = 0$

68. $\lim_{x\to\infty} \frac{\log x}{x} = \lim_{x\to\infty} \frac{\frac{1}{x}\log e}{1} = 0$

69. $\lim_{x\to\infty} \frac{(\ln x)^2}{x^{0.5}} = \lim_{x\to\infty} \frac{2(\ln x)\frac{1}{x}}{0.5x^{-0.5}} = \lim_{x\to\infty} \frac{4\ln x}{x^{0.5}} = \lim_{x\to\infty} \frac{4\frac{1}{x}}{0.5x^{-0.5}} = \lim_{x\to\infty} \frac{8}{x^{0.5}} = 0$

70. $[200 \log x] = [41 \ln x^2] < [\sqrt[4]{x}] < [420 x] < [17 x \log x] < [10x^2 - 3x + 5] < [2^x - x^2]$

71. $[\log x] = [\ln x] < [(\log x)2] < [\sqrt{x}] < [x] < [x \log x] < [x^3] = [2x^3 + x]$
$= [x^3 + \log x] < [e^x]$

EXERCISES 4.5

*1. 2, -4

2. x = 2, y = 4

*3. x = 1, y = 3, z = -2, w = 4

4. u = 1, v = -3, w = 7

5. *a. $\begin{bmatrix} 6 & -5 \\ 0 & 3 \\ 5 & 3 \end{bmatrix}$ b. $\begin{bmatrix} -2 & 7 \\ -2 & -3 \\ 1 & 5 \end{bmatrix}$ c. $\begin{bmatrix} 12 & 3 & 6 \\ 18 & -3 & 15 \\ 3 & 9 & 6 \end{bmatrix}$

d. $\begin{bmatrix} -4 & -8 \\ -12 & 2 \end{bmatrix}$ *e. $\begin{bmatrix} 14 & -17 \\ 2 & 9 \\ 9 & 1 \end{bmatrix}$ f. not possible

g. $\begin{bmatrix} 18 & -15 \\ 0 & 9 \\ 15 & 9 \end{bmatrix}$ h. $\begin{bmatrix} -12 & -24 \\ -36 & 6 \end{bmatrix}$ *i. $\begin{bmatrix} 21 & -23 \\ 33 & -44 \\ 11 & 1 \end{bmatrix}$

j. $\begin{bmatrix} -28 & 22 \\ 20 & 1 \\ -2 & 9 \end{bmatrix}$ k. $\begin{bmatrix} 10 & 7 \\ -2 & -4 \\ 30 & 8 \end{bmatrix}$ l. not possible

*m. $\begin{bmatrix} 28 & 4 \\ 6 & 25 \end{bmatrix}$ n. $\begin{bmatrix} 17 & 6 \\ 29 & 29 \\ 7 & 8 \end{bmatrix}$

6. a. $A \cdot B = \begin{bmatrix} 10 & 4 \\ 18 & -3 \end{bmatrix}$ $B \cdot A = \begin{bmatrix} 14 & 1 \\ 4 & -7 \end{bmatrix}$

b. $A(B \cdot C) = \begin{bmatrix} 3 & -1 \\ 2 & 5 \end{bmatrix}\begin{bmatrix} 26 & -22 \\ 10 & -8 \end{bmatrix} = \begin{bmatrix} 64 & -58 \\ 102 & -84 \end{bmatrix}$

$(A \cdot B)C = \begin{bmatrix} 10 & 4 \\ 18 & -3 \end{bmatrix}\begin{bmatrix} 6 & -5 \\ 2 & -2 \end{bmatrix} = \begin{bmatrix} 68 & -58 \\ 102 & -84 \end{bmatrix}$

*c. $A(B + C) = \begin{bmatrix} 3 & -1 \\ 2 & 5 \end{bmatrix}\begin{bmatrix} 10 & -4 \\ 4 & 3 \end{bmatrix} = \begin{bmatrix} 26 & -9 \\ 40 & -23 \end{bmatrix}$

$A \cdot B + A \cdot C = \begin{bmatrix} 10 & 4 \\ 18 & -3 \end{bmatrix} + \begin{bmatrix} 16 & -13 \\ 22 & -20 \end{bmatrix} = \begin{bmatrix} 26 & -9 \\ 40 & -23 \end{bmatrix}$

d. $(A + B)C = \begin{bmatrix} 7 & 0 \\ 4 & 4 \end{bmatrix}\begin{bmatrix} 6 & -5 \\ 2 & -2 \end{bmatrix} = \begin{bmatrix} 42 & -35 \\ 32 & -28 \end{bmatrix}$

$A \cdot C + B \cdot C = \begin{bmatrix} 16 & -13 \\ 22 & -20 \end{bmatrix} + \begin{bmatrix} 26 & -22 \\ 10 & -8 \end{bmatrix} = \begin{bmatrix} 42 & -35 \\ 32 & -28 \end{bmatrix}$

7. $x = 3, y = 4$

8. a. $I \cdot A = A$ for any $n \times n$ matrix A, in particular, if $A = I$, then $I^2 = I \cdot I = I$
b. $I^1 = I$. Assume $I^k = I$. Then $I^{k+1} = I^k \cdot I = I \cdot I = I$

9. a. Assume that row i of A is all 0s. Then for any j, the element in row i, column j of $A \cdot B$ is given by $\sum_{k=1}^{n} a_{ik} b_{kj}$. This sum is 0 because $a_{ik} = 0$ for all values of k.
b. Assume that column j of B is all 0s. Then for any i, the element in row i, column j of $A \cdot B$ is given by $\sum_{k=1}^{n} a_{ik} b_{kj}$. This sum is 0 because $b_{kj} = 0$ for all values of k.

10. a. Let $C = A + B$. Then $c_{ij} = a_{ij} + b_{ij}$. If $i \neq j$ then $a_{ij} = b_{ij} = 0$, so $c_{ij} = 0$.

b. Element i, j in rA is given by $r(a_{ij})$. If $a_{ij} = 0$, then $ra_{ij} = 0$, hence rA will be diagonal.

c. Let $C = A \cdot B$. Then c_{ij} is given by $\sum_{k=1}^{n} a_{ik} b_{kj}$. Assume that $i \neq j$. Then for any value of k, $1 \leq k \leq n$, k cannot equal both i and j, so that either a_{ik} or b_{kj} (or both) is 0. Therefore each term in the summation is 0 and $c_{ij} = 0$.

11.*a. $\begin{bmatrix} 1 & 3 \\ 2 & 2 \end{bmatrix}\begin{bmatrix} -1/2 & 3/4 \\ 1/2 & -1/4 \end{bmatrix} = \begin{bmatrix} 1 & 0 \\ 0 & 1 \end{bmatrix} = \begin{bmatrix} -1/2 & 3/4 \\ 1/2 & -1/4 \end{bmatrix}\begin{bmatrix} 1 & 3 \\ 2 & 2 \end{bmatrix}$

*b. For $\begin{bmatrix} 1 & 2 \\ 2 & 4 \end{bmatrix}\begin{bmatrix} b_{11} & b_{12} \\ b_{21} & b_{22} \end{bmatrix} = \begin{bmatrix} 1 & 0 \\ 0 & 1 \end{bmatrix}$

$$\begin{aligned} b_{11} + 2b_{21} &= 1 & \qquad b_{12} + 2b_{22} &= 0 \\ 2b_{11} + 4b_{21} &= 0 & \qquad 2b_{12} + 4b_{22} &= 1 \end{aligned}$$

which is an inconsistent system of equations with no solution.

c. $\begin{bmatrix} a_{11} & a_{12} \\ a_{21} & a_{22} \end{bmatrix}\begin{bmatrix} b_{11} & b_{12} \\ b_{21} & b_{22} \end{bmatrix} = \begin{bmatrix} 1 & 0 \\ 0 & 1 \end{bmatrix}$

implies

$$\begin{aligned} a_{11} b_{11} + a_{12} b_{21} &= 1 & \qquad a_{11} b_{12} + a_{12} b_{22} &= 0 \\ a_{21} b_{11} + a_{22} b_{21} &= 0 & \qquad a_{21} b_{12} + a_{22} b_{22} &= 1 \end{aligned}$$

Solving these systems of equations gives

$$b_{11} = \frac{a_{22}}{a_{11}a_{22} - a_{12}a_{21}} \qquad b_{12} = \frac{-a_{12}}{a_{11}a_{22} - a_{12}a_{21}}$$

$$b_{21} = \frac{-a_{21}}{a_{11}a_{22} - a_{12}a_{21}} \qquad b_{22} = \frac{a_{11}}{a_{11}a_{22} - a_{12}a_{21}}$$

These values can all be found if $a_{11}a_{22} - a_{12}a_{21} \neq 0$.

12. $(r\mathbf{A})(1/r)\mathbf{A}^{-1} = r(1/r)(\mathbf{A} \cdot \mathbf{A}^{-1}) = 1\mathbf{I} = \mathbf{I}$
$(1/r)\mathbf{A}^{-1}(r\mathbf{A}) = (1/r)(r)(\mathbf{A}^{-1} \cdot \mathbf{A}) = 1\mathbf{I} = \mathbf{I}$

13. If **A** is invertible, then $\mathbf{A}^{-1}$ exists, and

$$\begin{aligned} &\mathbf{A}^{-1}(\mathbf{A} \cdot \mathbf{B}) = \mathbf{A}^{-1}(\mathbf{A} \cdot \mathbf{C}) \\ &(\mathbf{A}^{-1} \cdot \mathbf{A})\mathbf{B} = (\mathbf{A}^{-1} \cdot \mathbf{A})\mathbf{C} \\ &\mathbf{IB} = \mathbf{IC} \\ &\mathbf{B} = \mathbf{C} \end{aligned}$$

14. a. Here are the steps:

$$A = \begin{bmatrix} 1 & 3 \\ 2 & 2 \end{bmatrix} \quad I = \begin{bmatrix} 1 & 0 \\ 0 & 1 \end{bmatrix}$$

Multiply row 1 by -2 and add to row 2:

$$\begin{bmatrix} 1 & 3 \\ 0 & -4 \end{bmatrix} \quad \begin{bmatrix} 1 & 0 \\ -2 & 1 \end{bmatrix}$$

Multiply row 2 by -1/4:

$$\begin{bmatrix} 1 & 3 \\ 0 & 1 \end{bmatrix} \quad \begin{bmatrix} 1 & 0 \\ 1/2 & -1/4 \end{bmatrix}$$

Multiply row 2 by -3 and add to row 1:

$$\begin{bmatrix} 1 & 0 \\ 0 & 1 \end{bmatrix} \quad \begin{bmatrix} -1/2 & 3/4 \\ 1/2 & -1/4 \end{bmatrix} = A^{-1}$$

b. Here are the steps:

$$\begin{bmatrix} -1 & 2 & -3 \\ 2 & 1 & 0 \\ 4 & -2 & 5 \end{bmatrix} \quad \begin{bmatrix} 1 & 0 & 0 \\ 0 & 1 & 0 \\ 0 & 0 & 1 \end{bmatrix}$$

Multiply row 1 by -1:

$$\begin{bmatrix} 1 & -2 & 3 \\ 2 & 1 & 0 \\ 4 & -2 & 5 \end{bmatrix} \quad \begin{bmatrix} -1 & 0 & 0 \\ 0 & 1 & 0 \\ 0 & 0 & 1 \end{bmatrix}$$

Multiply row 1 by -2 and add to row 2:

$$\begin{bmatrix} 1 & -2 & 3 \\ 0 & 5 & -6 \\ 4 & -2 & 5 \end{bmatrix} \quad \begin{bmatrix} -1 & 0 & 0 \\ 2 & 1 & 0 \\ 0 & 0 & 1 \end{bmatrix}$$

Multiply row 1 by -4 and add to row 3:

$$\begin{bmatrix} 1 & -2 & 3 \\ 0 & 5 & -6 \\ 0 & 6 & -7 \end{bmatrix} \begin{bmatrix} -1 & 0 & 0 \\ 2 & 1 & 0 \\ 4 & 0 & 1 \end{bmatrix}$$

Multiply row 2 by 1/5:

$$\begin{bmatrix} 1 & -2 & 3 \\ 0 & 1 & -6/5 \\ 0 & 6 & -7 \end{bmatrix} \begin{bmatrix} -1 & 0 & 0 \\ 2/5 & 1/5 & 0 \\ 4 & 0 & 1 \end{bmatrix}$$

Multiply row 2 by 2 and add to row 1:

$$\begin{bmatrix} 1 & 0 & 3/5 \\ 0 & 1 & -6/5 \\ 0 & 6 & -7 \end{bmatrix} \begin{bmatrix} -1/5 & 2/5 & 0 \\ 2/5 & 1/5 & 0 \\ 4 & 0 & 1 \end{bmatrix}$$

Multiply row 2 by -6 and add to row 3:

$$\begin{bmatrix} 1 & 0 & 3/5 \\ 0 & 1 & -6/5 \\ 0 & 0 & 1/5 \end{bmatrix} \begin{bmatrix} -1/5 & 2/5 & 0 \\ 2/5 & 1/5 & 0 \\ 8/5 & -6/5 & 1 \end{bmatrix}$$

Multiply row 3 by 5:

$$\begin{bmatrix} 1 & 0 & 3/5 \\ 0 & 1 & -6/5 \\ 0 & 0 & 1 \end{bmatrix} \begin{bmatrix} -1/5 & 2/5 & 0 \\ 2/5 & 1/5 & 0 \\ 8 & -6 & 5 \end{bmatrix}$$

Multiply row 3 by 6/5 and add to row 2:

$$\begin{bmatrix} 1 & 0 & 3/5 \\ 0 & 1 & 0 \\ 0 & 0 & 1 \end{bmatrix} \begin{bmatrix} -1/5 & 2/5 & 0 \\ 10 & -7 & 6 \\ 8 & -6 & 5 \end{bmatrix}$$

Multiply row 3 by -3/5 and add to row 1:

$$\begin{bmatrix} 1 & 0 & 0 \\ 0 & 1 & 0 \\ 0 & 0 & 1 \end{bmatrix} \begin{bmatrix} -5 & 4 & -3 \\ 10 & -7 & 6 \\ 8 & -6 & 5 \end{bmatrix} = A^{-1}$$

*15. First find A^{-1} by the method of Exercise 14:

$$A = \begin{bmatrix} 1 & 1 \\ 24 & 14 \end{bmatrix} \qquad I = \begin{bmatrix} 1 & 0 \\ 0 & 1 \end{bmatrix}$$

Multiply row 1 by -24 and add to row 2:

$$\begin{bmatrix} 1 & 1 \\ 0 & -10 \end{bmatrix} \begin{bmatrix} 1 & 0 \\ -24 & 1 \end{bmatrix}$$

Multiply row 2 by -1/10:

$$\begin{bmatrix} 1 & 1 \\ 0 & 1 \end{bmatrix} \begin{bmatrix} 1 & 0 \\ 24/10 & -1/10 \end{bmatrix}$$

Multiply row 2 by -1 and add to row 1:

$$\begin{bmatrix} 1 & 0 \\ 0 & 1 \end{bmatrix} \begin{bmatrix} -14/10 & 1/10 \\ 24/10 & -1/10 \end{bmatrix} = A^{-1}$$

Now multiply $A^{-1} \cdot B$:

$$\begin{bmatrix} -14/10 & 1/10 \\ 24/10 & -1/10 \end{bmatrix} \begin{bmatrix} 70 \\ 1180 \end{bmatrix} = \begin{bmatrix} 20 \\ 50 \end{bmatrix} = \begin{bmatrix} x \\ y \end{bmatrix}$$

so the solution is x = 20, y = 50.

16. First find the inverse of the matrix of coefficients:

$$\begin{bmatrix} 1 & 2 \\ 1 & 1 \end{bmatrix} \begin{bmatrix} 1 & 0 \\ 0 & 1 \end{bmatrix}$$

Multiply row 1 by -1 and add to row 2:

$$\begin{bmatrix} 1 & 2 \\ 0 & -1 \end{bmatrix} \begin{bmatrix} 1 & 0 \\ -1 & 1 \end{bmatrix}$$

Multiply row 2 by -1:

$$\begin{bmatrix} 1 & 2 \\ 0 & 1 \end{bmatrix} \begin{bmatrix} 1 & 0 \\ 1 & -1 \end{bmatrix}$$

Multiply row 2 by -2 and add to row 1:

$$\begin{bmatrix} 1 & 0 \\ 0 & 1 \end{bmatrix} \quad \begin{bmatrix} -1 & 2 \\ 1 & -1 \end{bmatrix}$$

Now multiply $A^{-1} \cdot B$:

$$\begin{bmatrix} -1 & 2 \\ 1 & -1 \end{bmatrix} \begin{bmatrix} -4 \\ 5 \end{bmatrix} = \begin{bmatrix} 14 \\ 9 \end{bmatrix}$$

so the solution is x = 14, y = -9.

17. a. $A^T = \begin{bmatrix} 1 & 6 \\ 3 & -2 \\ 4 & 1 \end{bmatrix}$

b. If A is symmetric then $a_{ij} = a_{ji}$ and $A^T(i, j) = A(j, i) = A(i, j)$. Therefore $A^T = A$. If $A^T = A$, then $A(i, j) = A^T(i, j) = A(j, i)$ and A is symmetric.

c. $(A^T)^T = A$ follows from the definition - two interchanges of row and column gets back to the original.

d. Let A + B = C. Then $C^T(i, j) = C(j, i) = A(j, i) + B(j, i) = A^T(i, j) + B^T(i, j)$ and $C^T = A^T + B^T$

e. Let A be an $n \times m$ matrix and B be an $m \times p$ matrix; then A^T is $m \times n$ and B^T is $p \times m$. Let $A \cdot B = C$. Then $C^T(i, j) = C(j, i) = \sum_{k=1}^{m} a_{jk} b_{ki} = \sum_{k=1}^{m} A^T(k, j) B^T(i, k) =$

$\sum_{k=1}^{m} B^T(i, k) A^T(k, j) = (B^T \cdot A^T)(i, j)$ and $C^T = B^T \cdot A^T$.

*18. For example, $\begin{bmatrix} 1 & 1 \\ -1 & -1 \end{bmatrix} \begin{bmatrix} 1 & 1 \\ -1 & -1 \end{bmatrix} = \begin{bmatrix} 0 & 0 \\ 0 & 0 \end{bmatrix}$

19. For example, $\begin{bmatrix} 1 & 2 \\ 2 & 1 \end{bmatrix} \begin{bmatrix} 1 & 2 \\ 1 & 2 \end{bmatrix} = \begin{bmatrix} 2 & 1 \\ 1 & 2 \end{bmatrix} \begin{bmatrix} 1 & 2 \\ 1 & 2 \end{bmatrix} = \begin{bmatrix} 3 & 6 \\ 3 & 6 \end{bmatrix}$

but $\begin{bmatrix} 1 & 2 \\ 2 & 1 \end{bmatrix} \neq \begin{bmatrix} 2 & 1 \\ 1 & 2 \end{bmatrix}$

20. This is not always true (for example, use the A and B of Practice 45). It is true if A = B = I, for example.

21. According to the Law of Cosines, in the following triangle

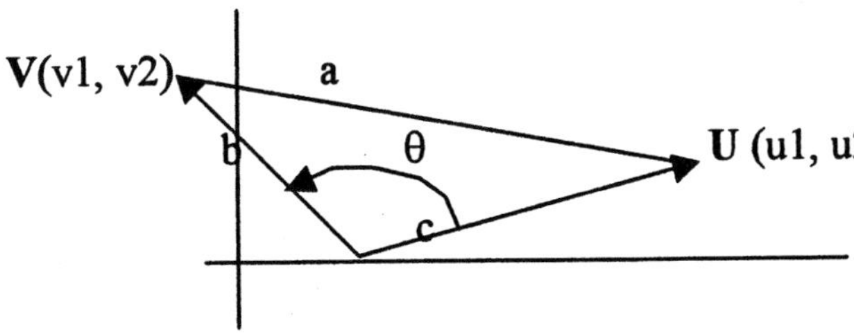

$a^2 = b^2 + c^2 - 2bc \cos\theta$. Also, $b = ||\mathbf{V}|| = \sqrt{v_1^2 + v_2^2}$, $c = ||\mathbf{U}|| = \sqrt{u_1^2 + u_2^2}$, and, by the distance formula, $a^2 = (u_1 - v_1)^2 + (u_2 - v_2)^2$. Therefore

$$(u_1 - v_1)^2 + (u_2 - v_2)^2 = v_1^2 + v_2^2 + u_1^2 + u_2^2 - 2\,||V|| \cdot ||U|| \cos\theta$$

$$\cos\theta = \frac{\mathbf{U} \bullet \mathbf{V}}{||\mathbf{U}|| \cdot ||\mathbf{V}||}$$

22.*a. The recurrence relation is in the form of Equation(1) of Section 2.5, where c = 7 and g(n) = 0; the solution is given by Equation (6) of Section 2.5 and is $M(n) = 7^{\log n}$.

b. A(1) = 0 because no additions are required to multiply two 1 × 1 matrices
To compute the product of two n × n matrices requires 7 multiplications of (n/2 × n/2) matrices, each of which requires $A\left(\frac{n}{2}\right)$ additions. The product also requires 18 additions of (n/2 × n/2) matrices, each of which requires $(n/2)^2$ additions.

c. Again using Equation (6) of Section 2.5, the solution is

$$A(n) = \sum_{i=1}^{\log n} 7^{\log n - i} 18\left(\frac{2^i}{2}\right)^2 = 7^{\log n}\frac{18}{4}\sum_{i=1}^{\log n} 7^{-i}2^{2i} = 7^{\log n}\frac{9}{2}\sum_{i=1}^{\log n}\frac{(2^2)^i}{7^i} = 7^{\log n}\frac{9}{2}\sum_{i=1}^{\log n}\left(\frac{4}{7}\right)^i =$$

$$7^{\log n}\frac{9}{2}\cdot\frac{4}{7}\left[1 + \frac{4}{7} + \left(\frac{4}{7}\right)^2 + \cdots + \left(\frac{4}{7}\right)^{\log n - 1}\right]$$ = (using formula for the sum of terms of a geometric sequence) $$7^{\log n}\frac{9}{2}\cdot\frac{4}{7}\cdot\frac{7}{3}\left[1 - \left(\frac{4}{7}\right)^{\log n}\right] = \Theta(7^{\log n})$$

d. log n log 7 = log 7 log n

e. The total work is $\Theta(7^{\log n}) + \Theta(7^{\log n}) = \Theta(7^{\log n}) = \Theta(n^{\log 7}) \cong \Theta(n^{2.8})$, which is better than the $\Theta(n^3)$ of the traditional algorithm.

*23. $A \wedge B = \begin{bmatrix} 1 & 0 & 0 \\ 0 & 1 & 0 \\ 0 & 1 & 1 \end{bmatrix}$ $A \vee B = \begin{bmatrix} 1 & 0 & 1 \\ 1 & 1 & 1 \\ 1 & 1 & 1 \end{bmatrix}$

$A \times B = \begin{bmatrix} 1 & 0 & 1 \\ 1 & 1 & 1 \\ 1 & 1 & 1 \end{bmatrix}$ $B \times A = \begin{bmatrix} 1 & 1 & 1 \\ 1 & 1 & 1 \\ 1 & 1 & 1 \end{bmatrix}$

24. $A \wedge B = \begin{bmatrix} 0 & 0 & 1 \\ 0 & 0 & 0 \\ 1 & 0 & 0 \end{bmatrix}$ $A \vee B = \begin{bmatrix} 0 & 1 & 1 \\ 1 & 1 & 0 \\ 1 & 0 & 0 \end{bmatrix}$

$A \times B = \begin{bmatrix} 1 & 0 & 0 \\ 0 & 1 & 1 \\ 0 & 1 & 1 \end{bmatrix}$ $B \times A = \begin{bmatrix} 1 & 1 & 0 \\ 0 & 0 & 0 \\ 0 & 0 & 1 \end{bmatrix}$

25. $A \wedge B = \begin{bmatrix} 0 & 1 & 0 \\ 0 & 0 & 1 \\ 0 & 0 & 0 \end{bmatrix}$ $A \vee B = \begin{bmatrix} 0 & 1 & 1 \\ 1 & 0 & 1 \\ 1 & 0 & 1 \end{bmatrix}$

$A \times B = \begin{bmatrix} 0 & 0 & 1 \\ 1 & 1 & 1 \\ 1 & 0 & 0 \end{bmatrix}$ $B \times A = \begin{bmatrix} 1 & 0 & 1 \\ 0 & 0 & 1 \\ 0 & 1 & 0 \end{bmatrix}$

26. In order for $A \vee B = A \wedge B$, it must be the case that $a_{ij} \vee b_{ij} = a_{ij} \wedge b_{ij}$ for all i, j. This is true if $a_{ij} = b_{ij} = 1$ or $a_{ij} = b_{ij} = 0$, therefore when A = B.

27. The i,j entry in $A \vee B$ is $\max(a_{ij}, b_{ij}) = \max(b_{ij}, a_{ij})$ = the i,j entry in $B \vee A$. A similar argument holds for $A \wedge B$.

28. Because the matrix is symmetric, only the entries on and below the main diagonal need be known. These entries will determine the entries above the main diagonal.

$$\begin{bmatrix} x & & & \\ x & x & & \\ x & x & x & \\ \cdots & & & \\ x & x & x \ldots & x \end{bmatrix}$$

There are $1 + 2 + 3 + \ldots + n = n(n + 1)/2$ such entries. Each entry can be either 0 or 1. The number of possibilities is therefore

$$2^{\frac{n(n+1)}{2}}$$

*29. The i,j entry of A^2 is $\sum_{k=1}^{n} a_{ik}a_{kj}$

The j,i entry of A^2 is $\sum_{k=1}^{n} a_{jk}a_{ki}$

But these are the same because $a_{ik} = a_{ki}$ and $a_{kj} = a_{jk}$ (A is symmetric)

30. $(A \cdot A^T)^T = (A^T)^T \cdot A^T$ by Exercise 17(e)
$= A \cdot A^T$ by Exercise 17(c)

Therefore $A \cdot A^T$ is its own transpose, so it is symmetric by Exercise 17(b).

31. For n = 1,

$$A = \begin{bmatrix} 1 & 1 \\ 1 & 0 \end{bmatrix} = \begin{bmatrix} F(2) & F(1) \\ F(1) & F(0) \end{bmatrix}$$

Assume that $A^k = \begin{bmatrix} F(k+1) & F(k) \\ F(k) & F(k-1) \end{bmatrix}$

Then $A^{k+1} = A^k \cdot A = \begin{bmatrix} F(k+1) & F(k) \\ F(k) & F(k-1) \end{bmatrix} \begin{bmatrix} 1 & 1 \\ 1 & 0 \end{bmatrix}$

$$= \begin{bmatrix} F(k+1)+F(k) & F(k+1) \\ F(k)+F(k-1) & F(k) \end{bmatrix} = \begin{bmatrix} F(k+2) & F(k+1) \\ F(k+1) & F(k) \end{bmatrix}$$

CHAPTER 5: Graphs and Trees

Students usually like to work with graphs because "it's fun to draw pictures." While the sheer volume of graph terminology at the beginning of the chapter is a little overwhelming, Section 5.1 on the whole goes rather easily. Graphs provide a nice way to introduce the concept of isomorphism, again because it is easy to "see" the correspondences, and check whether adjacency of nodes is preserved. This gentle beginning paves the way for later discussions on isomorphic Boolean algebras (Chapter 7) and isomorphic groups (Chapter 8). Graph planarity is an easily-understood idea that students can play with for a bit, and then Euler's formula gives some theoretical underpinnings; also the proof of Euler's formula provides a nice example of induction. The topic of graph coloring is treated in a series of exercises (68-76) at the end of Section 5.1.

It is worth pointing out that although a computer can store a picture of a graph as an image file, this is not a form that allows us to work with the graph data (note that the formal definition of a graph is not even given in visual terms). What the computer representation must capture is enough information to fulfill the definition of the graph, which would allow a visual representation to be reconstructed if desired. Adjacency matrix and linked-list graph representations are discussed in Section 5.1.

Trees and their representations are introduced in Section 5.2. Simple tree traversal algorithms - inorder, preorder, postorder - provide witness to the power of recursion, and results about trees provide still more practice with inductive proofs.

Applications of trees in Sections 5.3 and 5.4 include decision trees that give lower bounds on the work required for searching and sorting, binary search trees, and Huffman encoding. These topics are certainly optional, especially as the students may see them later in a data structures course.

EXERCISES 5.1

1. g(a) = (1, 2)
   g(b) = (1, 3)
   g(c) = (2, 3)
   g(d) = (2, 2)

2. a. yes b. no c. yes d. 3, a_5, 5, a_6, 6; 3, a_3, 4, a_4, 5, a_6, 6
   e. 3, a_3, 4, a_4, 5, a_5, 3 f. a_3, a_4, or a_5 g. a_1, a_2, a_6, or a_7

*3. a. 

b. For example, 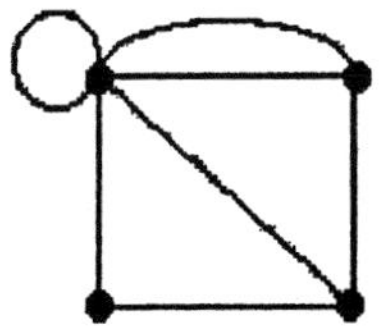

c. 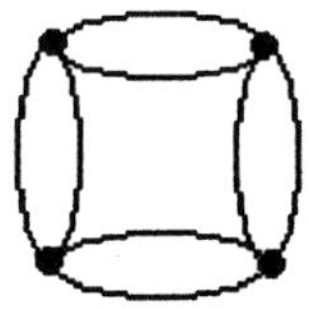

*4. a. 4, 5, 6
b. length 2
c. for example (naming the nodes), 1-2-1-2-2-1-4-5-6

5. a. 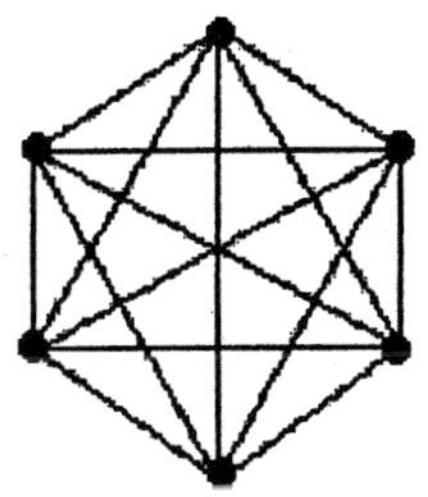

b. 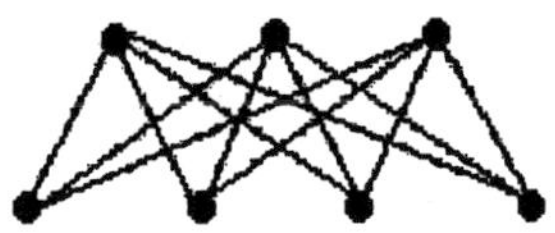

6. a. For example

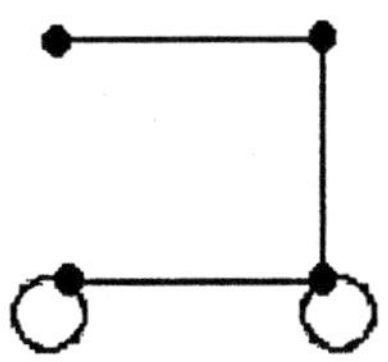

b. Does not exist; the node of degree 4 would have to have arcs going to 4 distinct other nodes, since no loops or parallel arcs are allowed, but there are not 4 distinct other nodes.

c.

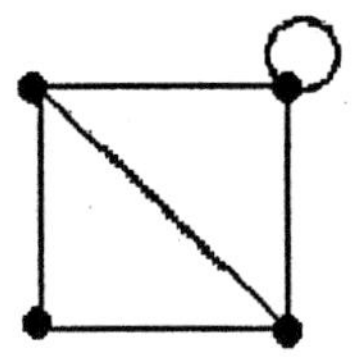

d. Does not exist; in such a graph, the sum of all the degrees would be 11, but the sum of all the degrees is the total number of arc ends, which must be twice the number of arcs, i.e., an even number.

*7. (b), because there is no node of degree 0.

8. (d), because the two nodes of degree 3 are not adjacent.

*9. f_1: 1 → a
2 → b
3 → c
4 → d

f_2: $a_1 \rightarrow e_2$
$a_2 \rightarrow e_7$
$a_3 \rightarrow e_6$
$a_4 \rightarrow e_1$
$a_5 \rightarrow e_3$
$a_6 \rightarrow e_4$
$a_7 \rightarrow e_5$

10. Not isomorphic; graph in (b) has a loop, graph in (a) does not.

11. f: 1 → a
2 → d
3 → b
4 → e
5 → c

12. f: 1 → b
2 → d
3 → c
4 → e
5 → f
6 → a

*13. Not isomorphic; graph in (b) has a node of degree 5, graph in (a) does not.

14. f: 1 → a
2 → c
3 → e
4 → g
5 → f
6 → h
7 → b
8 → d

15. a. There cannot be a bijection between the two node sets if they are not the same size.

b. For isomorphic graphs there is a bijection from one arc set to the other, either explicitly or, in the case of simple graphs, implicitly by means of the endpoints; this cannot happen if the arc sets are not the same size.

c. If arcs a_1 and a_2 in one graph both have endpoints x-y, then their image arcs in the second graph must have the same endpoints, which cannot happen if the second graph has no parallel arcs.

d. If an arc in one graph has endpoints x-x, then its image arc in the second graph must have endpoints f(x)-f(x), which is not possible if the second graph has no loops.

e. A node of degree k in one graph serves as an endpoint to k arcs; its image in the second graph must serve as an endpoint to the images of those k arcs, which implies it will have degree k also.

f. If there is a path n_1, a_1, n_2, a_2, ... , n_k between two nodes in one graph, then $f(n_1)$, $f(a_1)$, $f(n_2)$, $f(a_2)$, ... , $f(n_k)$ is a path in the second graph. Two nodes in the second graph are the images of nodes in the first graph; if the first graph is connected, there is a path between these nodes and hence there is a path between the two nodes in the second graph.

g. By the answer to part f, paths map to paths, so cycles map to cycles.

*16. 2 graphs

17. 4 graphs

18. 11 graphs

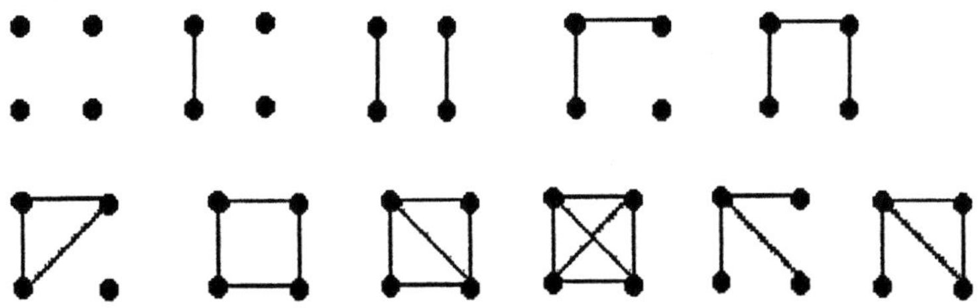

19. $\frac{n(n-1)}{2}$

The number of arcs is the number of ways to select 2 nodes out of n, C(n, 2) = $\frac{n(n-1)}{2}$. (Other proof methods include induction on the number of nodes.)

20. n = 13, a = 19, r = 8 and n - a + r = 2

*21.

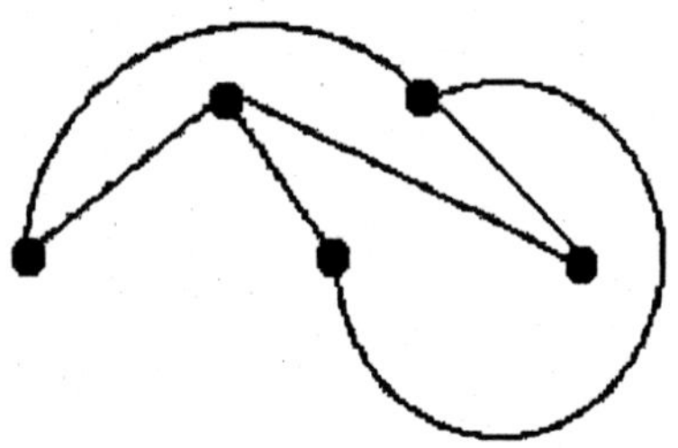

$K_{2,3}$

22.

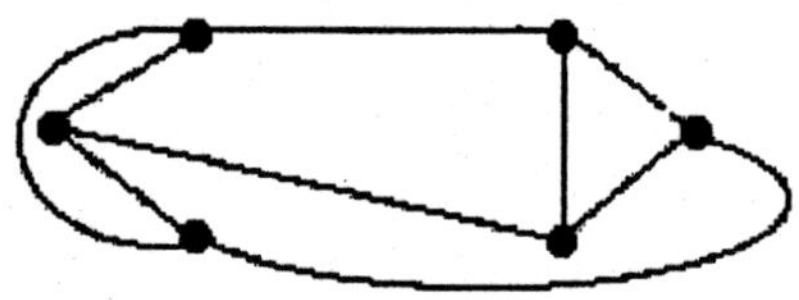

*23. 5 (by Euler's formula)

24. 8 (by Euler's formula)

25. The proof for Euler's formula does not depend on the graph being simple, so the result still holds for nonsimple graphs, but this is not true for inequalities (2) and (3).

26. In elementary subdivisions, the inserted node must be a new node, so once v is made a node from the first subdivision, it cannot be a node in the second subdivision.

*27. Planar

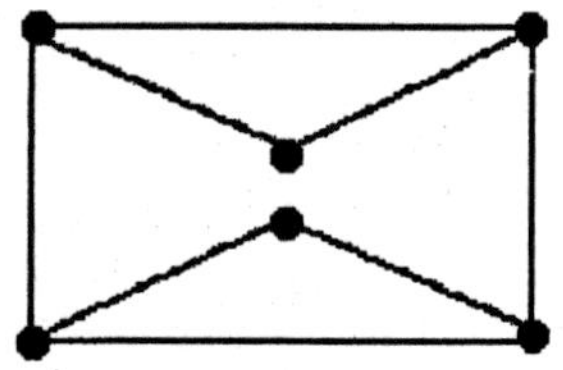

28. Planar

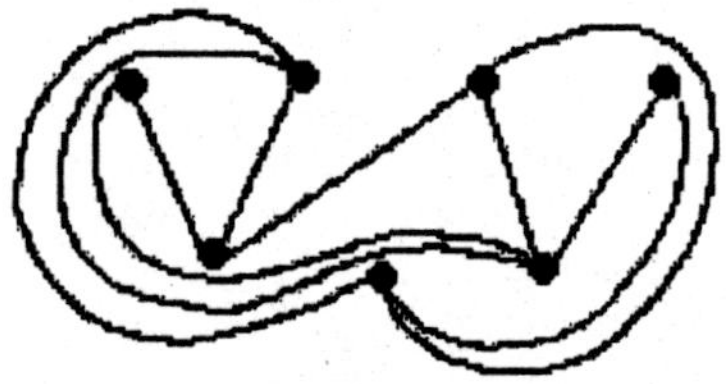

*29. Nonplanar - subgraph below can be obtained from $K_{3,3}$ by elementary subdivisions.

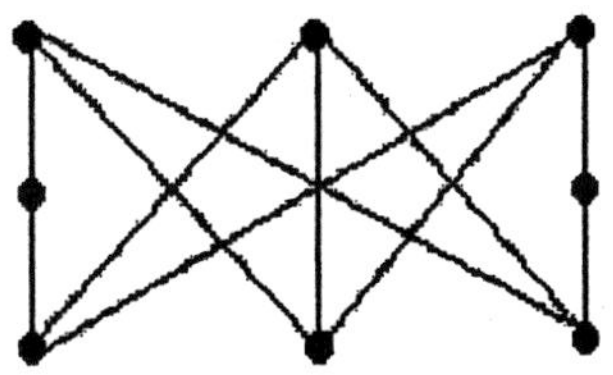

30. Nonplanar - the representation of the original graph shown below indicates that it can be obtained from K_5 by elementary subdivisions.

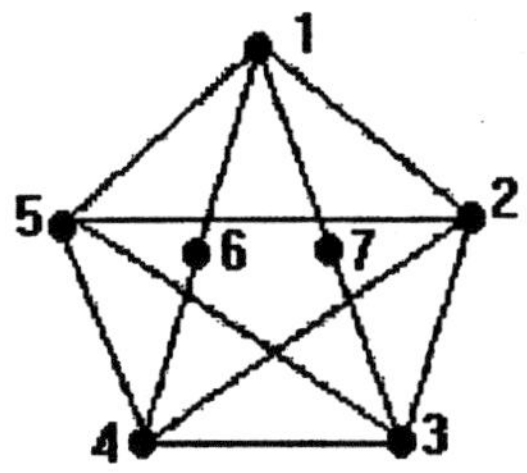

*31. $$\begin{bmatrix} 1 & 1 & 0 & 0 & 2 \\ 1 & 1 & 1 & 1 & 1 \\ 0 & 1 & 0 & 1 & 0 \\ 0 & 1 & 1 & 0 & 0 \\ 2 & 1 & 0 & 0 & 0 \end{bmatrix}$$

32. $$\begin{bmatrix} 1 & 0 & 1 & 0 & 0 \\ 0 & 0 & 1 & 1 & 1 \\ 1 & 1 & 0 & 1 & 0 \\ 0 & 1 & 1 & 0 & 1 \\ 0 & 1 & 0 & 1 & 0 \end{bmatrix}$$

33. $$\begin{bmatrix} 0 & 1 & 1 & 0 & 0 & 0 & 0 \\ 1 & 0 & 0 & 1 & 0 & 1 & 1 \\ 1 & 0 & 0 & 0 & 1 & 1 & 1 \\ 0 & 1 & 0 & 0 & 0 & 1 & 0 \\ 0 & 0 & 1 & 0 & 0 & 0 & 1 \\ 0 & 1 & 1 & 1 & 0 & 0 & 1 \\ 0 & 1 & 1 & 0 & 1 & 1 & 0 \end{bmatrix}$$

34. $$\begin{bmatrix} 0 & 1 & 1 & 0 & 0 & 0 \\ 1 & 0 & 1 & 0 & 0 & 0 \\ 1 & 1 & 0 & 0 & 0 & 0 \\ 0 & 0 & 0 & 0 & 1 & 1 \\ 0 & 0 & 0 & 1 & 0 & 1 \\ 0 & 0 & 0 & 1 & 1 & 0 \end{bmatrix}$$

*35. $$\begin{bmatrix} 0 & 1 & 0 & 0 \\ 0 & 0 & 1 & 1 \\ 0 & 0 & 0 & 1 \\ 0 & 0 & 1 & 0 \end{bmatrix}$$

36. $$\begin{bmatrix} 0 & 1 & 1 & 0 \\ 1 & 1 & 1 & 1 \\ 0 & 0 & 0 & 1 \\ 0 & 0 & 0 & 0 \end{bmatrix}$$

37.

38.

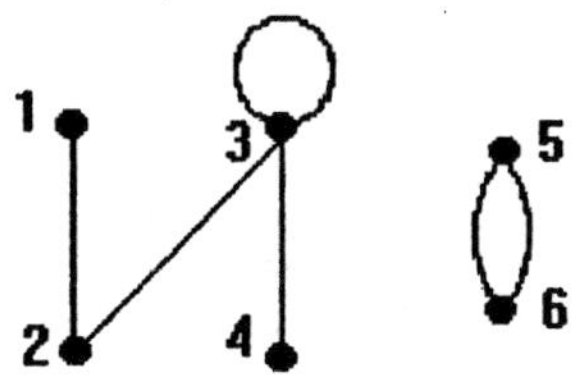

*39.

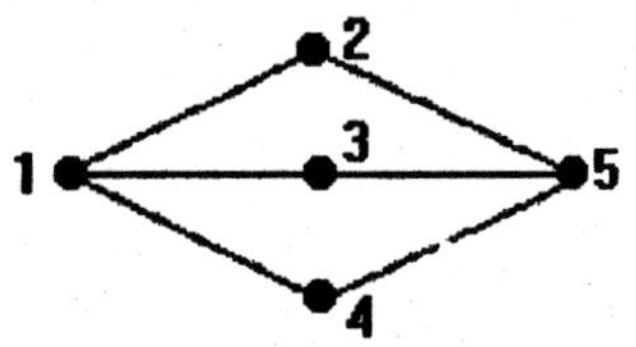

40.

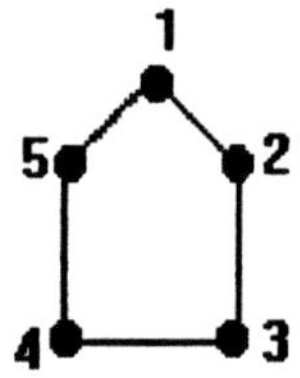

*41.

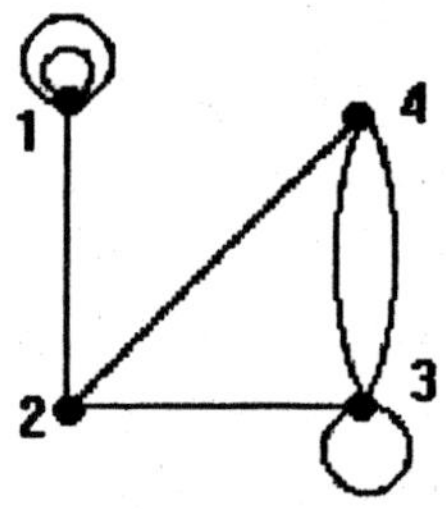

42.

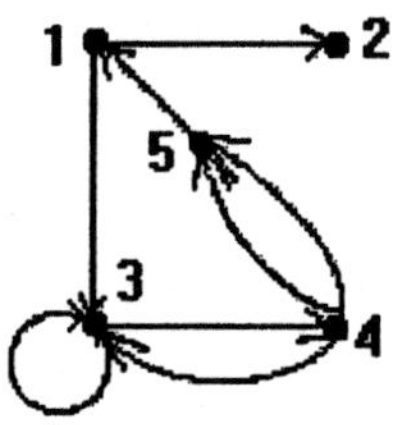

*43. The graph consists of n disconnected nodes with a loop at each node.

44. The n × n matrix with 0s down the main diagonal and 1s elsewhere.

45. The directed graph represented by A^T will look like the original graph with all the arrows reversed.

46.

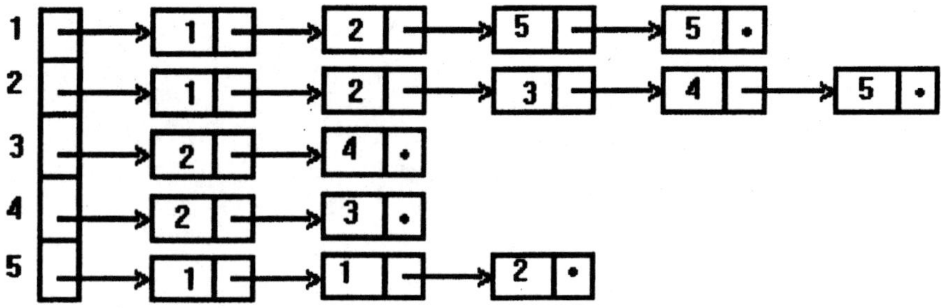

47.

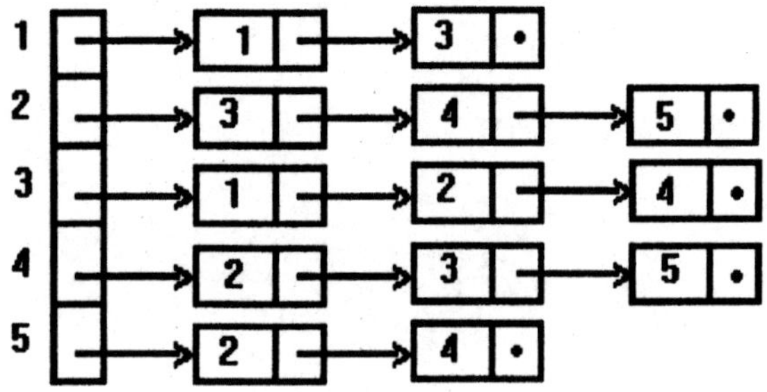

*48.

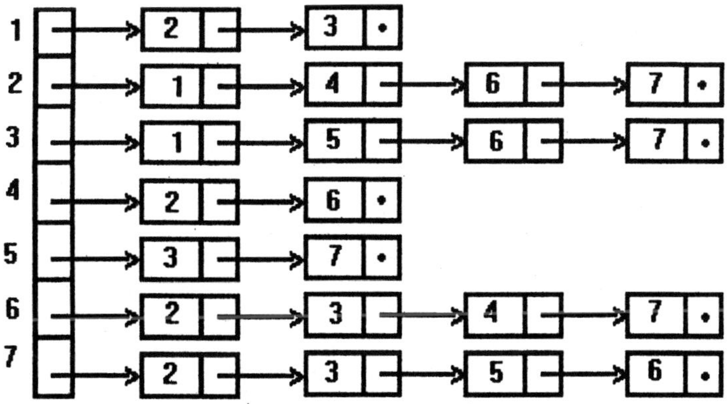
1
2
3
4
5
6
7
2
3
1
4
6
7
1
5
6
7
2
6
3
7
2
3
4
7
2
3
5
6

49.

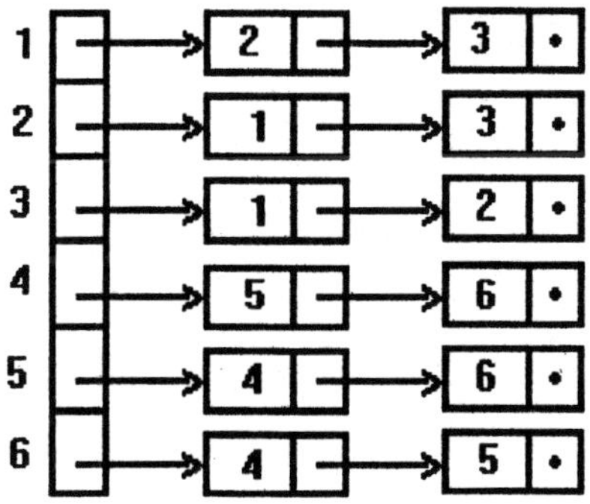
1
2
3
4
5
6
2
3
1
3
1
2
5
6
4
6
4
5

50.

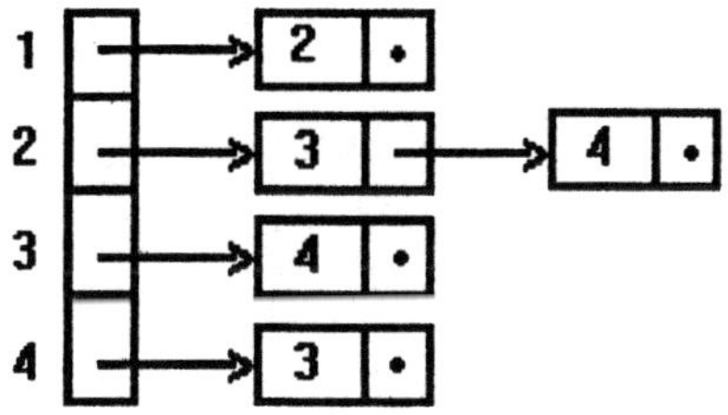
1
2
3
4
2
3
4
4
3

51.

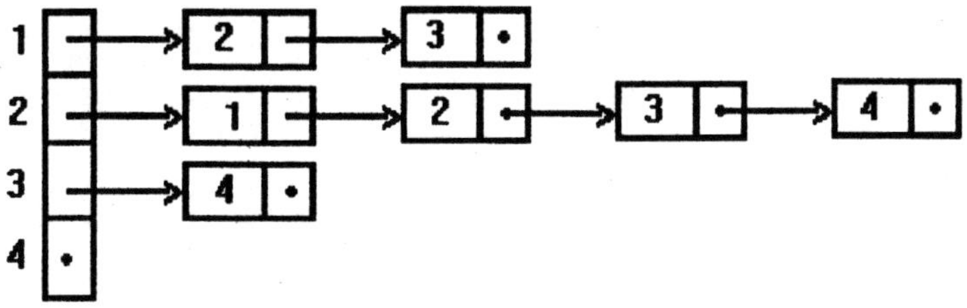
1
2
3
4
2
3
1
2
3
4
4

52. a.

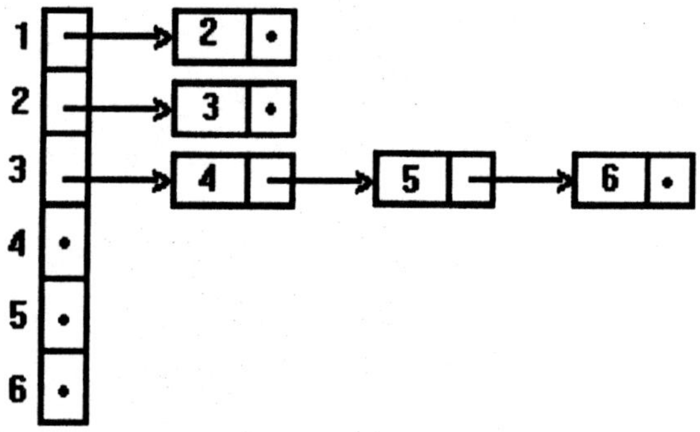

b. 16 c. 36

*53.

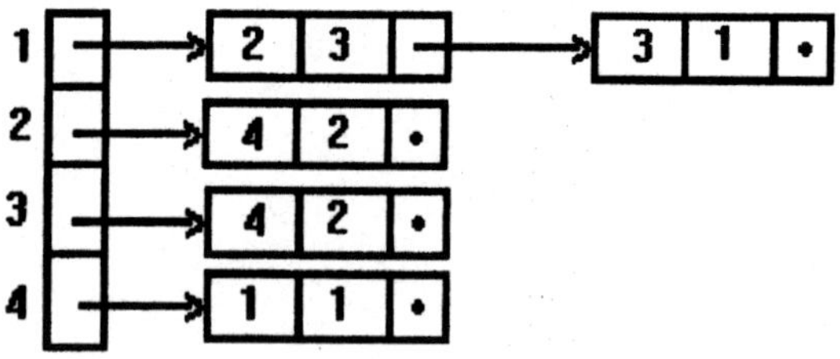

*54.

	Node	Pointer
1		5
2		7
3		11
4		0
5	2	6
6	3	0
7	1	8
8	2	9
9	3	10
10	4	0
11	4	0

55.

	Node	Weight	Pointer
1			5
2			7
3			8
4			9
5	2	3	6
6	3	1	0
7	4	2	0
8	4	2	0
9	1	1	0

56.

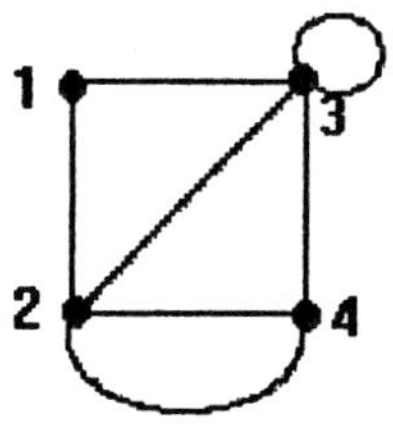

57.

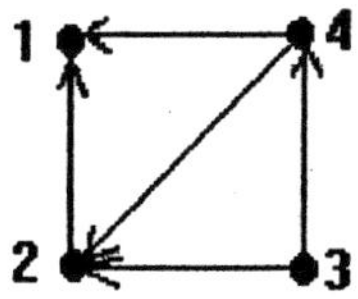

*58.

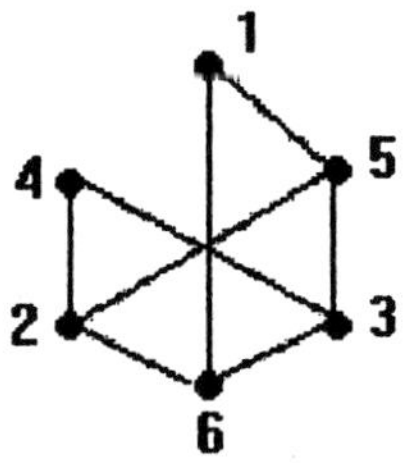

59.

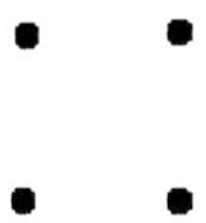

60. By the definition of isomorphic graphs, nodes x-y are adjacent in G_1 if and only if their images are adjacent in G_2. Therefore nodes are not adjacent in G_1 (and therefore are adjacent in G_1') if and only if their images are not adjacent in G_2 (and therefore are adjacent in G_2'). Thus the same function f makes the complement graphs isomorphic.

61. If G is isomorphic to G', then there is a bijection between the arcs of G and the arcs of G', so each has the same number of arcs. Thus considering all the arcs in the complete graph K_n, exactly half belong to G and half to G', so the number of arcs in K_n must be even, say 2m. There are n(n - 1)/2 arcs in K_n, so n(n - 1)/2 = 2m or n(n - 1) = 4m. In the product n(n - 1), one factor is even and the other odd, so the factor of 4 in the product must come either from n, in which case n = 4k for some k, or from n - 1, in which case n - 1 = 4k for some k, or n = 4k + 1.

*62. a. If G is not connected then G consists of two or more connected subgraphs that have no paths between them. Let x and y be distinct nodes. If x and y are in different subgraphs, there is no x-y arc in G, hence there is an x-y arc in G', and a path exists from x to y in G'. If x and y are in the same subgraph, then pick a node z in a different subgraph. There is an arc x-z in G' and an arc z-y in G', hence there is a path from x to y in G'.

b.

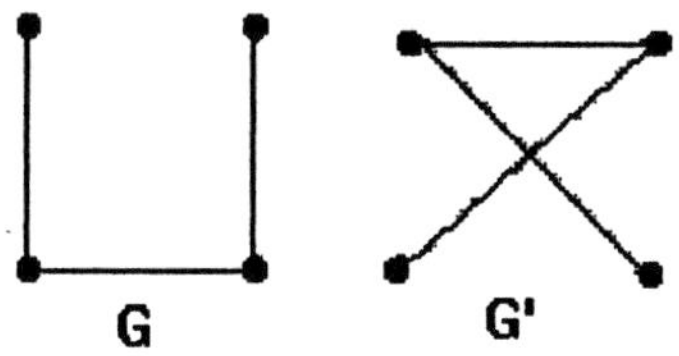

63. The matrix for G' will have 1s where A had 0s and 0s where A had 1s except for diagonal elements, which remain 0s.

64. In a simple connected planar graph G with n nodes, $n \geq 3$, and a arcs, $a \leq 3n - 6$ (Theorem on the Number of Nodes and Arcs). The number of arcs in G' is the number of arcs in the complete graph, $n(n - 1)/2$, minus a. If G' is also planar, then $n(n - 1)/2 - a \leq 3n - 6$, or $n(n - 1)/2 \leq 3n - 6 + a \leq 3n - 6 + 3n - 6 = 6n - 12$. This says that $n(n - 1) \leq 12n - 24$, or $n^2 - 13n + 24 \leq 0$. But a proof by induction shows that for $n \geq 11$, $n^2 - 13n + 24 > 0$, so not both G and G' can be planar.

65. The maximum number of arcs occurs in a complete graph; the maximum is $C(n,2) = n(n - 1)/2$, therefore $a \leq n(n - 1)/2$ or $2a \leq n^2 - n$.

*66. Let a simple connected graph have n nodes and m arcs. The original statement is $m \geq n - 1$, which says $n \leq m + 1$, or that the number of nodes is at most $m + 1$. For the base case, let $m = 0$. The only simple connected graph with 0 arcs consists of a single node, and the number of nodes, 1, is $\leq 0 + 1$. Now assume that any simple connected graph with r arcs, $0 \leq r \leq k$, has at most $r + 1$ nodes. Consider a simple connected graph with $k + 1$ arcs, and remove one arc. If the remaining graph is connected, it has k arcs and, by the inductive hypothesis, the number n of nodes satisfies $n \leq k + 1$. Therefore in the original graph (with the same nodes), $n \leq k + 1 < (k + 1) + 1$. If the remaining graph is not connected, it consists of two connected subgraphs with r_1 and r_2 arcs, $r_1 \leq k$ and $r_2 \leq k$, $r_1 + r_2 = k$. By the inductive hypothesis, one subgraph has at most $r_1 + 1$ nodes and the other has at most $r_2 + 1$ nodes. The original graph therefore had at most $r_1 + 1 + r_2 + 1 = (k + 1) + 1$ nodes.

67. Let G be a simple graph with n nodes, $n \geq 2$, and m arcs, $m > C(n-1,2) = (n - 1)(n - 2)/2$, and suppose that G is not connected. By Exercise 62, G' is connected. By Exercise 66, the number of arcs in G' is at least $n - 1$. Therefore the number m of arcs in G is $n(n - 1)/2$ - (the number of arcs in G') $\leq n(n - 1)/2 - (n - 1) = (n - 1)(n/2 - 1) = (n - 1)(n - 2)/2$, which is a contradiction.

68. At least three colors are required because of the overlapping boundaries. Once the following assignment has been made, the country marked A must be a third color.

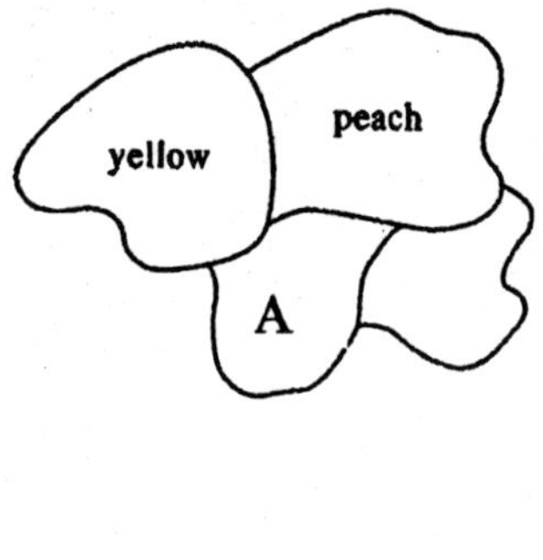

Three colors are sufficient:

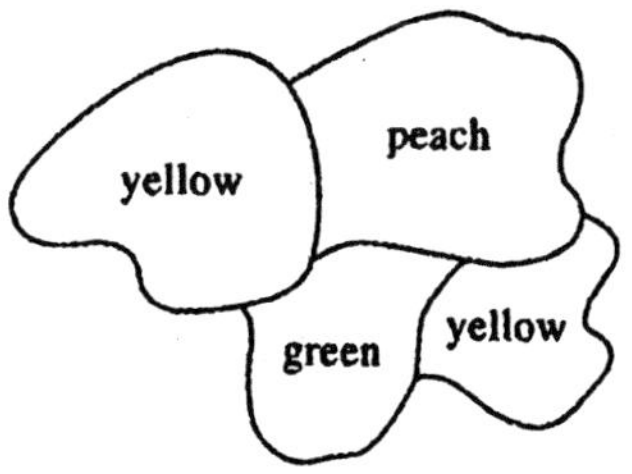

69. For example,

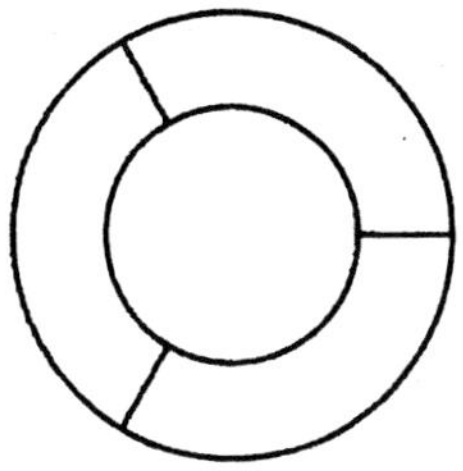

70. a.

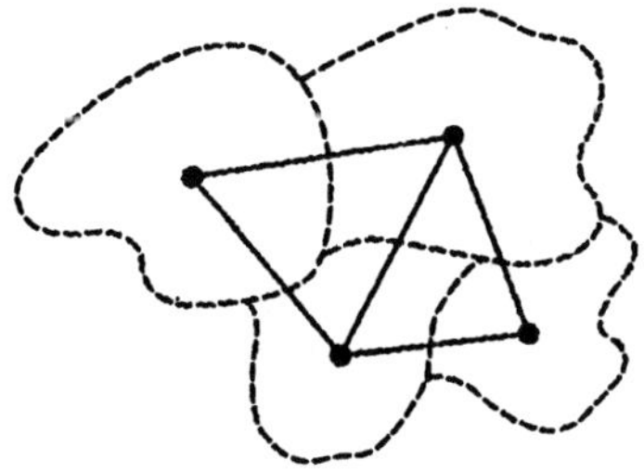

b.

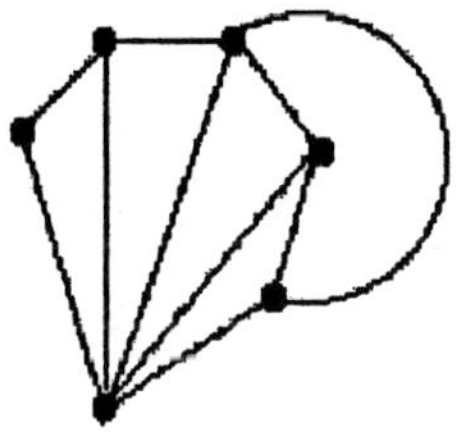

c.

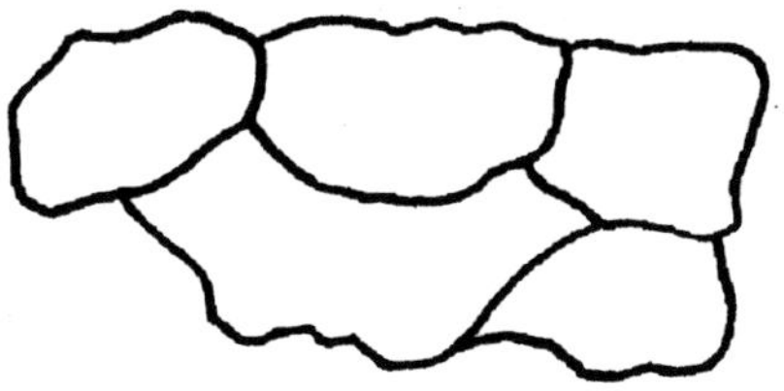

71. *a. 3 b. 4

72. The four-color conjecture is equivalent to the statement that the chromatic number for any simple, connected, planar graph is at most 4.

*73. If this result is not true, then every node in such a graph has degree greater than 5, that is, degree 6 or higher. The total number of arc ends in the graph is then at least 6n, where n is the number of nodes. But the number of arc ends is exactly twice the number a of arcs, so

$$6n \le 2a$$

By inequality (2) of the theorem on the number of nodes and arcs, $a \le 3n - 6$ or $2a \le 6n - 12$. Combining these inequalities, we obtain

$$6n \le 6n - 12$$

which is a contradiction.

74. The proof is by mathematical induction on the number of nodes in the graph. For the basis step of the induction process, it is clear that five colors are sufficient if the number of nodes is less than or equal to 5. Now assume that any simple, connected, planar graph with $\le k$ nodes can be colored with five colors, and consider such a graph with $k + 1$ nodes. We can assume that $k + 1$ is at least 6 because 5 or fewer nodes are taken care of. By Exercise 73, at least one node n of the graph has degree less than or equal to 5; temporarily removing n (and its adjoining arcs) from the graph will leave a collection of one or more simple, connected, planar subgraphs, each with no more than k nodes (Figures a and b). By the inductive hypothesis, each subgraph has a coloring with no more than five colors (use the same palette of five colors for each subgraph). Now look at the original graph again. If n has degree less than 5 or if the 5 nodes adjacent to n do not use five different colors, there is a fifth color left to use for n. Thus, we assume that n is adjacent to 5 nodes, n_1, n_2, n_3, n_4, and n_5, arranged clockwise around n and colored, respectively, colors 1, 2, 3, 4, and 5 (Figure c).

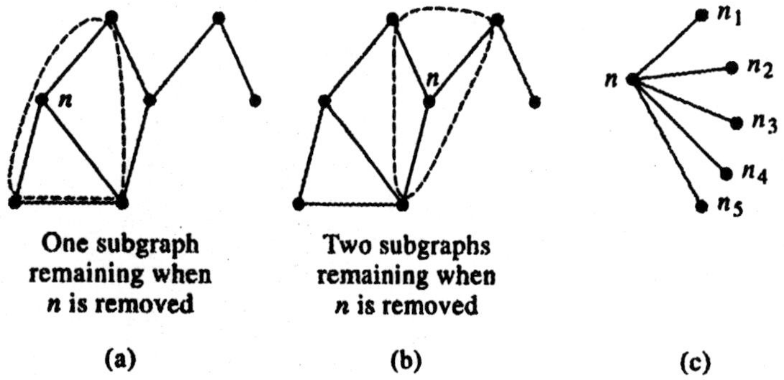

Now pick out all the nodes in the graph colored 1 or 3. Suppose there is no path, using just these nodes, between n_1 and n_3. Then, as far as nodes colored 1 and 3 are concerned, there are two separate sections of graph, one section containing n_1 and one containing n_3. In the section containing n_1 interchange colors 1 and 3 on all the nodes. Doing this does not violate the (proper) coloring of the subgraphs, it colors n_1 with 3, and it leaves color 1 for n.

Now suppose there is a path between n_1 and n_3 using only nodes colored 1 or 3. In this case we pick out all nodes in the original graph colored 2 or 4. Is there a path, using just these nodes, between n_2 and n_4? No, there is not. Because of the arrangement of nodes n_1, n_2, n_3, n_4, and n_5, such a path would have to cross the path connecting n_1 and n_3. Because the graph is planar, these two paths would have to meet at a node, which would then be colored 1 or 3 from the n_1-n_3 path and 2 or 4 from the n_2-n_4 path, an impossibility. Thus, there is no path using only nodes colored 2 or 4 between n2 and n4, and we can rearrange colors as in the previous case. This completes the proof.

75. *a. A graph with parallel arcs would contain

Temporarily create small countries at the nodes of parallel arcs, giving

which is a simple graph. Any coloring satisfactory for this graph is satisfactory for the original graph (let the new regions shrink to points).

b. By Euler's formula, in such a graph $n - a + r = 2$, but r is the total number of regions, including the one external region, so $r = R + 1$, therefore $n - a + R = 1$.

c. The total number of region edges in the graph is 2a, which is greater than or equal to the number of edges of enclosed regions, which in turn is greater than or equal to 6R. Therefore $2a \geq 6R = 6(1 - n + a)$ (from part (b)) $= 6 - 6n + 6a$, from which $2a \leq 3n - 3$.

d. Assume that every enclosed region has at least six adjacent edges. The total number of arc ends in the graph $= 2a \geq 3n > 3n - 3 \geq 2a$ (by part (b)). Contradiction.

e. Use induction on the number of enclosed regions. Six colors are sufficient for any map with ≤ 6 enclosed regions. Assume 6 colors are sufficient for any map with k enclosed regions, and consider a map with $k + 1$ enclosed regions. By part (d), there is at least one enclosed region with ≤ 5 edges. Shrink this region to a point. The remaining regions have a proper coloring by the inductive hypothesis; expand the region back and use the sixth color for it.

f. For countries with holes, put slits to open up the holes, making sure that a slit doesn't end at a boundary. Considering the slits as regions, the new map has no holes and can be colored by part (e). Then shrink the slits.

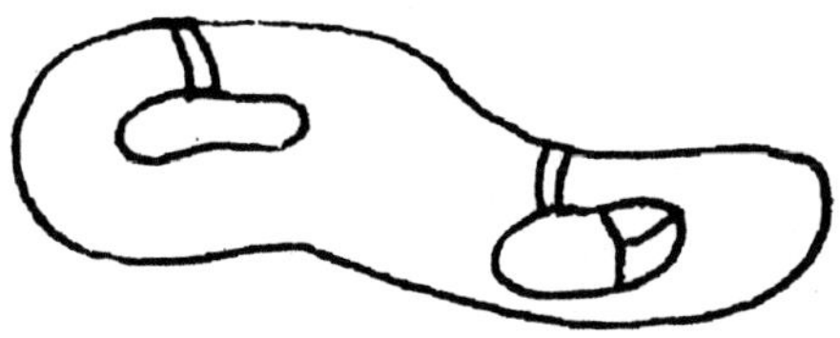

76. Let the members of Congress be the nodes of the graph, and include an arc x-y if any lobbyist must see both x and y. A time slot (instead of a color) is to be assigned to each node; adjacent nodes must have different time-slots. The resulting graph requires four time slots.

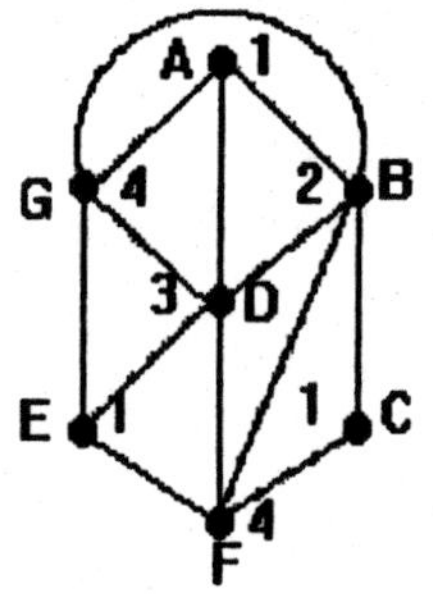

If lobbyist 3 does not need to see B and lobbyist 5 does not need to see D, the graph changes to one that requires only 3 time slots.

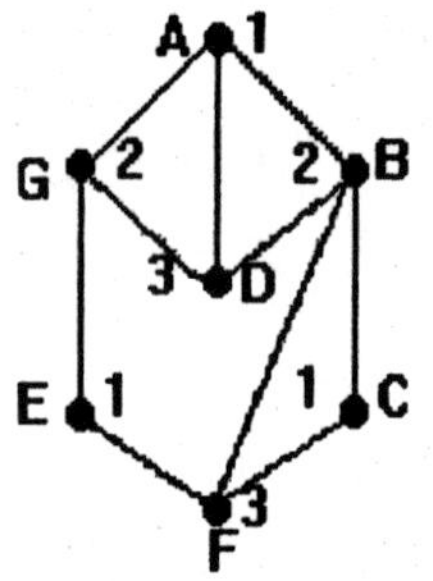

77. Let the processors be the nodes of a graph, and include an arc x-y if processors x and y both write to the data store at the same time. A data store block (instead of a color) is to be assigned to each node; adjacent nodes must use different blocks. The resulting graph requires three blocks.

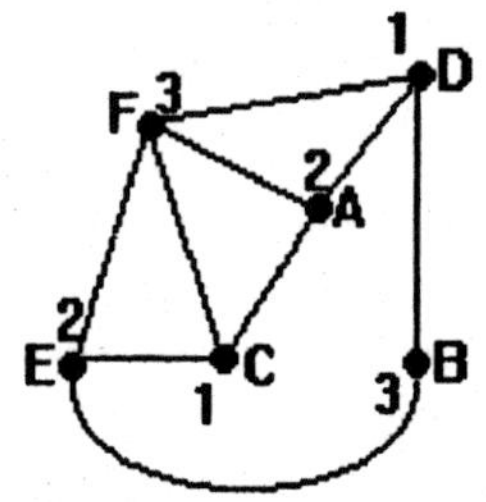

EXERCISES 5.2

*1. a.

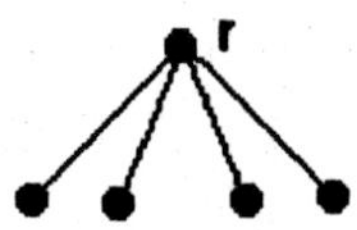

b.

c.

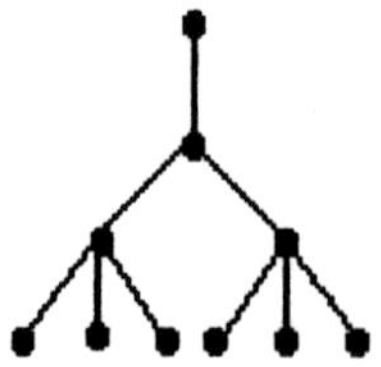

2. a. yes b. no c. yes d. b e. does not exist f. 2 g. 3

3.

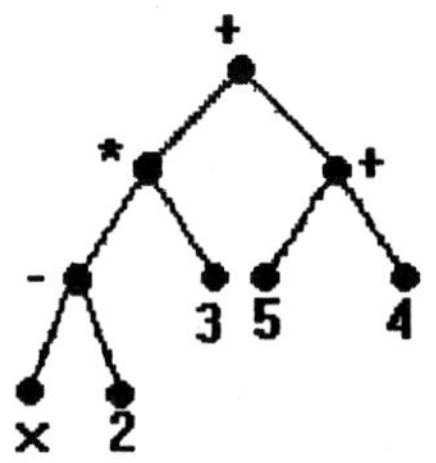

*4.

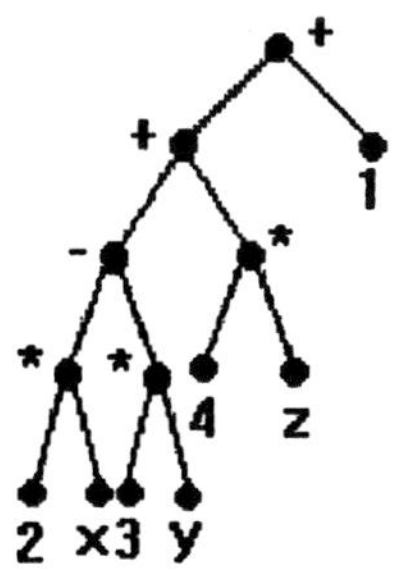

5.

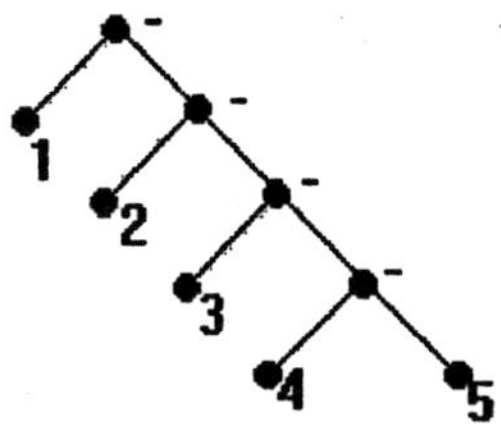

6.

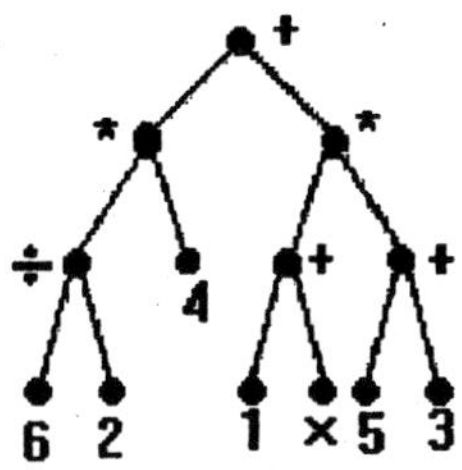

*7.

	Left child	Right child
1	2	3
2	0	4
3	5	6
4	7	0
5	0	0
6	0	0
7	0	0

8.

	Left child	Right child
1	2	3
2	4	5
3	6	7
4	8	9
5	10	11
6	12	13
7	14	15
8	0	0
9	0	0
10	0	0
11	0	0
12	0	0
13	0	0
14	0	0
15	0	0

*9.

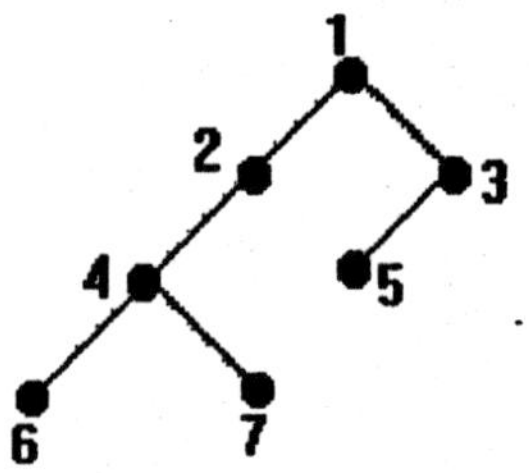

10.

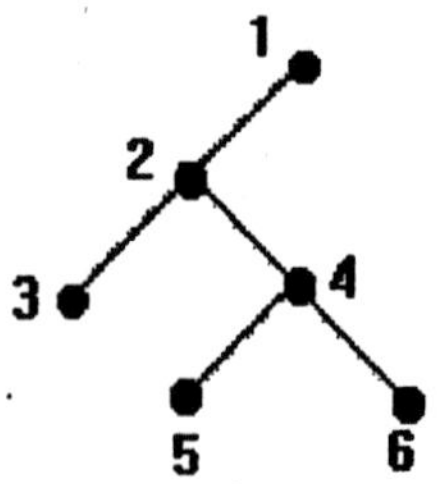

11.

	Name	Left child	Right child
1	All	0	2
2	Gaul	3	4
3	divided	0	0
4	is	5	6
5	into	0	0
6	three	7	0
7	parts	0	0

*12.

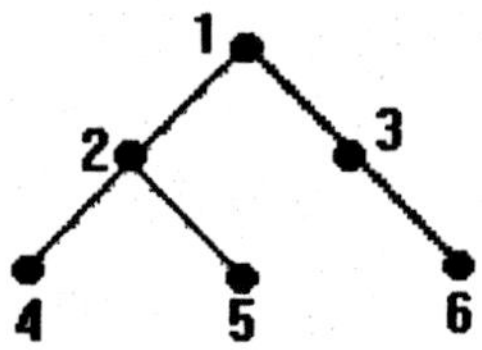

13.

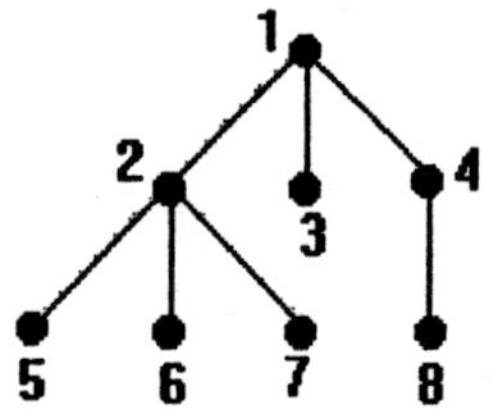

14. a.

	Left child	Right sibling
1	2	0
2	5	3
3	8	4
4	9	0
5	0	6
6	0	7
7	0	0
8	0	0
9	0	10
10	0	11
11	0	0

b.

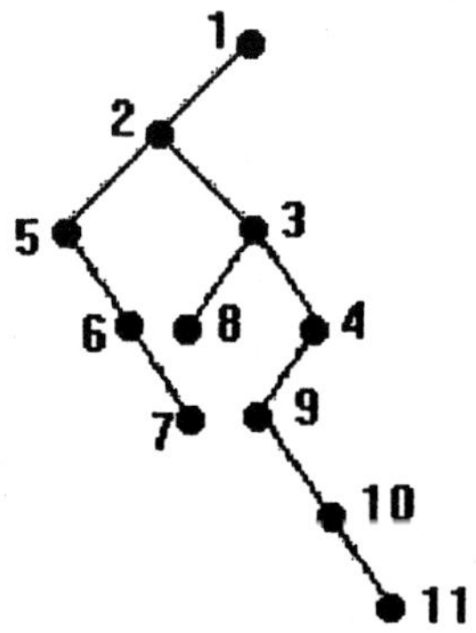

15.

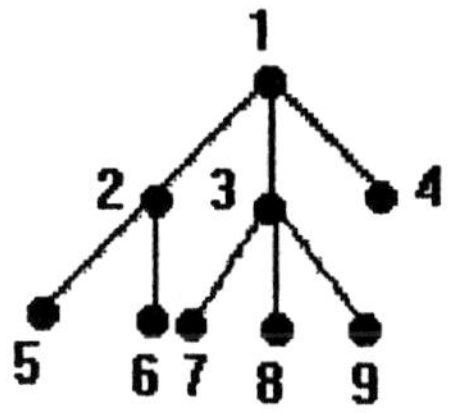

*16. preorder: a b d e h f c g
inorder: d b h e f a g c
postorder: d h e f b g c a

17. preorder: a b d g e c f h
inorder: g d b e a h f c
postorder: g d e b h f c a

18. preorder: a b e c f j g d h i
inorder: e b a j f c g h d i
postorder: e b j f g c h i d a

19. preorder: a b e f c g h d i
inorder: e b f a g c h i d
postorder: e f b g h c i d a

*20. preorder: a b c e f d g h
inorder: e c f b g d h a
postorder: e f c g h d b a

21. preorder: a b d h i c e f g
inorder: h d i b a e c f g
postorder: h i d b e f g c a

*22. prefix: + / 3 4 - 2 y
postfix: 3 4 / 2 y - +

23. prefix: * + * x y / 3 z 4
postfix: x y * 3 z / + 4 *

24. infix: ((2 + 3) * (6 * x)) - 7
postfix: 2 3 + 6 x * * 7 -

25. infix: ((x - y) + z) - w
postfix: x y - z + w -

*26. prefix: + * 4 - 7 x z
infix: (4 * (7 - x)) + z

27. prefix: / x - + 2 w * y z
infix: x / ((2 + w) - (y * z))

28.

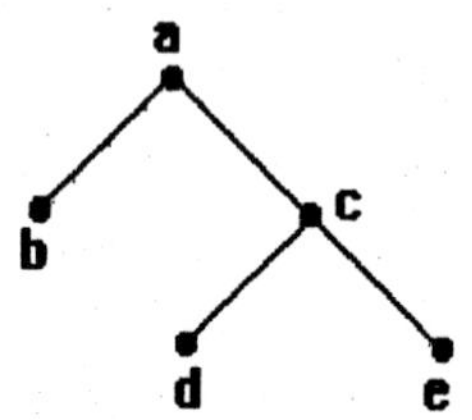

29.

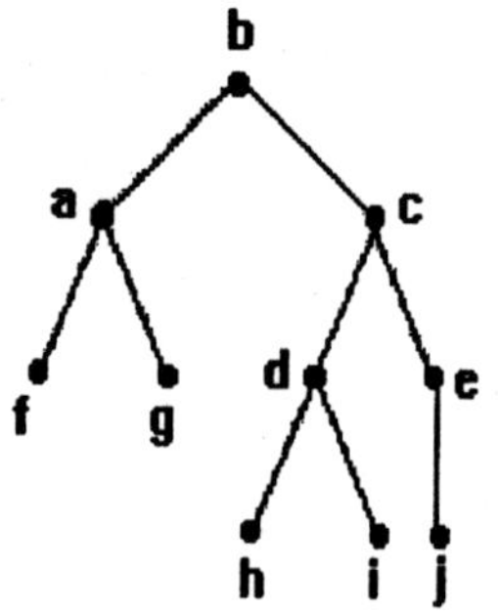

*30.

both inorder and postorder traversal give d c b a

31.

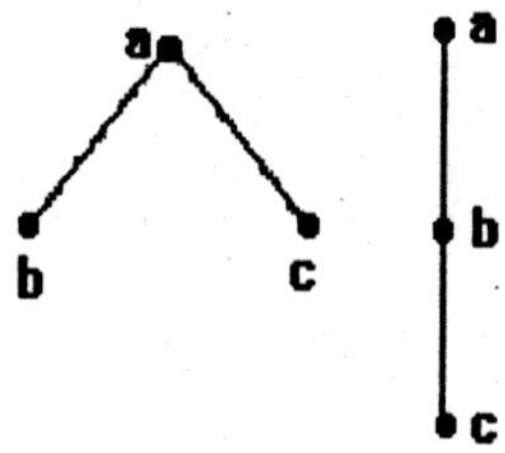

both preorder traversals give a b c

32. If the root has no left child and no right child, return 0 as the height, else invoke the algorithm on the left child if it exists, invoke the algorithm on the right child if it exists, return the maximum of those two values plus 1.

33. If the root has no left child and no right child, return 1, else invoke the algorithm on the left child if it exists, invoke the algorithm on the right child if it exists, return the sum of those two values plus 1.

*34. Consider a simple graph that is a nonrooted tree. A tree is an acyclic and connected graph, so for any two nodes x and y, a path from x to y exists. If the path is not unique, then the two paths diverge at some node n_1 and converge at some node n_2, and there is a cycle from n_1 through n_2 and back to n_1, which is a contradiction.

Now consider a simple graph that has a unique path between any two nodes. The graph is clearly connected. Also, there are no cycles because the presence of a cycle produces a nonunique path between two nodes on the cycle. The graph is thus acyclic and connected and is a nonrooted tree.

35. If G is a nonrooted tree, then G is connected. Suppose we remove an arc *a* between n_1 and n_2 and G remains connected. Then there is a path from n_1 to n_2. Adding *a* to this path results in a cycle from n_1 to n_1, which contradicts the definition of a tree. Suppose G is connected and removing any single arc makes G unconnected. Then there is a unique path between any two nodes and the graph is a nonrooted tree (Exercise 34).

36. Let G be a nonrooted tree and add an arc *a* between n_1 and n_2. Because G was originally connected, there was a path between n_1 and n_2; adding *a* to this path results in a cycle from n_1 to n_1. If adding one arc to G results in a graph with exactly one cycle, then the original graph was acyclic and connected, a nonrooted tree.

37. Using induction on n, a tree with 2 nodes has 2 nodes of degree one. Assume that a k-node tree has ≥ 2 nodes of degree one. Consider any tree with k + 1 nodes. Remove a leaf x (and its arc); the result is a tree with k nodes and, by the inductive hypothesis, ≥ 2 nodes of degree one. Putting node x back to obtain the original tree, node x has degree one but the parent of x may go from degree one to degree two; therefore the number of degree one nodes either increases by one or stays the same, so is still ≥ 2.

*38. Proof is by induction on d. For d = 0, the only node is the root, and $2^0 = 1$. Assume that there are at most 2^d nodes at depth d, and consider depth d + 1. There are at most two children for each node at depth d, so the maximum number of nodes at depth d + 1 is $2 \cdot 2^d = 2^{d+1}$.

39. a.

7 nodes

b.

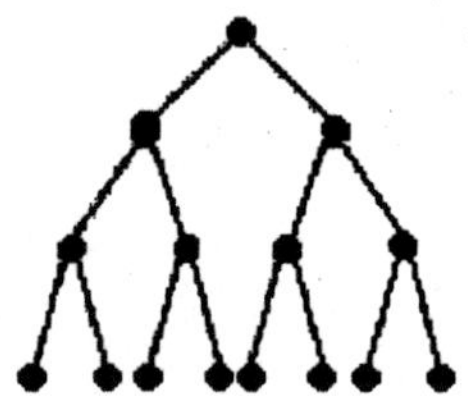

15 nodes

c. $2^{h+1} - 1$

d. Proof is by induction on h. For $h = 0$, the tree consists only of the root, so the number of nodes $= 1 = 2^{0+1} - 1$. Assume that a full binary tree of height h has $2^{h+1} - 1$ nodes. Consider a full binary tree of height $h + 1$. A full tree has the maximum number of leaves which, by Exercise 38, will be 2^{h+1}. Removing the leaves and associated arcs gives a full binary tree of height h, with $2^{h+1} - 1$ nodes by the inductive assumption. The total number of nodes in the original tree is

$$2^{h+1} + 2^{h+1} - 1 = 2 \cdot 2^{h+1} - 1 = 2^{h+2} - 1$$

Or, since by Exercise 36 there are 2^d nodes at each level, the total number of nodes is

$$1 + 2 + 2^2 + \ldots + 2^h = 2^{h+1} - 1$$

by Example 15 of Section 2.2. This approach does not avoid induction, however, because this result was obtained by induction.

40. a. In a full binary tree, all internal nodes have two children, so the total number of "children nodes" is 2x; the only "non-child" node is the root, so there is a total of $2x + 1$ nodes.

b. From part a, there are $2x + 1$ total nodes, x of which are internal, leaving $2x + 1 - x = x + 1$ leaves.

c. Consider a full binary tree with n nodes; let x be the number of internal nodes. From part a, n = 2x + 1. Therefore, x = (n - 1)/2. From part b, the number of leaves = x + 1 = (n - 1)/2 + 1 = (n + 1)/2.

41. Let x be the number of nodes with two children; we want to show there are x + 1 leaves; one proof is by induction on x. If x = 0, then the tree must be a "chain," i.e., it looks like this and there is only 1 leaf; 1 = x + 1. Assume in any binary tree with x nodes having two children that there are x + 1 leaves. Now consider a binary tree where x + 1 nodes have two children. Remove the subtree rooted by one of the children of a node with two children at the maximum depth at which such a node occurs, for example:

This subtree has no nodes with two children, and exactly one leaf. The remaining tree has x nodes with two children and, by the inductive hypothesis, x + 1 leaves. Thus the original tree had x + 2 leaves.

Or, do induction on the number of leaves and "grow" the tree. There are two cases: adding to a leaf produces no new leaves, and adding a second child increments both x and the number of leaves.

Still another proof parallels that of Exercise 40. Let x = the number of nodes with two children, let y = the number of nodes with one child. Then the total number of nodes is2x + y + 1, and the number of internal nodes is x + y, so the number of leaves is 2x + y + 1 - (x + y) = x + 1.

*42. By Exercise 39, a full tree of height h - 1 has $2^h - 1$ nodes. When $n = 2^h$, this is the beginning of level h. The height h remains the same until $n = 2^{h+1}$, when it increases by 1. Therefore for $2^h \le n < 2^{h+1}$, the height of the tree remains the same, and is given by $h = \lfloor \log n \rfloor$. For example, for $2^2 \le n < 2^3$, that is, n = 4, 5, 6, or 7, the tree height is 2, and $2 = \lfloor \log n \rfloor$.

43. Let x = the number of nodes with 2 children, let y = the number of nodes with 1 child. Then the total number n of nodes is 2x + y + 1 (1 is for the root, the only non-child node), and the number of leaves is x + 1 (Exercise 41). Each leaf contributes 2 null pointers and each node with one child contributes one. Therefore the total number of null pointers is 2(x + 1) + y = 2x + 2 + y = n + 1.

44. The proof is by induction on i, $i \ge 0$. For i = 0, a tree with no internal nodes consists only of the single root node. In this case E = 0, I = 0, and i = 0, so the equation E = I + 2i is true. Assume that in any tree with k internal nodes, all of which have two children, $E_k = I_k + 2k$.

Now consider a tree with k + 1 internal nodes, all of which have two children. We want to show that $E_{k+1} = I_{k+1} + 2(k + 1)$. In such a tree, pick an internal node x whose two children are leaves, and delete the two leaves(see figure). Now x is a leaf. Denote the path length to x by m. The new tree has k internal nodes (since we reduced

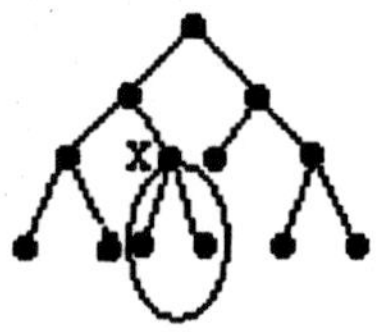

the number of internal nodes by 1). By the inductive hypothesis,

$E_k = I_k + 2k$ (1)

Also,

$I_k = I_{k+1} - m$ (subtract the path to x because x is no longer an internal node)

$E_k = E_{k+1} - 2(m + 1) + m$ (subtract the two paths to x's children, add the path to x)

Substituting in Equation (1),

$$E_{k+1} - 2(m + 1) + m = I_{k+1} - m + 2k$$
$$E_{k+1} - m - 2 = I_{k+1} - m + 2k$$
$$E_{k+1} = I_{k+1} + 2k + 2 = I_{k+1} + 2(k + 1)$$

45. a. There is only one binary tree with one node, so B(1) = 1. For a binary tree with n nodes, n > 1, the "shape" of the tree is determined by the "shape" of the left and right subtrees; the two subtrees have a total of n – 1 nodes. Let the left subtree have k nodes; the right subtree then has n – 1 – k nodes; k can range from 0 to n – 1. For each value of k, there are B(k) ways to form the left subtree, then B(n – 1 – k) ways to form the right subtree, so by the Multiplication Principle, there are B(k)B(n – 1 – k) different trees.

b. $B(0) = 1$
$B(1) = 1$

$$B(n) = = \sum_{k=0}^{n-1} B(k)B(n-1-k) = = \sum_{k=1}^{n} B(k-1)B(n-k)$$

which is the same as the Catalan sequence, so by Exercise 82 of Section 3.4,

$$B(n) = \frac{1}{n+1}C(2n, n)$$

c. $B(3) = \frac{1}{3+1}C(6,3) = 5$. The 5 distinct binary trees are:

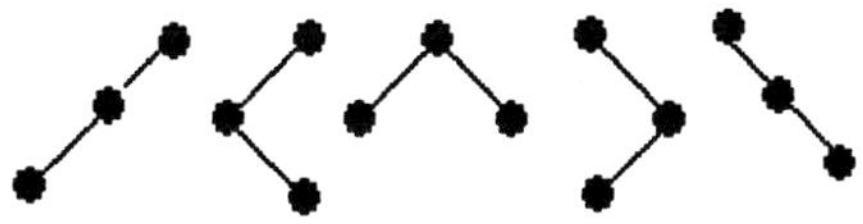

d. $B(6) = \frac{1}{6+1}C(12,6) = 132$

46. *a. 17. The growth of the tree proceeds as follows, building on the tree with 8 nodes:

Insert 9, 10, 11:

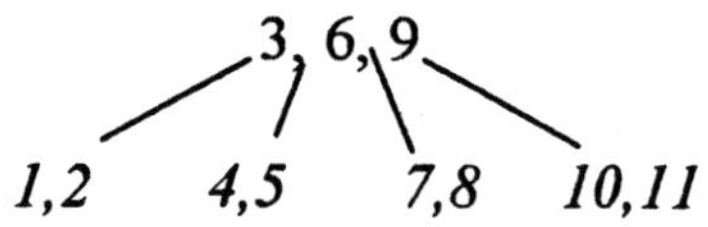

Insert 12, 13, 14:

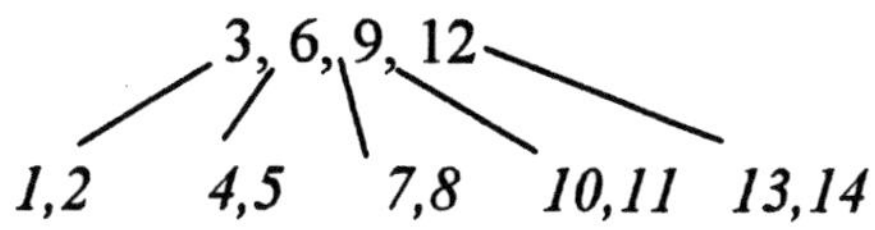

Insert 15, 16, 17:

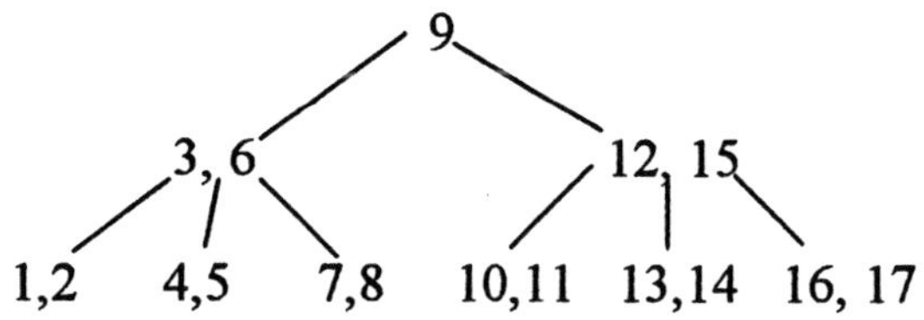

b. Proof is by induction on n. When n = 2, the tree is

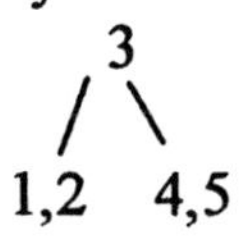

and the bottom level has 2 nodes. $2 = 2\cdot3^{2-2}$, true. Assume that when the tree has just achieved k levels, the bottom level has $2\cdot3^{k-2}$ nodes. Now let the tree grow just to achieve level k + 1. By the inductive assumption, the level above the new bottom level has $2\cdot3^{k-2}$ nodes, each of which has 3 links to child nodes. Hence the number of nodes in the bottom level is $2\cdot3^{k-2}\cdot3 = 2\cdot3^{k-1}$.

c. From part (b), the number of nodes when the tree has attained n levels is

1 (the root) + $2\cdot3^0 + 2\cdot3^1 + \ldots + 2\cdot3^{n-2}$

The root node contains one data value; all other nodes contain two data values, so the total number of nodes is

$1 + 4(3^0 + 3^1 + \ldots + 3^{n-2}) = 1 + 4\left(\frac{3^n - 3}{6}\right)$ (this equality can be proved by induction)

This expression gives 5 when n = 2 and 17 when n = 3, which agrees with earlier results.

47. The chromatic number of a tree is 2; all the nodes at an odd level can be one color, and all the nodes at an even level can be the other color.

*48.

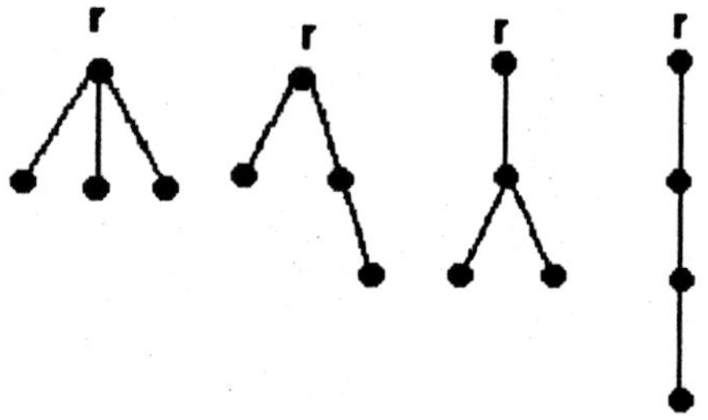

49.

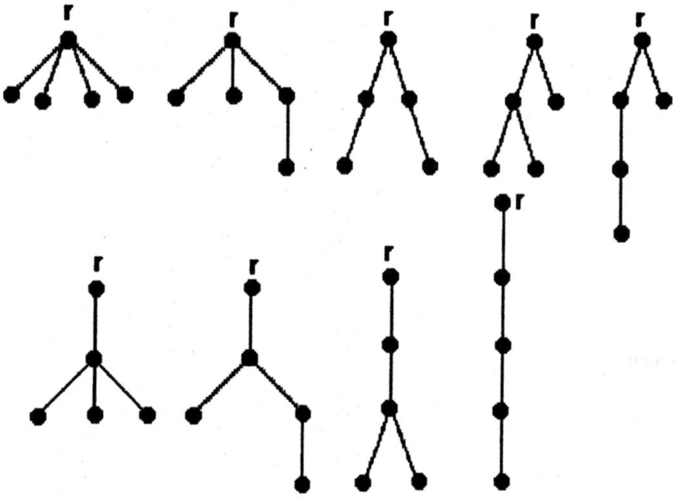

50.

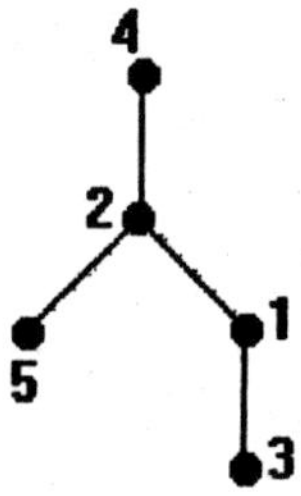

51.

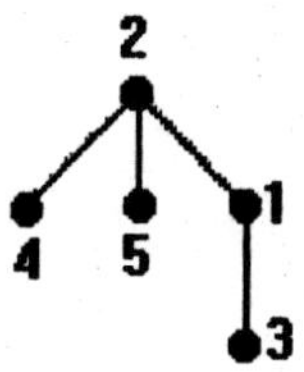

EXERCISES 5.3

*1.

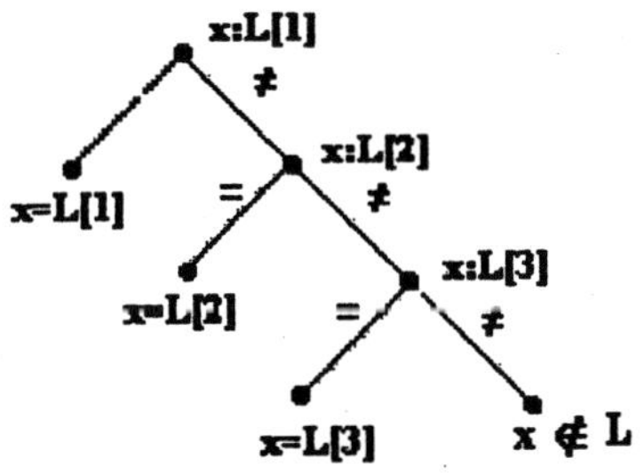

2.

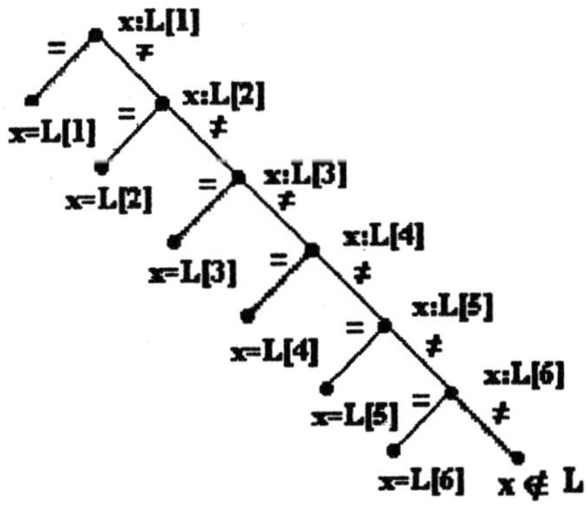

3.

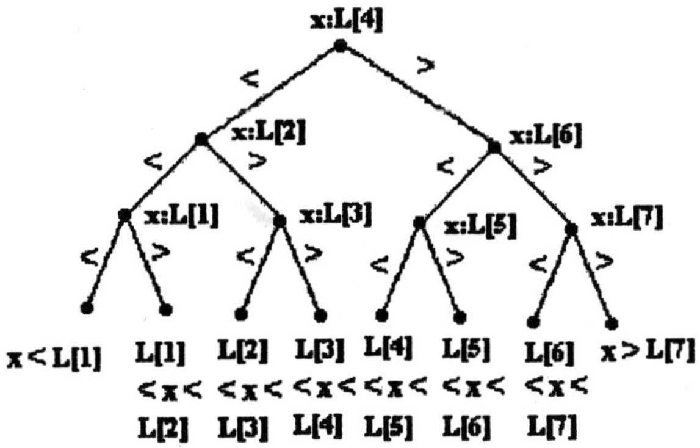

depth = 3 = 1 + $\lfloor \log 7 \rfloor$

4.

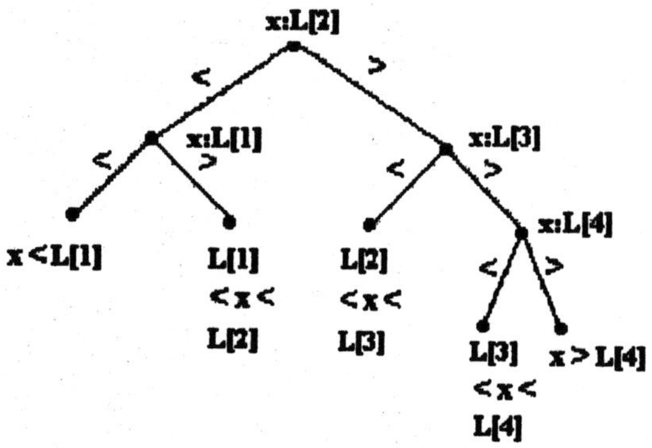

depth = 3 = 1 + $\lfloor \log 4 \rfloor$

*5.

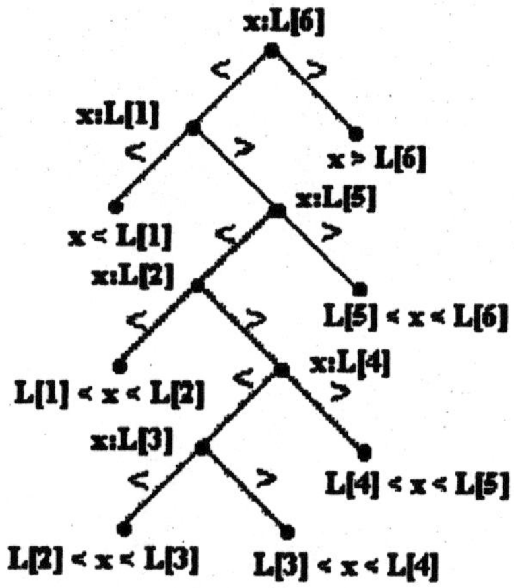

depth = 6; algorithm is not optimal

6.

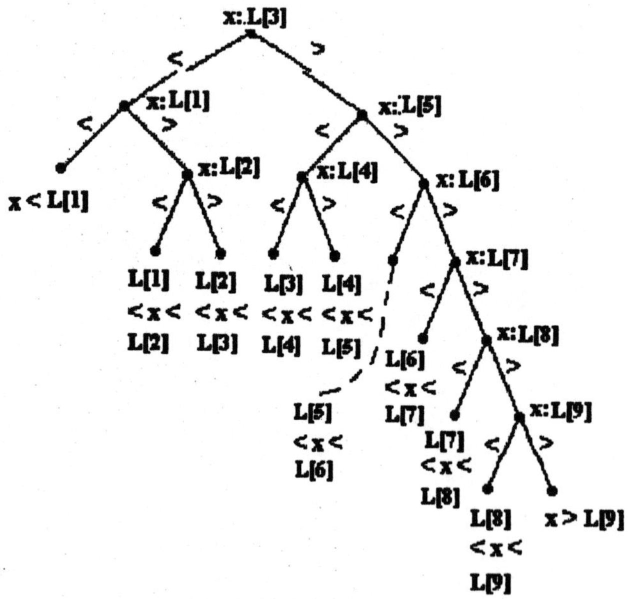

depth = 6; algorithm is not optimal

*7. a.

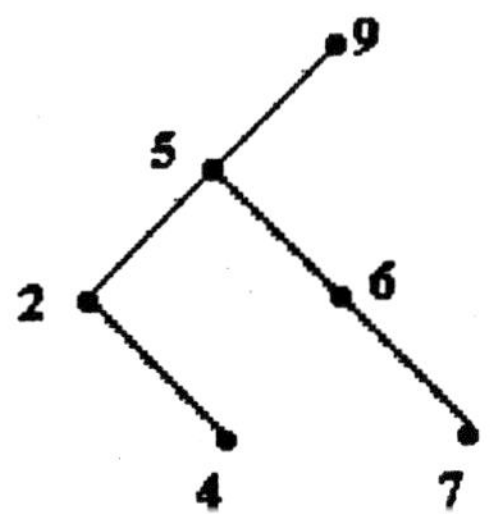

depth = 3

b. 2.83

8. a.

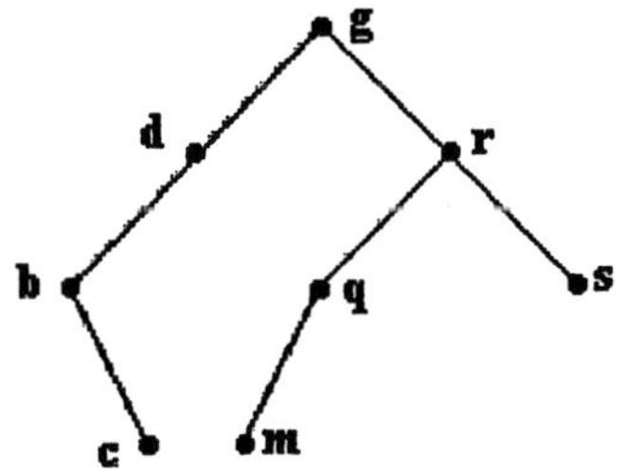

depth = 3

b. 2.75

9. a. $\lfloor \log 6 \rfloor + 1 = 3$

b. For example: g, d, a, k, i, s

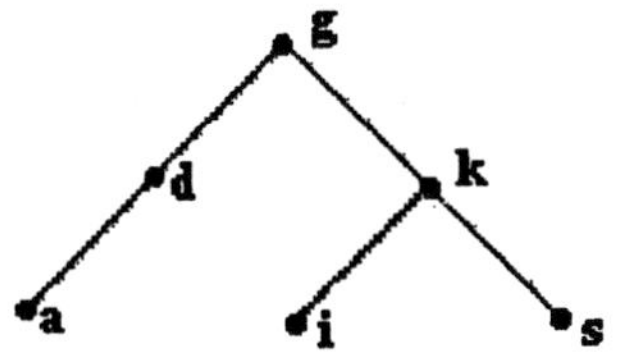

depth = 2 (Note that a binary search tree of depth 2 implies 3 compares in the worst case.)

10. a. $\lfloor \log 9 \rfloor + 1 = 4$

b. For example: 12, 8, 10, 15, 18, 19, 7, 4

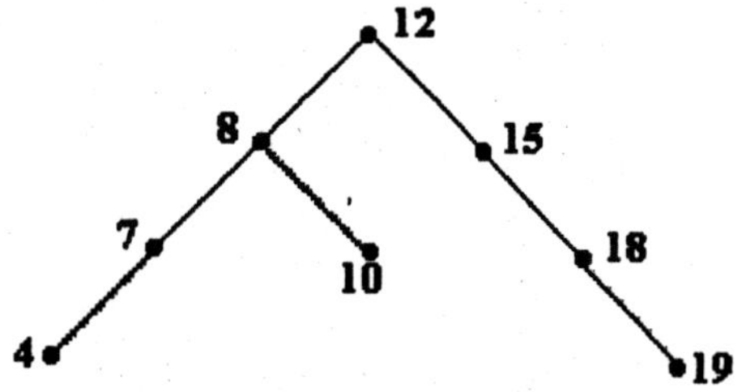

depth = 3 (Note that a binary search tree of depth 3 implies 4 compares in the worst case.)

*11.

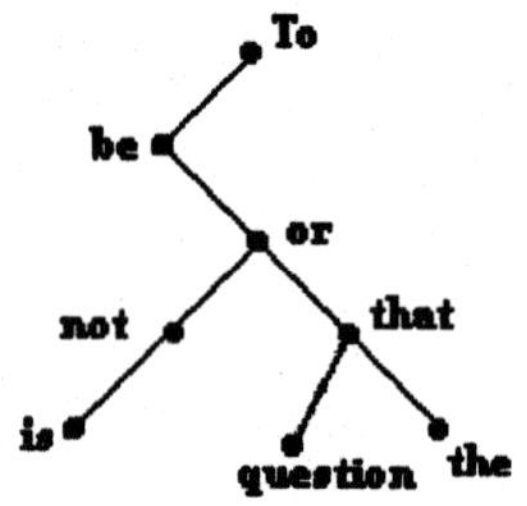

be is not or question that the To

12.

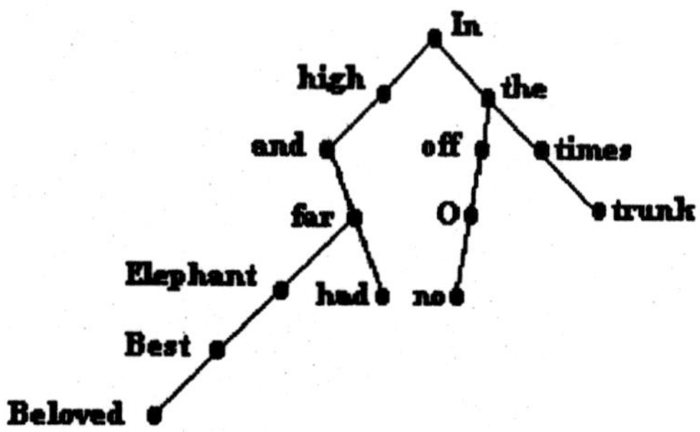

and Beloved Best Elephant far had high in no O off the times trunk

*13. a. $\lceil \log 4! \rceil = \lceil \log 24 \rceil = 5$

b. $\lceil \log 8! \rceil = \lceil \log 40320 \rceil = 16$

c. $\lceil \log 16! \rceil = \lceil \log 2.09 \times 10^{13} \rceil = 45$

14.

n	Exer. 13	Sel. sort	Mergesort
4	5	6	5
8	16	28	17
16	45	120	49

Mergesort is close to optimal.

15. a. 5 (any of thc five coins can be the light one)
    b. 2 (the minimum depth for a ternary tree with 5 leaves)
    c.

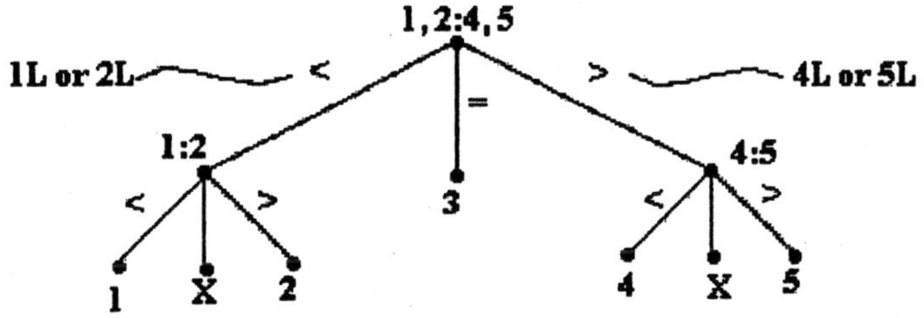

*16. a. 10 (any of the five coins can be heavy or light)
    b. 3 (the minimum depth for a ternary tree with 10 leaves)
    c.

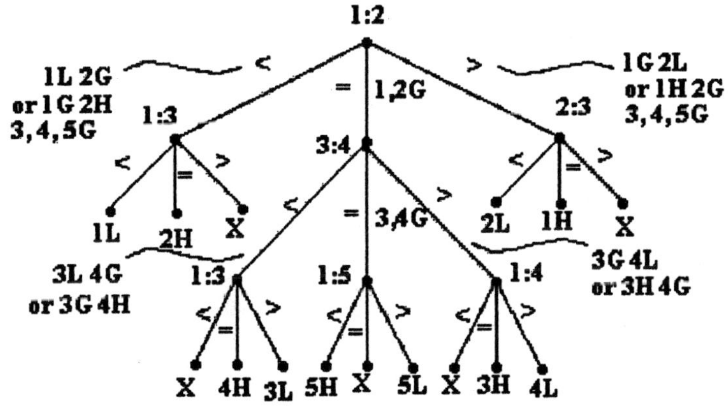

17. a. 4 (any of the four coins can be the counterfeit one)
    b. 2 (the minimum depth for a ternary tree with 4 leaves)
    c.

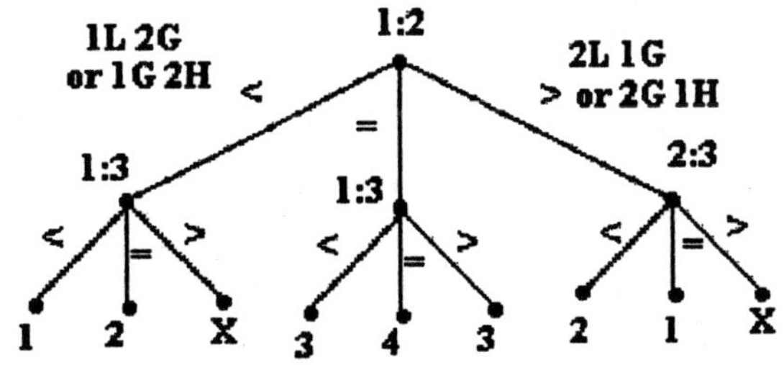

18. a. 8 (any of the four coins can be heavy or light)
    b. 2 (the minimum depth for a ternary tree with 8 leaves)
    c. Suppose that a decision tree of depth 2 can solve the problem.

    Case 1: first compare involves 2 coins. Then the decision tree has the form shown:

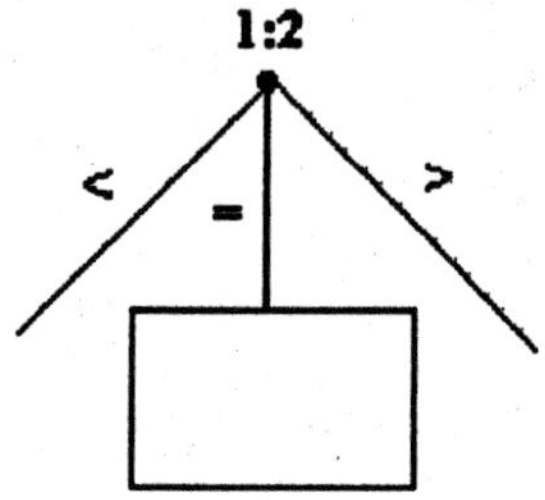

    If coins 1 and 2 balance, then within the box, the four possible outcomes of 3L, 3H, 4L, 4H must be determined, but there can be at most 3 leaves produced here.

    Case 2: first compare involves 4 coins. Then the decision tree has the form shown (the four coins cannot balance because one of them is bad):

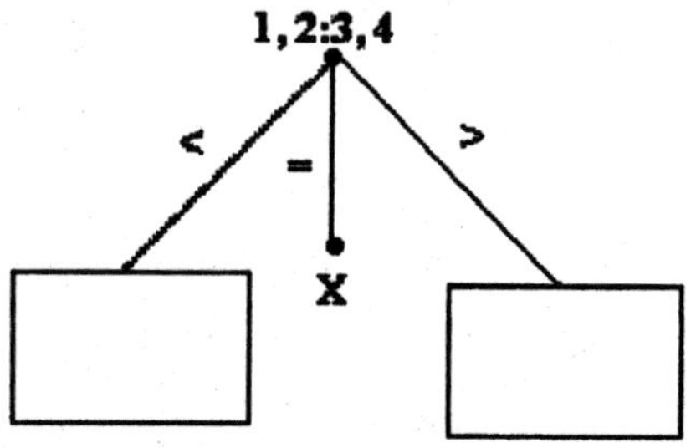

    Then within the two boxes, the eight possible outcomes must be determined, but there can be at most 6 leaves produced here.

19.

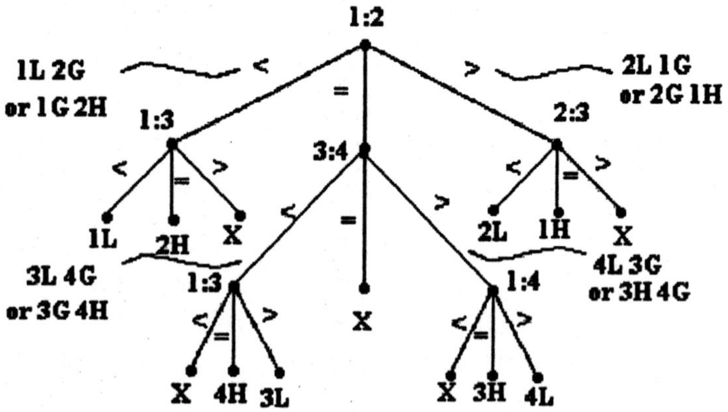

20. a. 16 (any of the eight coins can be heavy or light)
    b. 3 (the minimum depth for a ternary tree with 16 leaves)

c.

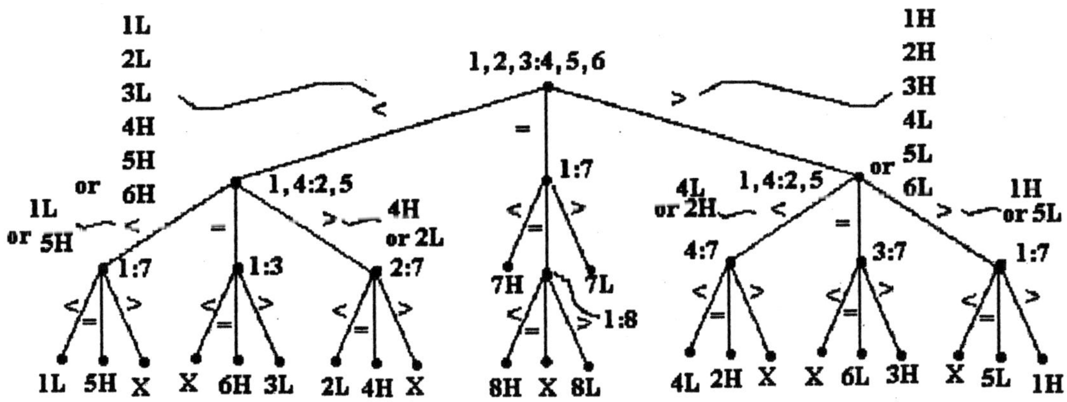

*21. The three-way comparison would be done something like

if (x = node element)
    write "found";
else
    if (x < node element)
        ....
    else
        ...

so that in the worst case, where x is not equal to the node element, 2 comparisons are done at the node. The number of comparisons in the worst case for binary search of n elements is therefore 2 times the depth of the tree, or $2*(1 + \lfloor \log n \rfloor)$.

22. a. (For simplicity, not all leaves are labeled)

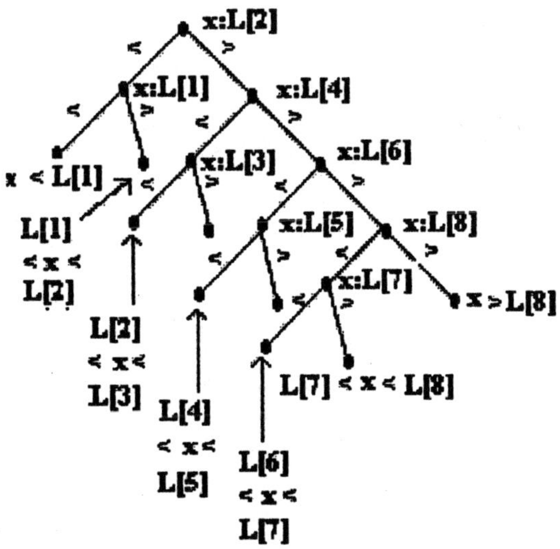

b. 10 comparisons (for example, if $L(7) < x < L(8)$). As explained in the solution to Exercise 21, two comparisons are performed at each internal node.

c. The longest branch of the decision tree is approximately n/2 nodes in length, with 2 comparisons at each node, so the search is $\Theta(n)$. This is a higher order of magnitude than the original binary search algorithm, and is equivalent to sequential search.

23\. a.
$$\begin{aligned}\log n! &= \log[(n)(n-1)(n-2)\ldots(2)(1)]\\ &= \log n + \log(n-1) + \log(n-2) + \ldots + \log 2 + \log 1\\ &\le \log n + \log n + \log n + \ldots + \log n \quad \text{for } n \ge 1\\ &= n \log n\end{aligned}$$

b.
$$\begin{aligned}\log n! &= \log[(n)(n-1)(n-2)\ldots(2)(1)]\\ &= \log n + \log(n-1) + \log(n-2) + \ldots + \log 2 + \log 1\\ &\ge \log n + \log(n-1) + \ldots + \log\lceil n/2\rceil\\ &\ge \log\lceil n/2\rceil + \log\lceil n/2\rceil + \ldots + \log\lceil n/2\rceil\\ &\ge \lceil n/2\rceil \log\lceil n/2\rceil\\ &\ge \left(\frac{n}{2}\right)\log\left(\frac{n}{2}\right) = \left(\frac{n}{2}\right)(\log n - \log 2) = \left(\frac{n}{2}\right)(\log n - 1)\\ &= \left(\frac{n}{2}\right)\log n - \left(\frac{n}{2}\right) = \left(\frac{n}{4}\right)\log n + \left(\frac{n}{4}\right)\log n - \left(\frac{n}{2}\right) = \left(\frac{n}{4}\right)\log n + \left(\frac{n}{4}\right)(\log n - 2)\\ &\ge \left(\frac{n}{4}\right)\log n \quad \text{because } \log n \ge 2 \text{ for } n \ge 4\end{aligned}$$

EXERCISES 5.4

*1. a. ooue b. iaou c. eee

2\. a. bh%% b. wwq c. qhwb

3\. a. (pw)a b. paw c. ((a))

4\. 319
31771
3175
11119
1111771
111175
139
13771
1375

5. a - 0101
   b - 011
   c - 10
   d - 11

6. r - 000
   s - 001
   t - 01
   u - 1

*7. a.

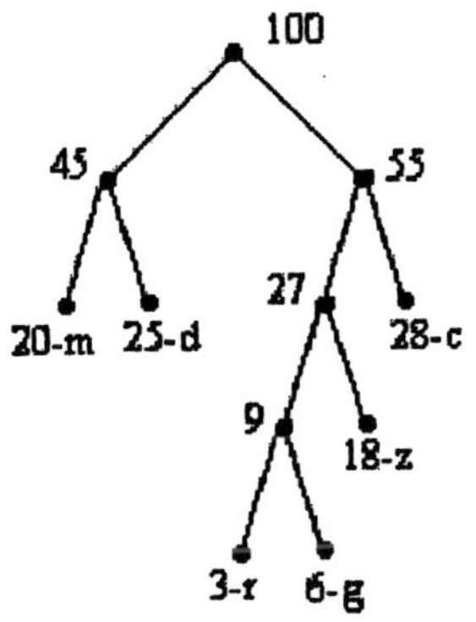

b. c - 11
   d - 01
   g - 1001
   m - 00
   r – 1000
   z - 101

8. a.

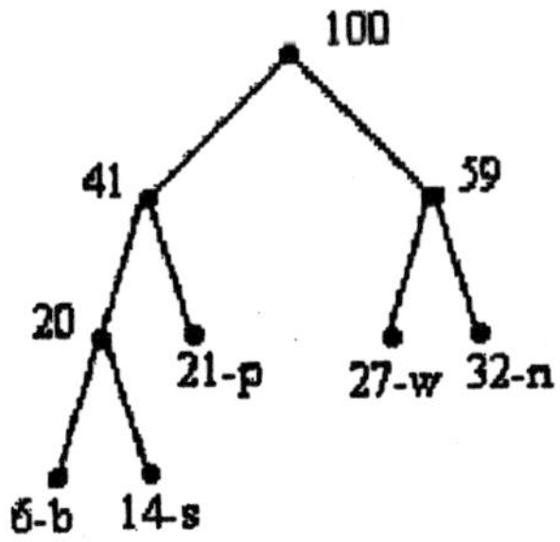

b.
   b – 000
   n - 11
   p - 01
   s - 001
   w - 10

9. a.

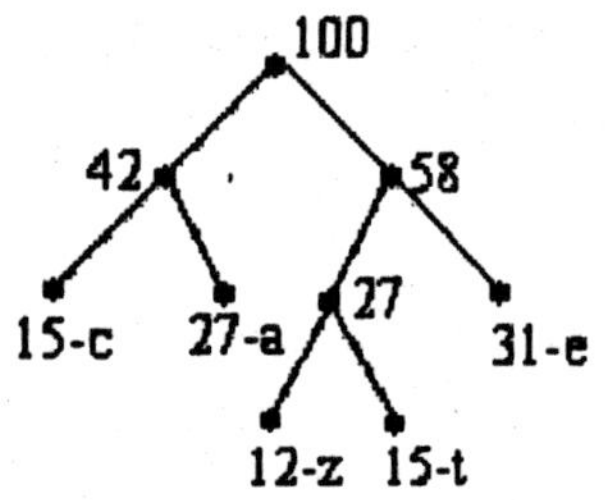

or

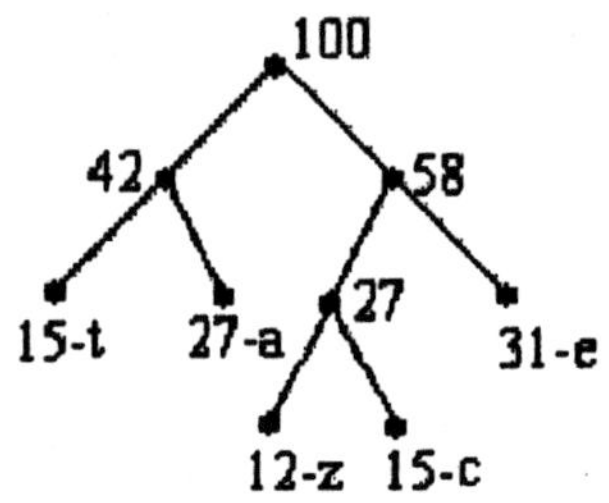

b.

a - 01		a - 01
z - 100		z - 100
t - 101	or	t - 00
e - 11		e - 11
c - 00		c - 101

*10. a.

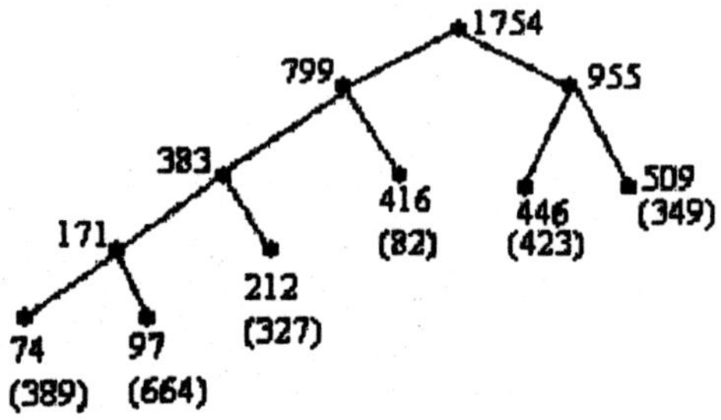

b. 82 - 01
664 - 0001
327 - 001
349 - 11
423 - 10
389 – 0000

11. a.

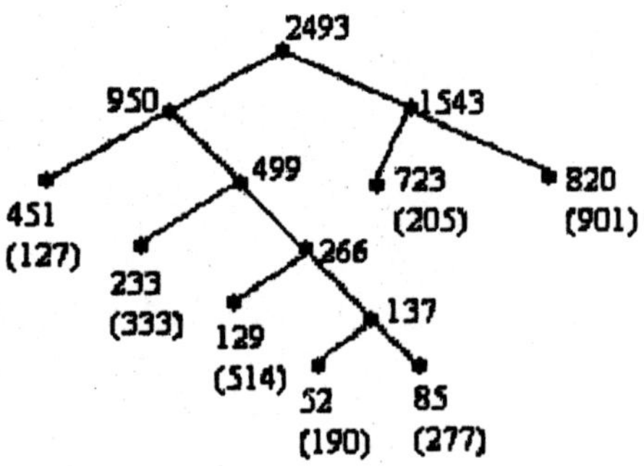

b. 190 – 01110
205 – 10
514 – 0110
333 – 010
127 – 00
901 – 11
277 – 01111

12. Greyscale images contain only luminance information with no color components; the luminance information is more accurately preserved through the various preprocessing steps.

*13. Every single character z has been replaced by the two-character string sh. The frequencies of occurrence of all the characters must be recomputed. For example, if there were 100 characters in the original file, there are now 112. If there were 27 a's per 100 characters in the original file, there are now 27 a's per 112 characters, so the frequency of occurrence of an a is (27/112)*100 = 24.

The new frequencies are:

a	s	h	t	e	c
24	11	11	13	28	13

The new Huffman tree is

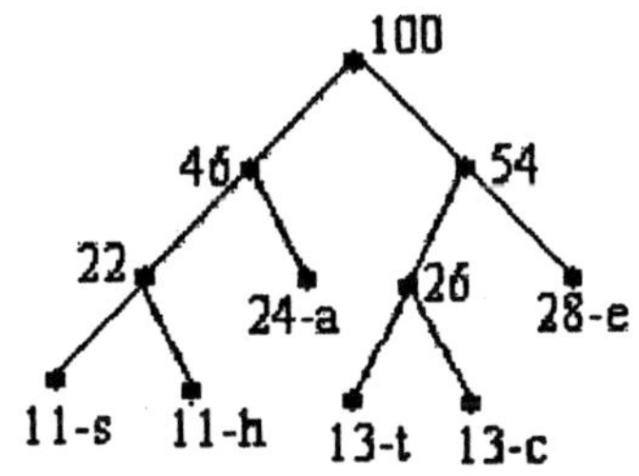

and a new encoding (one of several possibilities) is

s - 000
h - 001
a - 01
t - 100
c - 101
e - 11

14. Aside from punctuation or uppercase characters, the single character that appears least frequently (1 time) is k, so it would have one of the longest codes. The single character that appears most frequently is the blank character separating words, so it would have one of the shortest codes. The next most frequent character (27 times) is t, so it would also have a short code.

15. a. A possible Huffman tree is:

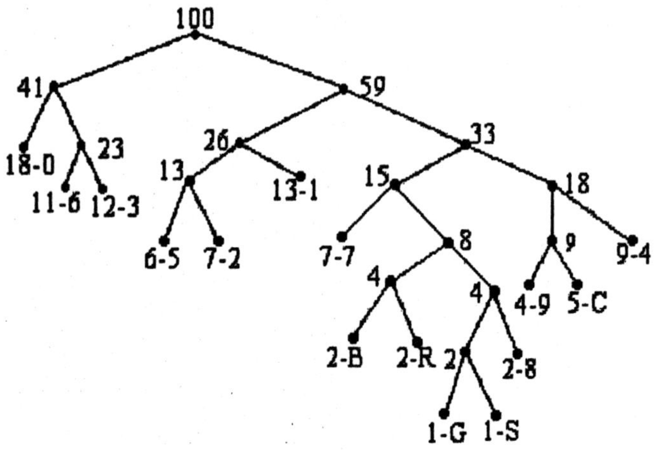

which would lead to the codes:

B - 110100	0 - 01	5 - 1000
C - 11101	1 - 101	6 - 010
G - 1101100	2 - 1001	7 - 1100
R - 110101	3 - 011	8 - 110111
S - 1101101	4 - 1111	9 - 11100

b. The space required is

$9 \times 5 \times 10^8(.02*6 + .05*5 + .01*7 + .02*6 + .01*7 + .18*2 + .13*3 + .07*4$
$+ .12*3 + .09*4 + .06*4 + .11*3 + .07*4 + .02*6 + .04*5)$

$= 9 \times 5 \times 10^8(3.55)$

as compared to $9 \times 5 \times 10^8(8)$. The new file takes 3.55/8 = 44 percent of the space of the original file.

*16. a. Because we are assuming that f(x) < f(p), we can write f(x) + j = f(p) for some positive quantity j. Because x is above p in tree T (Figure 5.58a), we can write d(x) + k = d(p) for some positive integer k. The contributions to E(T) from nodes x and p are given by

$$f(x)d(x) + f(p)d(p) = f(x)d(x) + [(f(x) + j)(d(x) + k)]$$
$$= 2f(x)d(x) + jd(x) + kf(x) + jk$$

In tree T' (Figure 5.58b), the contributions to E(T') from nodes x and p are given by the following (using the original d(x) and d(p) values):

$$f(x)d(p) + f(p)d(x) = f(x)(d(x) + k) + (f(x) + j)d(x)$$
$$= 2f(x)d(x) + kf(x) + jd(x)$$

which is jk smaller than the previous expression.

b. In Figure 5.59d, the contribution to E(B) from the node with frequency f(x) + f(y) is

[f(x) + f(y)]*r

where r is the depth of that node. The corresponding contribution to E(B') (Figure 5.59c) is

$$\begin{aligned} f(x)(r + 1) + f(y)*(r + 1) &= f(x)*r + f(y)*r + f(x) + f(y) \\ &= [f(x) + f(y)]*r + f(x) + f(y) \end{aligned}$$

so E(B') exceeds E(B) by f(x) + f(y).

CHAPTER 6: Graph Algorithms

This chapter contains (naturally) a number of different graph algorithms. Section 6.1 explores the relationship between directed graphs, associated binary relations, and adjacency matrices. It is seen that the earlier topic of the transitive closure of a binary relation relates to reachability. The two different algorithms for computing the reachability matrix - a "brute force" approach and Warshall's algorithm - provide a good illustration of an order of magnitude improvement obtained by use of a clever algorithm.

The Euler path algorithm (Section 6.2) is interesting because it answers an old children's game that the students have all seen. The contrast of the Euler path problem with the seemingly-similar Hamiltonian circuit problem is well worth mentioning, especially as a pointer to the brief discussion of computational complexity in Chapter 8. Shortest-path and minimal spanning tree algorithms (Section 6.3) introduce the idea of a greedy algorithm; additional shortest-path and minimal spanning tree algorithms appear in the exercises. These algorithms, as well as those for depth-first and breadth-first search (Section 6.4), are also subjected to (somewhat informal) analysis of algorithm techniques.

Students can readily appreciate the applicability of the shortest-path algorithm for routing in computer networks and the minimal spanning tree algorithm for network design Applications of depth-first search to reachability in a directed graph and to topological sorting are also discussed.

Despite these motivating applications, there may be good reason to pick and choose what you cover in this chapter, or even to skip it entirely. A data structures course will probably cover many of these algorithms.

EXERCISES 6.1

*1. $A = \begin{bmatrix} 1 & 0 & 0 \\ 1 & 0 & 1 \\ 0 & 1 & 0 \end{bmatrix}$ $\rho = \{(1, 1), (2, 1), (2, 3), (3, 2)\}$

2. $A = \begin{bmatrix} 0 & 0 & 1 & 0 \\ 0 & 0 & 1 & 0 \\ 0 & 0 & 0 & 1 \\ 0 & 0 & 0 & 1 \end{bmatrix}$ $\rho = \{(1, 3), (2, 3), (3, 4), (4, 4)\}$

*3.

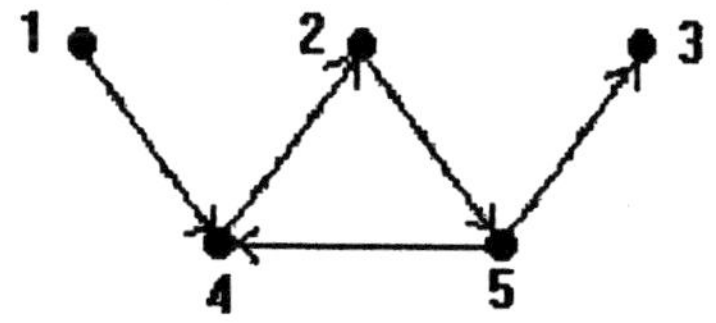

$\rho = \{(1, 4), (2, 5), (4, 2), (5, 3), (5, 4)\}$

4.

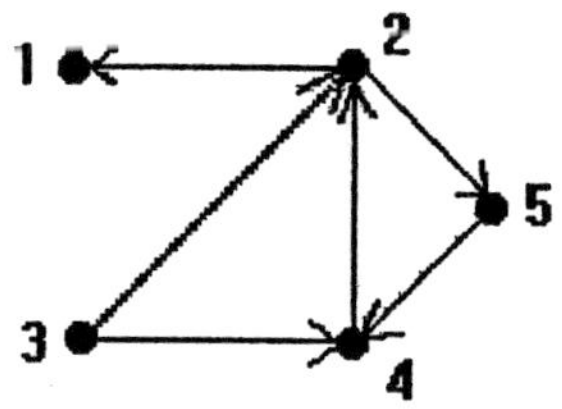

$\rho = \{(2, 1), (2, 5), (3, 2), (3, 4), (4, 2), (5, 4)\}$

5.

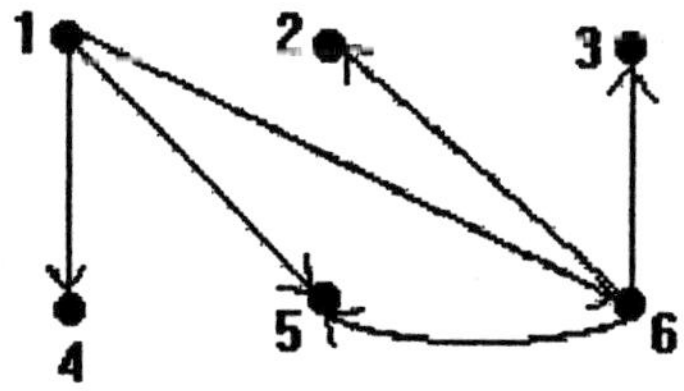

$$A = \begin{bmatrix} 0 & 0 & 0 & 1 & 1 & 1 \\ 0 & 0 & 0 & 0 & 0 & 0 \\ 0 & 0 & 0 & 0 & 0 & 0 \\ 0 & 0 & 0 & 0 & 0 & 0 \\ 0 & 0 & 0 & 0 & 0 & 0 \\ 0 & 1 & 1 & 0 & 1 & 0 \end{bmatrix}$$

6.

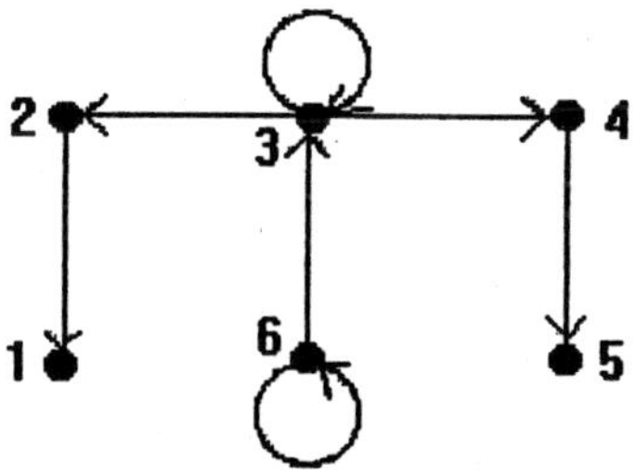

$$A = \begin{bmatrix} 0 & 0 & 0 & 0 & 0 & 0 \\ 1 & 0 & 0 & 0 & 0 & 0 \\ 0 & 1 & 1 & 1 & 0 & 0 \\ 0 & 0 & 0 & 0 & 1 & 0 \\ 0 & 0 & 0 & 0 & 0 & 0 \\ 0 & 0 & 1 & 0 & 0 & 1 \end{bmatrix}$$

7. For every pair of nodes a and b, if there is an arc from a to b, then there is also an arc from b to a.

*8. The graph is a "star" with node 1 at the center, i.e., 1 is adjacent to every node and every node is adjacent to 1, but no other nodes are adjacent.

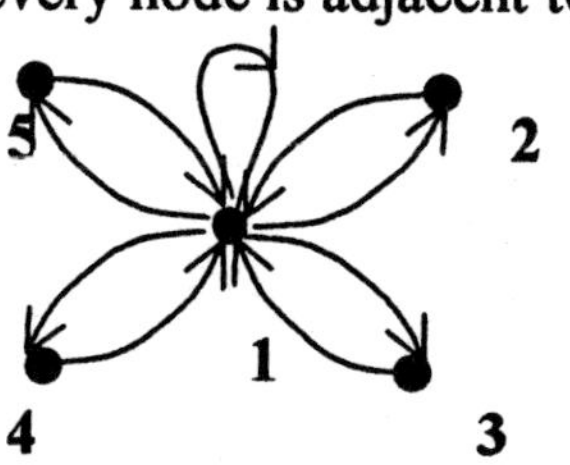

9. The graph is a cycle through all the nodes.

10. If entry i,j = 1, then entry j,i = 0 unless i = j.

11. The adjacency matrix for $\rho \cup \sigma$ is

$$\begin{bmatrix} 0 & 1 & 1 & 0 \\ 0 & 0 & 1 & 1 \\ 1 & 1 & 0 & 1 \\ 1 & 0 & 0 & 1 \end{bmatrix}$$

The adjacency matrix for $\rho \cap \sigma$ is

$$\begin{bmatrix} 0 & 1 & 0 & 0 \\ 0 & 0 & 0 & 0 \\ 1 & 0 & 0 & 0 \\ 1 & 0 & 0 & 0 \end{bmatrix}$$

*12.

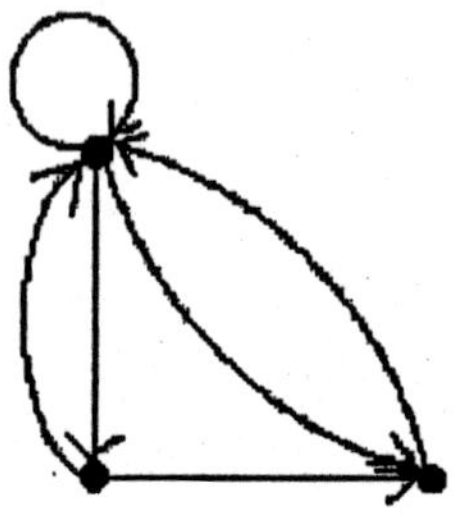

$\rho \cup \sigma$

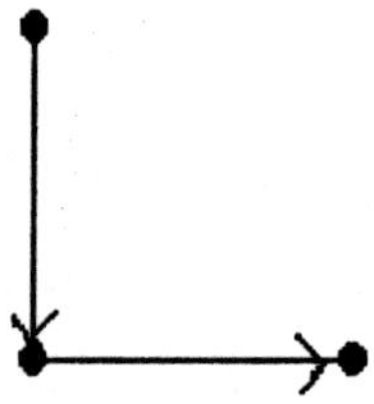

$\rho \cap \sigma$

13. $A^2 = \begin{bmatrix} 2 & 1 & 2 & 1 \\ 1 & 0 & 0 & 1 \\ 2 & 1 & 1 & 1 \\ 2 & 1 & 0 & 2 \end{bmatrix}$ $\quad A^{(2)} = \begin{bmatrix} 1 & 1 & 1 & 1 \\ 1 & 0 & 0 & 1 \\ 1 & 1 & 1 & 1 \\ 1 & 1 & 0 & 1 \end{bmatrix}$

14. Because there is a path from every node to every other node, the matrix R will have all 1 entries.

*15. $R = \begin{bmatrix} 1 & 1 & 1 & 1 \\ 1 & 1 & 1 & 1 \\ 1 & 1 & 1 & 1 \\ 0 & 0 & 0 & 0 \end{bmatrix}$

16. $R = \begin{bmatrix} 0 & 1 & 1 & 1 & 1 \\ 0 & 0 & 1 & 1 & 1 \\ 0 & 0 & 0 & 1 & 1 \\ 0 & 0 & 0 & 0 & 1 \\ 0 & 0 & 0 & 0 & 0 \end{bmatrix}$

17. $A = \begin{bmatrix} 1 & 0 & 0 \\ 1 & 0 & 1 \\ 0 & 1 & 0 \end{bmatrix}$ $A^{(2)} = \begin{bmatrix} 1 & 0 & 0 \\ 1 & 1 & 1 \\ 1 & 0 & 1 \end{bmatrix}$ $A^{(3)} = \begin{bmatrix} 1 & 0 & 0 \\ 1 & 0 & 1 \\ 1 & 1 & 1 \end{bmatrix}$

$$R = A \vee A^{(2)} \vee A^{(3)} = \begin{bmatrix} 1 & 0 & 0 \\ 1 & 1 & 1 \\ 1 & 1 & 1 \end{bmatrix}$$

18. $R = \begin{bmatrix} 0 & 0 & 1 & 1 \\ 0 & 0 & 1 & 1 \\ 0 & 0 & 0 & 1 \\ 0 & 0 & 0 & 1 \end{bmatrix}$

*19. $R = \begin{bmatrix} 0 & 1 & 1 & 1 & 1 \\ 0 & 1 & 1 & 1 & 1 \\ 0 & 0 & 0 & 0 & 0 \\ 0 & 1 & 1 & 1 & 1 \\ 0 & 1 & 1 & 1 & 1 \end{bmatrix}$

20. $R = \begin{bmatrix} 0 & 0 & 0 & 0 & 0 \\ 1 & 1 & 0 & 1 & 1 \\ 1 & 1 & 0 & 1 & 1 \\ 1 & 1 & 0 & 1 & 1 \\ 1 & 1 & 0 & 1 & 1 \end{bmatrix}$

21. $R = \begin{bmatrix} 0 & 1 & 1 & 1 & 1 & 1 \\ 0 & 0 & 0 & 0 & 0 & 0 \\ 0 & 0 & 0 & 0 & 0 & 0 \\ 0 & 0 & 0 & 0 & 0 & 0 \\ 0 & 0 & 0 & 0 & 0 & 0 \\ 0 & 1 & 1 & 0 & 1 & 0 \end{bmatrix}$

22. $R = \begin{bmatrix} 0 & 0 & 0 & 0 & 0 & 0 \\ 1 & 0 & 0 & 0 & 0 & 0 \\ 1 & 1 & 1 & 1 & 1 & 0 \\ 0 & 0 & 0 & 0 & 1 & 0 \\ 0 & 0 & 0 & 0 & 0 & 0 \\ 1 & 1 & 1 & 1 & 1 & 1 \end{bmatrix}$

*23. $A = M_0 = \begin{bmatrix} 1 & 0 & 0 \\ 1 & 0 & 1 \\ 0 & 1 & 0 \end{bmatrix}$ $M_1 = \begin{bmatrix} 1 & 0 & 0 \\ 1 & 0 & 1 \\ 0 & 1 & 0 \end{bmatrix}$ $M_2 = \begin{bmatrix} 1 & 0 & 0 \\ 1 & 0 & 1 \\ 1 & 1 & 1 \end{bmatrix}$ $M_3 = \begin{bmatrix} 1 & 0 & 0 \\ 1 & 1 & 1 \\ 1 & 1 & 1 \end{bmatrix} = R$

24. $R = \begin{bmatrix} 0 & 0 & 1 & 1 \\ 0 & 0 & 1 & 1 \\ 0 & 0 & 0 & 1 \\ 0 & 0 & 0 & 1 \end{bmatrix}$

25. $R = \begin{bmatrix} 0 & 1 & 1 & 1 & 1 \\ 0 & 1 & 1 & 1 & 1 \\ 0 & 0 & 0 & 0 & 0 \\ 0 & 1 & 1 & 1 & 1 \\ 0 & 1 & 1 & 1 & 1 \end{bmatrix}$

26. $R = \begin{bmatrix} 0 & 0 & 0 & 0 & 0 \\ 1 & 1 & 0 & 1 & 1 \\ 1 & 1 & 0 & 1 & 1 \\ 1 & 1 & 0 & 1 & 1 \\ 1 & 1 & 0 & 1 & 1 \end{bmatrix}$

*27. $R = \begin{bmatrix} 0 & 1 & 1 & 1 & 1 & 1 \\ 0 & 0 & 0 & 0 & 0 & 0 \\ 0 & 0 & 0 & 0 & 0 & 0 \\ 0 & 0 & 0 & 0 & 0 & 0 \\ 0 & 0 & 0 & 0 & 0 & 0 \\ 0 & 1 & 1 & 0 & 1 & 0 \end{bmatrix}$

28. $R = \begin{bmatrix} 0 & 0 & 0 & 0 & 0 & 0 \\ 1 & 0 & 0 & 0 & 0 & 0 \\ 1 & 1 & 1 & 1 & 1 & 0 \\ 0 & 0 & 0 & 0 & 1 & 0 \\ 0 & 0 & 0 & 0 & 0 & 0 \\ 1 & 1 & 1 & 1 & 1 & 1 \end{bmatrix}$

*29. $A^2[i,j] = \sum_{k=1}^{n} a_{ik}a_{kj}$

If a term such as $a_{i2}a_{2j}$ in this sum is 0, then either $a_{i2} = 0$ or $a_{2j} = 0$ (or both) and there is either no path of length 1 from n_i to n_2 or no path of length 1 from n_2 to n_j (or both). Thus there are no paths of length 2 from n_i to n_j passing through n_2. If $a_{i2}a_{2j} \neq 0$, then $a_{i2} = p$ and $a_{2j} = q$, where p and q are positive integers. Then there are p paths of length 1 from n_i to n_2 and q paths of length 1 from n_2 to n_j. By the Multiplication Principle, there are pq possible paths of length 2 from n_i to n_j through n_2. By the Addition Principle, the sum of all such terms gives all possible paths of length 2 from n_i to n_j.

30. The proof is by induction on n. The result for n = 1 is true by the definition of A. (The result for n = 2 is true by Exercise 29 above.) Assume that $A^p[i,j]$ equals the number of paths of length p from n_i to n_j. Then

$$A^{p+1}[i,j] = \sum_{k=1}^{n} A^p[i,k]a_{kj}$$

A term such as $A^p[i,q]a_{qj}$ is the product of the number of paths of length p from n_i to n_q and the number of paths of length 1 from n_q to n_j, which is the number of paths of length p + 1 from n_i to n_j through n_q. The sum of all such terms gives all possible paths of length p + 1 from n_i to n_j.

31. 3; $A^2 = \begin{bmatrix} 0 & 0 & 3 & 0 \\ 0 & 0 & 0 & 0 \\ 0 & 0 & 0 & 0 \\ 0 & 0 & 0 & 0 \end{bmatrix}$

32. 6; $A^4 = \begin{bmatrix} 1 & 0 & 4 & 2 & 6 \\ 0 & 1 & 4 & 2 & 6 \\ 0 & 0 & 3 & 2 & 6 \\ 0 & 0 & 2 & 1 & 4 \\ 0 & 0 & 4 & 2 & 7 \end{bmatrix}$

33.

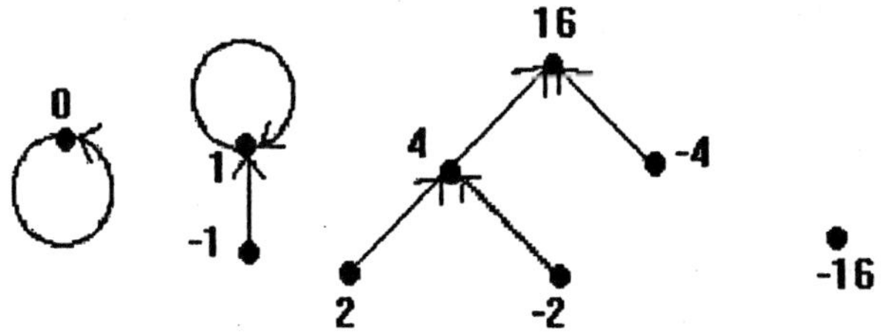

EXERCISES 6.2

1. The graph of Example 3, Chapter 2, has 4 nodes of degree 3, hence, by the theorem on Euler paths, it has no Euler path.

*2. yes 3. yes 4. no 5. no

6. yes *7. no 8. yes 9. no 10. yes

*11. $\begin{bmatrix} 0 & 1 & 1 & 1 & 1 & 0 \\ 1 & 0 & 0 & 0 & 0 & 1 \\ 1 & 0 & 0 & 0 & 0 & 1 \\ 1 & 0 & 0 & 0 & 0 & 1 \\ 1 & 0 & 0 & 0 & 0 & 1 \\ 0 & 1 & 1 & 1 & 1 & 0 \end{bmatrix}$

total after row 2 is 0

12. $\begin{bmatrix} 0 & 0 & 0 & 1 & 1 & 0 \\ 0 & 0 & 0 & 1 & 1 & 1 \\ 0 & 0 & 0 & 0 & 1 & 1 \\ 1 & 1 & 0 & 0 & 1 & 0 \\ 1 & 1 & 1 & 1 & 0 & 1 \\ 0 & 1 & 1 & 0 & 1 & 0 \end{bmatrix}$

total after row 4 is 2

*13. $\begin{bmatrix} 0 & 1 & 0 & 1 & 1 & 0 & 0 \\ 1 & 0 & 1 & 0 & 1 & 1 & 0 \\ 0 & 1 & 0 & 0 & 0 & 1 & 1 \\ 1 & 0 & 0 & 0 & 1 & 0 & 0 \\ 1 & 1 & 0 & 1 & 0 & 1 & 0 \\ 0 & 1 & 1 & 0 & 1 & 0 & 1 \\ 0 & 0 & 1 & 0 & 0 & 1 & 0 \end{bmatrix}$

$i = 8$

14. $\begin{bmatrix} 0 & 1 & 0 & 1 & 0 & 0 & 0 & 0 \\ 1 & 0 & 1 & 0 & 0 & 1 & 0 & 0 \\ 0 & 1 & 0 & 1 & 0 & 0 & 1 & 0 \\ 1 & 0 & 1 & 0 & 0 & 0 & 0 & 1 \\ 0 & 0 & 0 & 0 & 0 & 1 & 0 & 1 \\ 0 & 1 & 0 & 0 & 1 & 0 & 1 & 0 \\ 0 & 0 & 1 & 0 & 0 & 1 & 0 & 1 \\ 0 & 0 & 0 & 1 & 1 & 0 & 1 & 0 \end{bmatrix}$

$i = 5$

*15. no 16. yes 17. yes 18. no

19. yes *20. yes 21. yes 22. no

23.

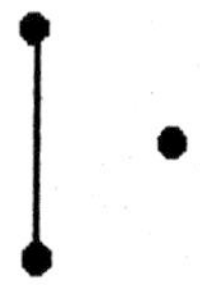

A path through the graph exists that uses each arc exactly once; there is no requirement that each node be visited.

*24. Any two nodes must be part of the Hamiltonian circuit, therefore there is a path between them, namely that part of the circuit that is between them.

25. a. $(n - 1)^n$ (At each of the n nodes that begins an arc of the path, there are n - 1 possible arcs to use.)

b. $(n - 1)(n - 2)^{n-1}$ (There are n - 1 choices for the first arc of the path, but only n - 2 choices for each arc after that because the same arc cannot be repeated.)

c. $(n - 1)!$ (There are n - 1 choices for the first arc, n - 2 for the second, etc.)

26. No; using rooms as nodes (plus an outside node), and doorways as arcs, the resulting graph has 4 odd vertices, hence no Euler path.

*27. a. $n = 2$ or $n =$ any odd number
b. $n > 2$

28. a. Euler paths exist for $m = n = 1$, for $m = 2$ and $n =$ any odd number (or vice versa), or for m even and n even.
b. Hamiltonian circuits exist for $m = n \geq 2$

29. Each odd vertex is the beginning or end of such a path, so there exist at least n such paths. If each pair of odd vertices is connected by a temporary arc, the resulting graph has no odd vertices and has an Euler cycle. The removal of the temporary arcs leaves $\leq n$ disjoint Euler paths which traverse the original graph. Therefore n paths are necessary and sufficient.

*30. Begin at any node and take one of the arcs out from that node. Each time a new node is entered on an arc, there is exactly one unused arc on which to exit that node; because the arc is unused, it will lead to a new node or to the initial node. Upon return to the initial node, if all nodes have been used, we are done. If there is an unused node, because the graph is connected, there is an unused path from that node to a used node, which means the used node has degree ≥ 3, a contradiction.

EXERCISES 6.3

*1. IN = {2}

	1	2	3	4	5	6	7	8
d	3	0	2	∞	∞	∞	1	∞
s	2	-	2	2	2	2	2	2

p = 7, IN = {2,7}

	1	2	3	4	5	6	7	8
d	3	0	2	∞	∞	6	1	2
s	2	-	2	2	2	7	2	7

p = 3, IN = {2,7,3}

	1	2	3	4	5	6	7	8
d	3	0	2	3	∞	6	1	2
s	2	-	2	3	2	7	2	7

p = 8, IN = {2,7,3,8}

	1	2	3	4	5	6	7	8
d	3	0	2	3	3	6	1	2
s	2	-	2	3	8	7	2	7

p = 5, IN = {2,7,3,8,5}

	1	2	3	4	5	6	7	8
d	3	0	2	3	3	6	1	2
s	2	-	2	3	8	7	2	7

path: 2,7,8,5 distance = 3

2. IN = {3}

	1	2	3	4	5	6	7	8
d	5	2	0	1	∞	∞	∞	2
s	3	3	-	3	3	3	3	3

p = 4, IN = {3,4}

	1	2	3	4	5	6	7	8
d	5	2	0	1	5	∞	∞	2
s	3	3	-	3	4	3	3	3

p = 2, IN = {3,4,2}

	1	2	3	4	5	6	7	8
d	5	2	0	1	5	∞	4	2
s	3	3	-	3	4	3	2	3

p = 8, IN = {3,4,2,8}

	1	2	3	4	5	6	7	8
d	5	2	0	1	3	∞	3	2
s	3	3	-	3	8	3	8	3

p = 5, IN = {3,4,2,8,5}

	1	2	3	4	5	6	7	8
d	5	2	0	1	3	9	3	2
s	3	3	-	3	8	5	8	3

p = 7, IN = {3,4,2,8,5,7}

	1	2	3	4	5	6	7	8
d	5	2	0	1	3	8	3	2
s	3	3	-	3	8	7	8	3

p = 1, IN = {3,4,2,8,5,7,1}

	1	2	3	4	5	6	7	8
d	5	2	0	1	3	6	3	2
s	3	3	-	3	8	1	8	3

p = 6, IN = {3,4,2,8,5,7,1,6}

	1	2	3	4	5	6	7	8
d	5	2	0	1	3	6	3	2
s	3	3	-	3	8	1	8	3

path: 3,1,6 (alternate
path: 3,2,1,6) distance = 6

3. IN = {1}

	1	2	3	4	5	6	7	8
d	0	3	5	∞	8	1	∞	∞
s	-	1	1	1	1	1	1	1

p = 6, IN = {1,6}

	1	2	3	4	5	6	7	8
d	0	3	5	∞	7	1	6	∞
s	-	1	1	1	6	1	6	1

p = 2, IN = {1,6,2}

	1	2	3	4	5	6	7	8
d	0	3	5	∞	7	1	4	∞
s	-	1	1	1	6	1	2	1

p = 7, IN = {1,6,2,7}

	1	2	3	4	5	6	7	8
d	0	3	5	∞	7	1	4	5
s	-	1	1	1	6	1	2	7

p = 3, IN = {1,6,2,7,3}

	1	2	3	4	5	6	7	8
d	0	3	5	6	7	1	4	5
s	-	1	1	3	6	1	2	7

p = 8, IN = {1,6,2,7,3,8}

	1	2	3	4	5	6	7	8
d	0	3	5	6	6	1	4	5
s	-	1	1	3	8	1	2	7

p =5, IN = {1,6,2,7,3,8,5}

	1	2	3	4	5	6	7	8
d	0	3	5	6	6	1	4	5
s	-	1	1	3	8	1	2	7

path: 1,2,7,8,5 distance = 6

4.

IN = {4}

	1	2	3	4	5	6	7	8
d	∞	∞	1	0	4	∞	∞	∞
s	4	4	4	-	4	4	4	4

p = 3, IN = {4,3}

	1	2	3	4	5	6	7	8
d	6	3	1	0	4	∞	∞	3
s	3	3	4	-	4	4	4	3

p = 2, IN = {4,3,2}

	1	2	3	4	5	6	7	8
d	6	3	1	0	4	∞	4	3
s	3	3	4	-	4	4	2	3

p = 8, IN = {4,3,2,8}

	1	2	3	4	5	6	7	8
d	6	3	1	0	4	∞	4	3
s	3	3	4	-	4	4	2	3

p = 7, IN = {4,3,2,8,7}

	1	2	3	4	5	6	7	8
d	6	3	1	0	4	9	4	3
s	3	3	4	-	4	7	2	3

path: 4,3,2,7 distance = 4

*5.

IN = {a}

	a	b	c	d	e	f
d	0	1	3	∞	∞	∞
s	-	a	a	a	a	a

p = b, IN = {a,b}

	a	b	c	d	e	f
d	0	1	2	∞	∞	2
s	-	a	b	a	a	b

p = c, IN = {a,b,c}

	a	b	c	d	e	f
d	0	1	2	4	6	2
s	-	a	b	c	c	b

p = f, IN = {a,b,c,f}

	a	b	c	d	e	f
d	0	1	2	4	3	2
s	-	a	b	c	f	b

p = e, IN = {a,b,c,f,e}

	a	b	c	d	e	f
d	0	1	2	4	3	2
s	-	a	b	c	f	b

path: a,b,f,e distance = 3

6.

IN = {d}

	a	b	c	d	e	f
d	∞	∞	2	0	1	2
s	d	d	d	-	d	d

p = e, IN = {d,e}

	a	b	c	d	e	f
d	∞	∞	2	0	1	2
s	d	d	d	-	d	d

p = c, IN = {d,e,c}

	a	b	c	d	e	f
d	5	3	2	0	1	2
s	c	c	d	-	d	d

p = f, IN = {d,e,c,f}

	a	b	c	d	e	f
d	5	3	2	0	1	2
s	c	c	d	-	d	d

p = b, IN = {d,e,c,f,b}

	a	b	c	d	e	f
d	4	3	2	0	1	2
s	b	c	d	-	d	d

p = a, IN = {d,e,c,f,b,a}

	a	b	c	d	e	f
d	4	3	2	0	1	2
s	b	c	d	-	d	d

path: d,c,b,a distance = 4

*7.

IN = {1}

	1	2	3	4	5	6	7
d	0	2	∞	∞	3	2	∞
s	-	1	1	1	1	1	1

p = 2, IN = {1,2}

	1	2	3	4	5	6	7
d	0	2	3	∞	3	2	∞
s	-	1	2	1	1	1	1

p = 6, IN = {1,2,6}

	1	2	3	4	5	6	7
d	0	2	3	∞	3	2	5
s	-	1	2	1	1	1	6

p = 3, IN = {1,2,6,3}

	1	2	3	4	5	6	7
d	0	2	3	4	3	2	5
s	-	1	2	3	1	1	6

p = 5, IN = {1,2,6,3,5}

	1	2	3	4	5	6	7
d	0	2	3	4	3	2	5
s	-	1	2	3	1	1	6

p = 4, IN = {1,2,6,3,5,4}

	1	2	3	4	5	6	7
d	0	2	3	4	3	2	5
s	-	1	2	3	1	1	6

p = 7, IN = {1,2,6,3,5,4,7}

	1	2	3	4	5	6	7
d	0	2	3	4	3	2	5
s	-	1	2	3	1	1	6

path: 1,6,7 distance = 5

8.

IN = {3}

	1	2	3	4	5	6	7
d	∞	∞	0	1	∞	∞	∞
s	3	3	-	3	3	3	3

p = 4, IN = {3,4}

	1	2	3	4	5	6	7
d	∞	∞	0	1	2	∞	2
s	3	3	-	3	4	3	4

p = 5, IN = {3,4,5}

	1	2	3	4	5	6	7
d	∞	∞	0	1	2	3	2
s	3	3	-	3	4	5	4

p = 7, IN = {3,4,5,7}

	1	2	3	4	5	6	7
d	∞	∞	0	1	2	3	2
s	3	3	-	3	4	5	4

p = 6, IN = {3,4,5,7,6}

	1	2	3	4	5	6	7
d	∞	∞	0	1	2	3	2
s	3	3	-	3	4	5	4

p = 1, IN = {3,4,5,7,6,1}

	1	2	3	4	5	6	7
d	∞	∞	0	1	2	3	2
s	3	3	-	3	4	5	4

No path from 3 to 1

9. a. Change the condition on the **while** loop to continue until all nodes are in IN. Also, rather than writing out a particular shortest path, make *d* and *s* output parameters that carry the information about shortest paths and their distances.
   b. No, the worst case was already computed assuming the destination node is the last node brought into IN.

10.

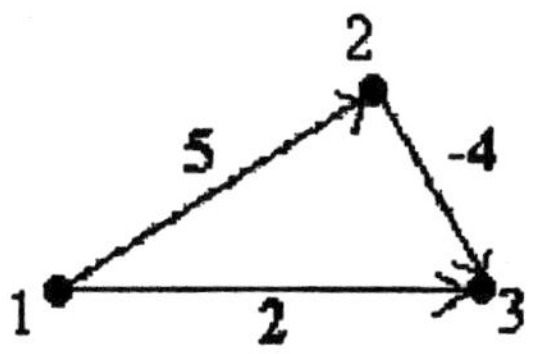

To find the shortest path from 1 to 3, the algorithm proceeds as follows:

I = {1}

	1	2	3
d	0	5	2
s	-	1	1

p = 3, IN = {1,3}

	1	2	3
d	0	5	2
s	-	1	1

Thus the algorithm will select the path 1-3 with distance 2, although the shortest path is 1-2-3 with distance 1. Allowing negative weights makes the greedy property insufficient for success, because a path with an initially high weight can later have its weight reduced by negative values, but this cannot be seen locally.

*11.

	1	2	3	4	5	6	7	8
d	3	0	2	∞	∞	∞	1	∞
s	2	-	2	2	2	2	2	2

(1)

	1	2	3	4	5	6	7	8
d	3	0	2	3	11	4	1	2
s	2	-	2	3	1	1	2	7

(2)

	1	2	3	4	5	6	7	8
d	3	0	2	3	3	4	1	2
s	2	-	2	3	8	1	2	7

(3)

No further changes in *d* or *s*.

12.

	1	2	3	4	5	6	7	8
d	0	3	5	∞	8	1	∞	∞
s	-	1	1	1	1	1	1	1

(1)

	1	2	3	4	5	6	7	8
d	0	3	5	6	7	1	4	7
s	-	1	1	3	6	1	2	3

(2)

	1	2	3	4	5	6	7	8
d	0	3	5	6	7	1	4	5
s	-	1	1	3	6	1	2	7

(3)

	1	2	3	4	5	6	7	8
d	0	3	5	6	6	1	4	5
s	-	1	1	3	8	1	2	7

(4)

No further changes in *d* or *s*.

13.

	1	2	3	4	5	6	7
d	0	2	∞	∞	3	2	∞
s	-	1	1	1	1	1	1

(1)

	1	2	3	4	5	6	7
d	0	2	3	∞	3	2	5
s	-	1	2	1	1	1	6

(2)

	1	2	3	4	5	6	7
d	0	2	3	4	3	2	5
s	-	1	2	3	1	1	6

(3)

No further changes in *d* or *s*.

14.

	1	2	3
d	0	5	2
s	-	1	1

(1)

	1	2	3
d	0	5	1
s	-	1	2

(2)

*15. Initial A and after k = x:

	x	1	2	3	y
x	0	1	∞	4	∞
1	1	0	3	1	5
2	∞	3	0	2	2
3	4	1	2	0	3
y	∞	5	2	3	0

after k = 1 and k = 2:

	x	1	2	3	y
x	0	1	4	2	6
1	1	0	3	1	5
2	4	3	0	2	2
3	2	1	2	0	3
y	6	5	2	3	0

after k = 3 and k = y:

	x	1	2	3	y
x	0	1	4	2	5
1	1	0	3	1	4
2	4	3	0	2	2
3	2	1	2	0	3
y	5	4	2	3	0

16. Initial A:

	1	2	3	4	5	6	7	8
1	0	3	5	∞	8	1	∞	∞
2	3	0	2	∞	∞	∞	1	∞
3	5	2	0	1	∞	∞	∞	2
4	∞	∞	1	0	4	∞	∞	∞
5	8	∞	∞	4	0	6	∞	1
6	1	∞	∞	∞	6	0	5	∞
7	∞	1	∞	∞	∞	5	0	1
8	∞	∞	2	∞	1	∞	1	0

after k = 1:

	1	2	3	4	5	6	7	8
1	0	3	5	∞	8	1	∞	∞
2	3	0	2	∞	11	4	1	∞
3	5	2	0	1	13	6	∞	2
4	∞	∞	1	0	4	∞	∞	∞
5	8	11	13	4	0	6	∞	1
6	1	4	6	∞	6	0	5	∞
7	∞	1	∞	∞	∞	5	0	1
8	∞	∞	2	∞	1	∞	1	0

after k = 2:

	1	2	3	4	5	6	7	8
1	0	3	5	∞	8	1	4	∞
2	3	0	2	∞	11	4	1	∞
3	5	2	0	1	13	6	3	2
4	∞	∞	1	0	4	∞	∞	∞
5	8	11	13	4	0	6	12	1
6	1	4	6	∞	6	0	5	∞
7	4	1	3	∞	12	5	0	1
8	∞	∞	2	∞	1	∞	1	0

after k = 3:

	1	2	3	4	5	6	7	8
1	0	3	5	6	8	1	4	7
2	3	0	2	3	11	4	1	4
3	5	2	0	1	13	6	3	2
4	6	3	1	0	4	7	4	3
5	8	11	13	4	0	6	12	1
6	1	4	6	7	6	0	5	8
7	4	1	3	4	12	5	0	1
8	7	4	2	3	1	8	1	0

after k = 4:

	1	2	3	4	5	6	7	8
1	0	3	5	6	8	1	4	7
2	3	0	2	3	7	4	1	4
3	5	2	0	1	5	6	3	2
4	6	3	1	0	4	7	4	3
5	8	7	5	4	0	6	8	1
6	1	4	6	7	6	0	5	8
7	4	1	3	4	8	5	0	1
8	7	4	2	3	1	8	1	0

after k = 5:

	1	2	3	4	5	6	7	8
1	0	3	5	6	8	1	4	7
2	3	0	2	3	7	4	1	4
3	5	2	0	1	5	6	3	2
4	6	3	1	0	4	7	4	3
5	8	7	5	4	0	6	8	1
6	1	4	6	7	6	0	5	7
7	4	1	3	4	8	5	0	1
8	7	4	2	3	1	7	1	0

after k = 6:

	1	2	3	4	5	6	7	8
1	0	3	5	6	7	1	4	7
2	3	0	2	3	7	4	1	4
3	5	2	0	1	5	6	3	2
4	6	3	1	0	4	7	4	3
5	7	7	5	4	0	6	8	1
6	1	4	6	7	6	0	5	7
7	4	1	3	4	8	5	0	1
8	7	4	2	3	1	7	1	0

after k = 7:

	1	2	3	4	5	6	7	8
1	0	3	5	6	7	1	4	5
2	3	0	2	3	7	4	1	2
3	5	2	0	1	5	6	3	2
4	6	3	1	0	4	7	4	3
5	7	7	5	4	0	6	8	1
6	1	4	6	7	6	0	5	6
7	4	1	3	4	8	5	0	1
8	5	2	2	3	1	6	1	0

after k = 8:

	1	2	3	4	5	6	7	8
1	0	3	5	6	6	1	4	5
2	3	0	2	3	3	4	1	2
3	5	2	0	1	3	6	3	2
4	6	3	1	0	4	7	4	3
5	6	3	3	4	0	6	2	1
6	1	4	6	7	6	0	5	6
7	4	1	3	4	2	5	0	1
8	5	2	2	3	1	6	1	0

17. IN = {1,6,2,7,8,5,3,4}

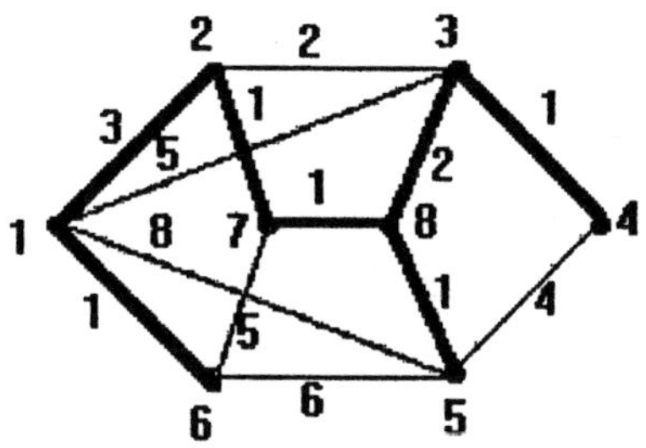

18. IN = {1,5,2,6,7,8,4,3}

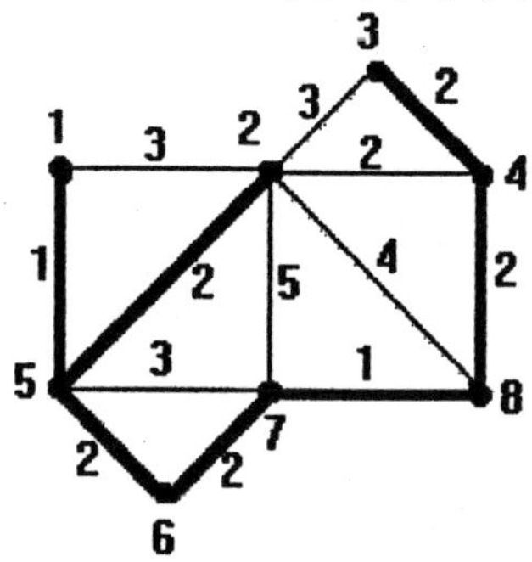

*19. IN = {1,8,5,6,2,7,4,3}

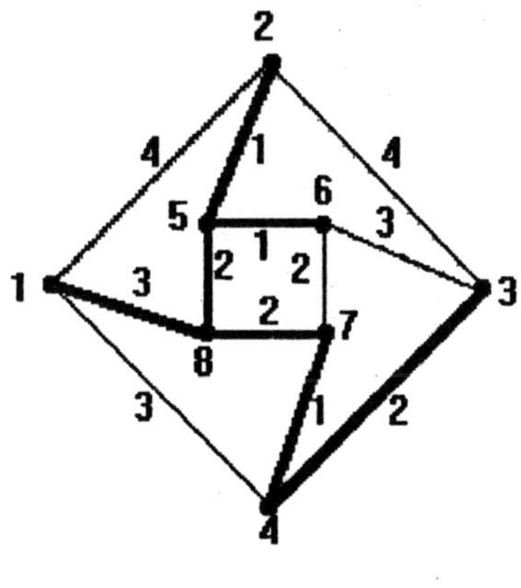

20. IN = {1,3,4,2,5,7,9,8,6}

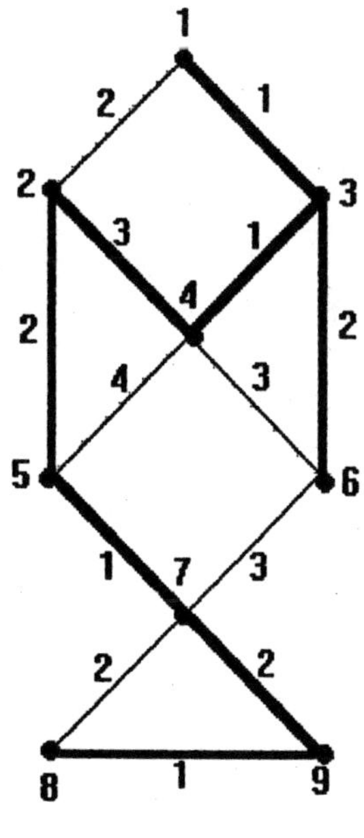

*21. Kruskal's algorithm develops a minimal spanning tree as follows.

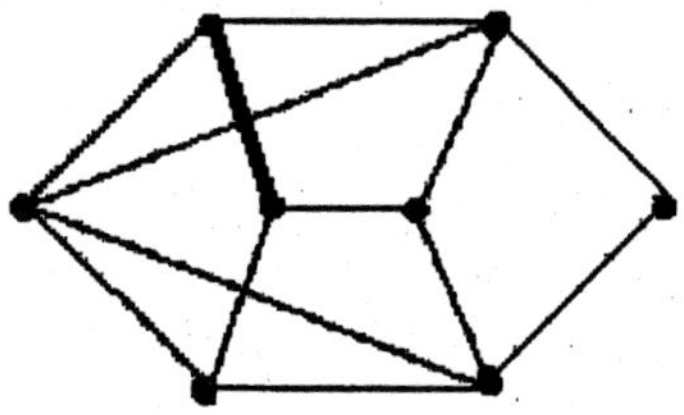

(1)

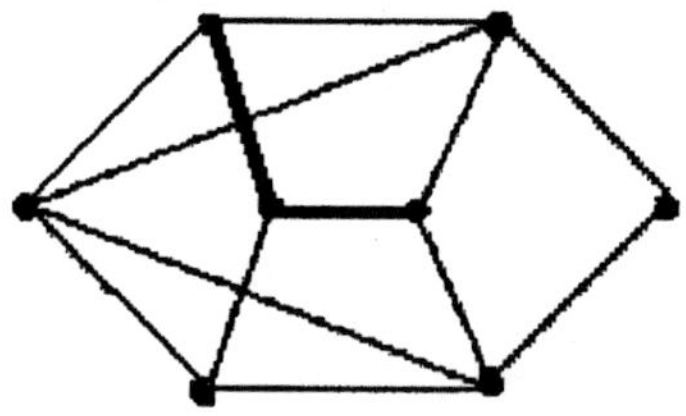

(2)

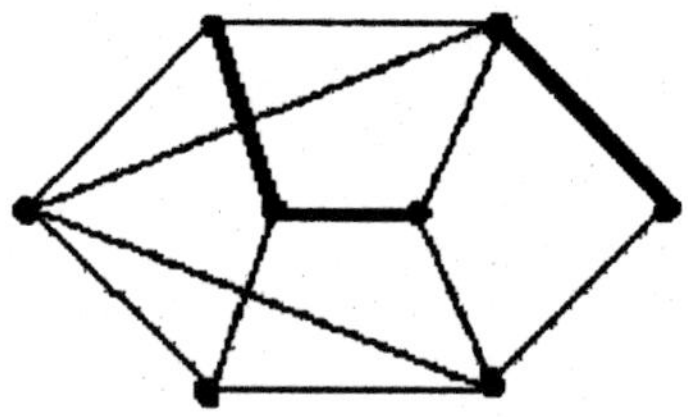

(3)

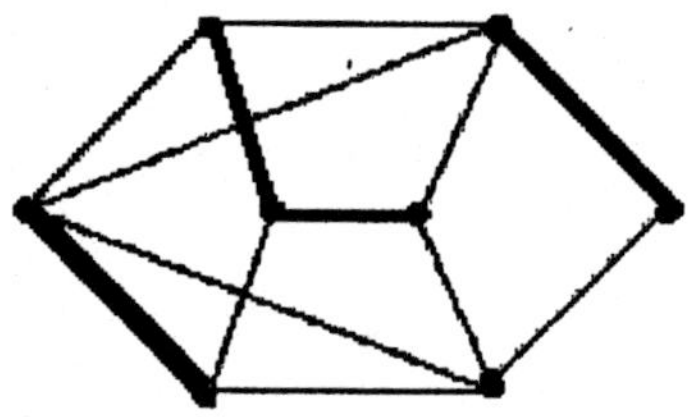

(4)

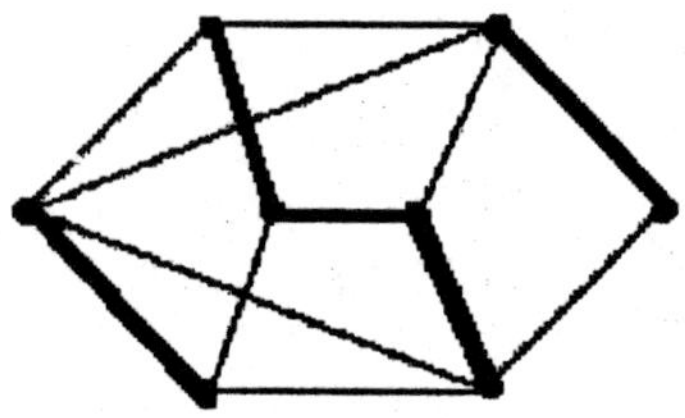

(5)

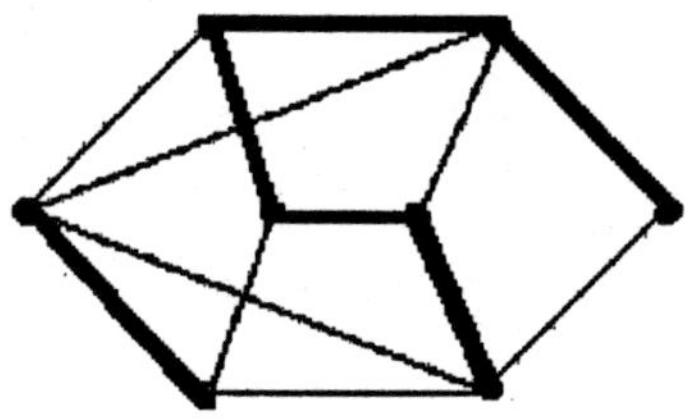

(6)

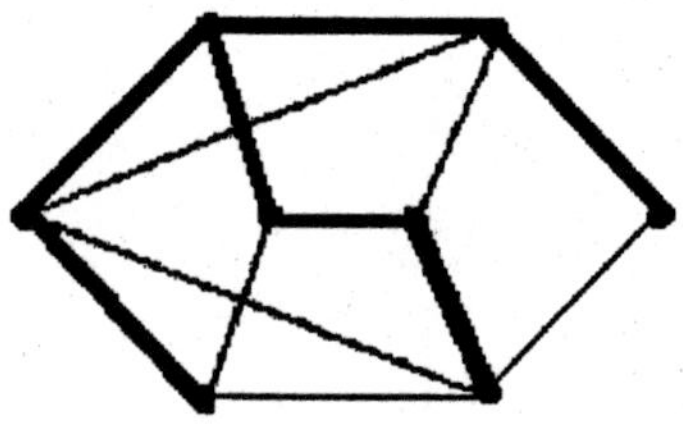

(7)

22. Result can agree with Exercise 16.

23. Result can agree with Exercise 17.

24. Result can agree with Exercise 18.

25.

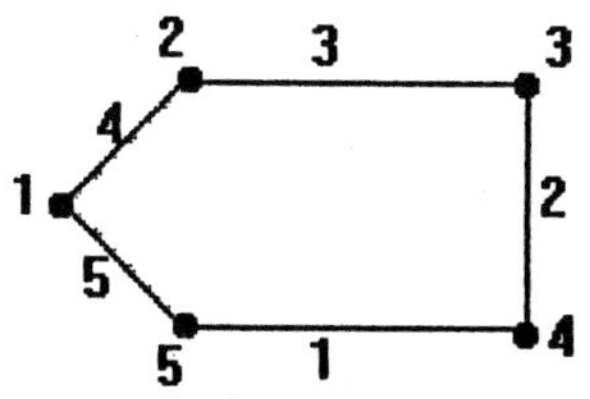

The shortest path from 1 to 5 is 1-5 with distance 5. If the algorithm added the node closest to IN at each step, it would choose the path 1-2-3-4-5 with distance 10.

26. The solution is to find a minimal spanning tree for the graph, as shown here.

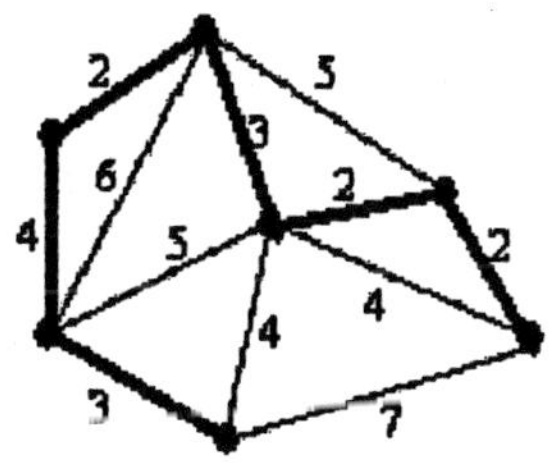

*27. a. weight = 300

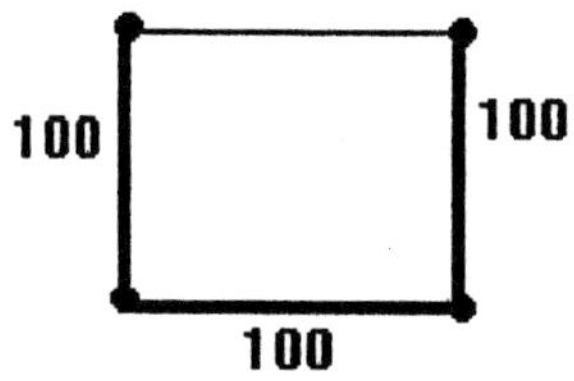

b. $d = 50\sqrt{2} = 70.750$ $4d = 282.8$

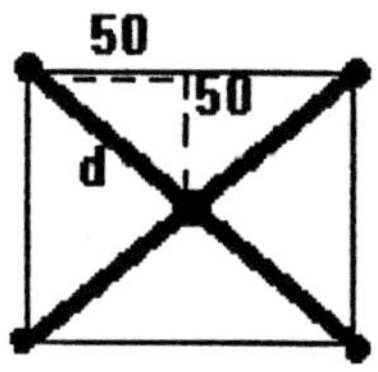

28. Dijkstra's algorithm was $\Theta(n^2)$ in the worst case, which was when all nodes are brought into IN. This is the situation to find the distance from the start node to any other node. Repeating this process with all n nodes, in turn, as the start node would result in an algorithm of order $n\Theta(n^2) = \Theta(n^3)$. Algorithm AllPairsShortestPath is clearly $\Theta(n^3)$ because of the nested **for** loops. Therefore the algorithms are the same order of magnitude (although AllPairsShortestPath has the advantage of simplicity).

EXERCISES 6.4

*1. a b c e f d h g j i

2. c a b e f d h g j i

3. d a b c e f h g j i

4. g b a c e f d h i j

*5. e b a c f d h g j i

6. h f c a b e g j i d

*7. a b c f j g d e h k i

8. e b a c f j g d h k i

*9 f c a b d e h k i g j

10. h e b a c f j g d i k

*11. a b c d e g f h j i

12. c a b e f d g h j i

13. d a f b c e h g i j

14. g b e h j a c f i d

15. e b c f g a d h j i

16. h f g i c d e b j a

*17. a b c d e f g h i j k

18. e b d h i a c k f g j

19. f c j a b g d e h i k

20. h e k b d i a c f g j

*21. a b c e g d f h

22. g d a b c e h f

23. f b

*24. a b c d e g h f

25. g d f h a e b c

26. f b

27. Begin a dfs at node C, using only the subgraph involving nodes C, A, E, F, and G. The arcs to previously unvisited nodes form the spanning tree.

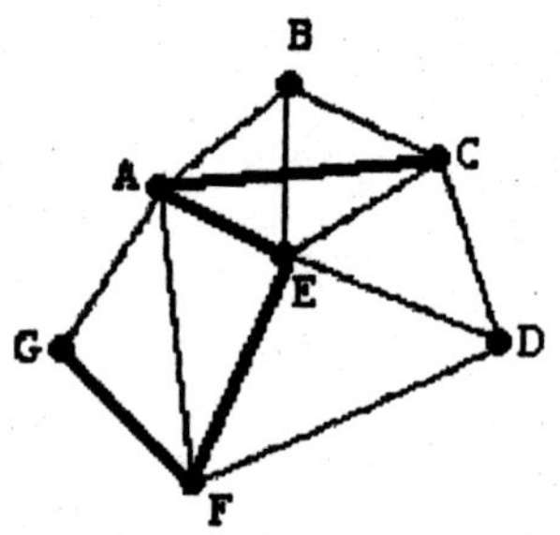

*28. Begin a dfs at node a: a c f g e b d

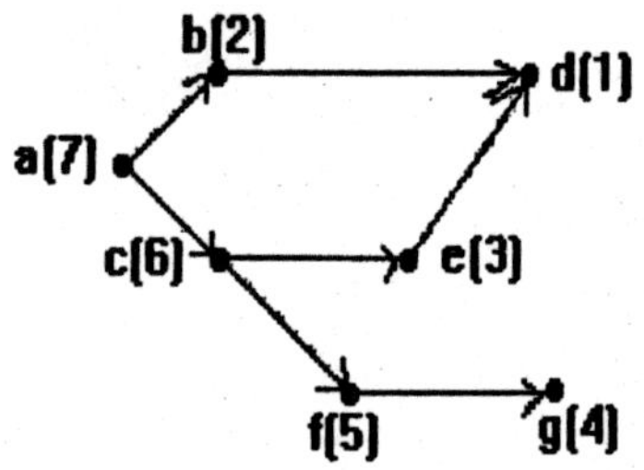

29. Begin a dfs at node d, and then at node a: a d c e f b

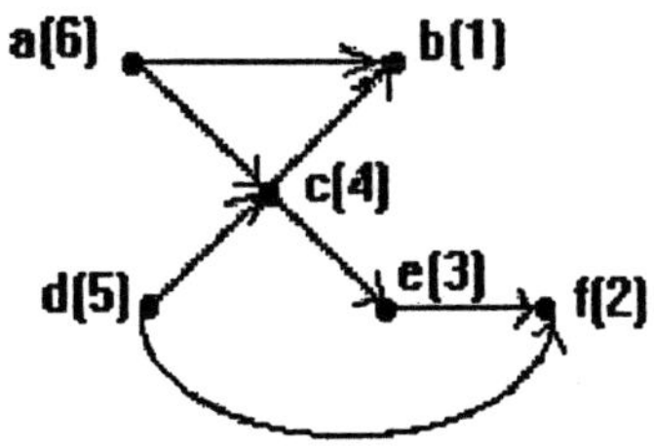

30. Do a breadth-first search starting from the root and visiting adjacent nodes left to right.

CHAPTER 7: Boolean Algebra and Computer Logic

The Boolean algebra structure is presented in this chapter as an abstraction of the common properties of propositional logic and set theory. It is the first example of a "metastructure" or a model. The theme of modeling runs throughout Chapter 8 as well.

Abstraction in notation is also used here, and it is worth pointing out (several times) that the "+" in a particular instance of a Boolean algebra may not be addition, and the "·" may not be multiplication. Because of this tendency to see + as addition and · as multiplication, and because some of the properties of a Boolean algebra agree with those of ordinary arithmetic, I like to highlight those properties that do NOT agree with ordinary arithmetic, such as property (3a) or the idempotent property.

To transform a Boolean expression into an equivalent form requires that one be able to recognize when a particular Boolean algebra property is applicable. Thus the process is somewhat an exercise in pattern recognition, and therefore is similar to constructing proofs in propositional logic. By this time, however, the students are a bit more adept at this sort of thing.

Graph isomorphism was defined in Section 5.1, where there was an easy geometric interpretation. An isomorphism between two Boolean algebras requires that the effects of operations be preserved, which is expressed as "operate and map equals map and operate." Here, and in the group theory section in Chapter 8, I draw many pictures of two circles and say "we're here, what is the operation? now we've mapped over here, what is the operation?"

The translations between Boolean expressions, truth functions, and logic networks generally go smoothly. Building circuitry for binary addition is fun because it is such a practical application. Students cope well with Karnaugh maps, but find the Quine-McCluskey procedure tedious.

EXERCISES 7.1

*1.

+	0	1	a	a'
0	0	1	a	a'
1	1	1	1	1
a	a	1	a	1
a'	a'	1	1	a'

·	0	1	a	a'
0	0	0	0	0
1	0	1	a	a'
a	0	a	a	0
a'	0	a'	0	a'

2. a. max(x, y) = max(y, x)   min(x, y) = min(y, x)
max(max(x, y), z) = max(x, max(y, z)) = max(x, y, z)
min(min(x, y), z) = min(x, min(y, z)) = min(x, y, z)
max(x, min(y, z)) = min(max(x, y),max(x, z))
min(x, max(y, z)) = max(min(x, y), min(x, z))
(The last two can be shown by taking cases: $x < y < z$, $y < x < z$, etc.)

b. Let m be the 0 element. Then we must have max(x, m) = x for all $x \in Z$. But for $x = m - 1$, max(m - 1, m) = m.

3. a. $2^4 = 16$

b.

$(f_1 + f_2)(0,0) = 1$	$(f_1 \cdot f_2)(0,0) = 1$	$f_1'(0,0) = 0$
$(f_1 + f_2)(0,1) = 1$	$(f_1 \cdot f_2)(0,1) = 0$	$f_1'(0,1) = 1$
$(f_1 + f_2)(1,0) = 1$	$(f_1 \cdot f_2)(1,0) = 0$	$f_1'(1,0) = 0$
$(f_1 + f_2)(1,1) = 0$	$(f_1 \cdot f_2)(1,1) = 0$	$f_1'(1,1) = 1$

c. + and $\cdot$ are binary operations on B, ' is a unary operation on B. Max and min are commutative and associative operations; the distributive laws follow by considering different cases for the values of $f_1(x, y)$, $f_2(x, y)$, and $f_3(x, y)$ for a fixed $(x, y) \in S^2$. For any f in B and (x,y) in S^2, (f + 0)(x, y) = max(f(x, y), 0(x, y)) = max(f(x, y), 0) = f(x, y), and (f $\cdot$ 1)(x, y) = min(f(x, y), 1(x, y)) = min(f(x, y), 1) = f(x, y). Also (f + f')(x, y) = max(f(x, y), f '(x, y)) = 1 and (f $\cdot$ f ')(x, y) = min(f(x, y), f '(x, y)) = 0 since one value of the pair (f(x, y), f '(x, y)) is 1 and the other 0.

4. *a.

$x' + x = x + x'$	(1a)
$= 1$	(5a)

and

$x' \cdot x = x \cdot x'$	(1b)
$= 0$	(5b)

Therefore $x = (x')'$ by the theorem on the uniqueness of complements.

b.

$(x + y) + (x' \cdot y') = ((x + y) + x') \cdot ((x + y) + y')$	(3a)
$= (x + (y + x')) \cdot (x + (y + y'))$	(2a)
$= (x + (x' + y)) \cdot (x + 1)$	(1a, 5a)
$= ((x + x') + y) \cdot 1$	(2a, Practice 3a)
$= (1 + y) \cdot 1$	(5a)
$= (y + 1) \cdot 1$	(1a)
$= 1 \cdot 1$	(Practice 3a)
$= 1$	(4b)

and

$$
\begin{aligned}
(x+y)\cdot(x'\cdot y') &= (x'\cdot y')\cdot(x+y) && \text{(1b)}\\
&= (x'\cdot y')\cdot x + (x'\cdot y')\cdot y && \text{(3b)}\\
&= x\cdot(x'\cdot y') + (x'\cdot y')\cdot y && \text{(1b)}\\
&= (x\cdot x')\cdot y' + x'\cdot(y'\cdot y) && \text{(2b)}\\
&= 0\cdot y' + x'\cdot(y\cdot y') && \text{(5b, 1b)}\\
&= y'\cdot 0 + x'\cdot 0 && \text{(1b, 5b)}\\
&= 0+0 && \text{(Practice 3b)}\\
&= 0 && \text{(4a)}
\end{aligned}
$$

Therefore $x'\cdot y' = (x+y)'$ by the theorem on the uniqueness of complements.
$(x\cdot y)' = x' + y'$ follows by duality.

5. *a.

$$
\begin{aligned}
x+(x\cdot y) &= x\cdot 1 + x\cdot y && \text{(4b)}\\
&= x(1+y) && \text{(3b)}\\
&= x(y+1) && \text{(1a)}\\
&= x\cdot 1 && \text{(Practice 3a)}\\
&= x && \text{(4b)}
\end{aligned}
$$

$x\cdot(x+y) = x$ follows by duality

b.

$$
\begin{aligned}
x\cdot[y+(x\cdot z)] &= x\cdot y + x\cdot(x\cdot z) && \text{(3b)}\\
&= x\cdot y + (x\cdot x)\cdot z && \text{(2b)}\\
&= x\cdot y + x\cdot z && \text{(dual of idempotent property)}
\end{aligned}
$$

$x+[y\cdot(x+z)] = (x+y)\cdot(x+z)$ follows by duality

c.

$$
\begin{aligned}
(x+y)\cdot(x'+y) &= (y+x)\cdot(y+x') && \text{(1a)}\\
&= y+(x\cdot x') && \text{(3a)}\\
&= y+0 && \text{(5b)}\\
&= y && \text{(4a)}
\end{aligned}
$$

$(x\cdot y)+(x'\cdot y) = y$ follows by duality

d.

$$
\begin{aligned}
(x+(y\cdot z))' &= x'\cdot(y\cdot z)' && \text{(De Morgan's Laws)}\\
&= x'\cdot(y'+z') && \text{(De Morgan's Laws)}\\
&= (x'\cdot y')+(x'\cdot z') && \text{(3b)}
\end{aligned}
$$

$(x\cdot(y+z))' = (x'+y')\cdot(x'+z')$ follows by duality

e.

$$
\begin{aligned}
(x+y)\cdot(x+1) &= (x+y)\cdot x + (x+y)\cdot 1 && \text{(3b)}\\
&= x\cdot(x+y) + (x+y)\cdot 1 && \text{(1b)}\\
&= (x\cdot x)+(x\cdot y)+(x+y)\cdot 1 && \text{(3b)}\\
&= x+(x\cdot y)+(x+y)\cdot 1 && \text{(dual of idempotent property)}\\
&= x+(x\cdot y)+(x+y) && \text{(4b)}\\
&= (x\cdot y)+x+(x+y) && \text{(1a)}\\
&= (x\cdot y)+(x+x)+y && \text{(2a)}\\
&= (x\cdot y)+x+y && \text{(idempotent property)}\\
&= x+(x\cdot y)+y && \text{(1a)}
\end{aligned}
$$

$(x\cdot y)+(x\cdot 0) = x\cdot(x+y)\cdot y$ follows by duality

f. $(x + y) + (y \cdot x') = (y + x) + (y \cdot x')$ (1a)
$= y + (x + (y \cdot x'))$ (2a)
$= y + (x + y) \cdot (x + x')$ (3a)
$= y + (x + y) \cdot 1$ (5a)
$= y + (x + y)$ (4b)
$= (x + y) + y$ (1a)
$= x + (y + y)$ (2a)
$= x + y$ (idempotent property)
$(x \cdot y) \cdot (y + x') = x \cdot y$ follows by duality

6. *a. $x + (x' \cdot y + x \cdot y)' = x + (y \cdot x' + y \cdot x)'$ (1b)
$= x + (y \cdot (x' + x))'$ (3b)
$= x + (y \cdot (x + x'))'$ (1a)
$= x + (y \cdot 1)'$ (5a)
$= x + y'$ (4b)

b. $((x \cdot y) \cdot z) + (y \cdot z) = (x \cdot (y \cdot z)) + (y \cdot z)$ (2b)
$= ((y \cdot z) \cdot x) + y \cdot z$ (1b)
$= ((y \cdot z) \cdot x) + (y \cdot z) \cdot 1$ (4b)
$= (y \cdot z) \cdot (x + 1)$ (3b)
$= (y \cdot z) \cdot 1$ (Practice 3a)
$= y \cdot z$ (4b)

c. $(y' \cdot x) + x + (y + x) \cdot y' = (y' \cdot x) + x + y' \cdot (y + x)$ (1b)
$= (y' \cdot x) + x + (y' \cdot y) + (y' \cdot x)$ (3b)
$= (y' \cdot x) + x + (y \cdot y') + (y' \cdot x)$ (1b)
$= (y' \cdot x) + x + 0 + (y' \cdot x)$ (5b)
$= (y' \cdot x) + x + (y' \cdot x)$ (4a)
$= (y' \cdot x) + (y' \cdot x) + x$ (1a)
$= (y' \cdot x) + x$ (idempotent property)
$= x + (y' \cdot x)$ (1a)

d. $((x' + z') \cdot (y + z'))' = (x' + z')' + (y + z')'$ (De Morgan's Laws)
$= ((x')' \cdot (z')') + y' \cdot (z')'$ (De Morgan's Laws)
$= x \cdot z + y' \cdot z$ (Exercise 4a)
$= z \cdot x + z \cdot y'$ (1b)
$= z \cdot (x + y')$ (3b)
$= (x + y') \cdot z$ (1b)

e. $(x \cdot y) + (x' \cdot z) + (x' \cdot y \cdot z') = (x \cdot y) + (x' \cdot z) \cdot 1 + (x' \cdot y \cdot z')$ (4b)
$= (x \cdot y) + (x' \cdot z) \cdot (y + y') + (x' \cdot y \cdot z')$ (5a)
$= (x \cdot y) + (x' \cdot z \cdot y) + (x' \cdot z \cdot y') + (x' \cdot y \cdot z')$ (3b)
$= (x \cdot y) + (x' \cdot z \cdot y) + (x' \cdot z \cdot y) + (x' \cdot z \cdot y') + (x' \cdot y \cdot z')$ (idempotent)
$= (x \cdot y) + (x' \cdot z \cdot y) + (x' \cdot y \cdot z') + (x' \cdot z \cdot y) + (x' \cdot z \cdot y')$ (1a)
$= (x \cdot y) + (x' \cdot y \cdot z) + (x' \cdot y \cdot z') + (x' \cdot z \cdot y) + (x' \cdot z \cdot y')$ (1b)
$= (x \cdot y) + (x' \cdot y) \cdot (z + z') + (x' \cdot z) \cdot (y + y')$ (3b)

$= (x \cdot y) + (x' \cdot y) \cdot 1 + (x' \cdot z) \cdot 1$ (5a)
$= (x \cdot y) + (x' \cdot y) + (x' \cdot z)$ (4b)
$= (y \cdot x) + (y \cdot x') + (x' \cdot z)$ (1b)
$= y \cdot (x + x') + (x' \cdot z)$ (3b)
$= y \cdot 1 + (x' \cdot z)$ (5a)
$= y + (x' \cdot z)$ (4b)

f. $(x \cdot y') + (y \cdot z') + (x' \cdot z) = (x \cdot y') \cdot 1 + (y \cdot z') \cdot 1 + (x' \cdot z) \cdot 1$ (4b)
$= (x \cdot y') \cdot (z + z') + (y \cdot z') \cdot (x + x') + (x' \cdot z) \cdot (y + y')$ (5a)
$= x \cdot y' \cdot z + x \cdot y' \cdot z' + y \cdot z' \cdot x + y \cdot z' \cdot x' + x' \cdot z \cdot y + x' \cdot z \cdot y'$ (3b)
$= x \cdot y' \cdot z + x' \cdot z \cdot y' + y \cdot z' \cdot x + x \cdot y' \cdot z' + x' \cdot z \cdot y + y \cdot z' \cdot x'$ (1a)
$= y' \cdot z \cdot x + y' \cdot z \cdot x' + x \cdot z' \cdot y + x \cdot z' \cdot y' + x' \cdot y \cdot z + x' \cdot y \cdot z'$ (1b)
$= (y' \cdot z) \cdot (x + x') + (x \cdot z') \cdot (y + y') + (x' \cdot y) \cdot (z + z')$ (3b)
$= (y' \cdot z) \cdot 1 + (x \cdot z') \cdot 1 + (x' \cdot y) \cdot 1$ (5a)
$= (y' \cdot z) + (x \cdot z') + (x' \cdot y)$ (4b)
$= (x' \cdot y) + (y' \cdot z) + (x \cdot z')$ (1a)

7. a. $(x + y \cdot x)' = (x \cdot 1 + y \cdot x)'$ (4b)
$= (x \cdot 1 + x \cdot y)'$ (1b)
$= (x \cdot (1 + y))'$ (3b)
$= (x \cdot (y + 1))'$ (1a)
$= (x \cdot 1)'$ (Practice 3a)
$= x'$ (4b)

b. $x \cdot (z + y) + (x' + y)' = x \cdot (z + y) + ((x')' \cdot y')$ (De Morgan's Laws)
$= x \cdot (z + y) + (x \cdot y')$ (double negation)
$= x \cdot ((z + y) + y')$ (3b)
$= x \cdot (z + (y + y'))$ (2a)
$= x \cdot (z + 1)$ (5a)
$= x \cdot 1$ (Practice 3a)
$= x$ (4b)

c. $(x \cdot y)' + x' \cdot z + y' \cdot z = (x' + y') + x' \cdot z + y' \cdot z$ (DeMorgan's Laws)
$= (x' + y') + z \cdot x' + z \cdot y'$ (1b)
$= (x' + y') + z \cdot (x' + y')$ (3b)
$= (x' + y') \cdot 1 + z \cdot (x' + y')$ (4b)
$= (x' + y') \cdot 1 + (x' + y') \cdot z$ (1b)
$= (x' + y') \cdot (1 + z)$ (3b)
$= (x' + y') \cdot (z + 1)$ (1a)
$= (x' + y') \cdot 1$ (Practice 3a)
$= x' + y'$ (4b)

d. $x \cdot y + x' = x \cdot y + x' \cdot 1$ (4b)
$= x \cdot y + x' \cdot (y + y')$ (5a)
$= x \cdot y + x' \cdot y + x' \cdot y'$ (3b)
$= y \cdot x + y \cdot x' + x' \cdot y'$ (1b)
$= y \cdot (x + x') + x' \cdot y'$ (3b)
$= y \cdot 1 + x' \cdot y'$ (5a)
$= y + x' \cdot y'$ (4b)

8. a. Let $x \cdot y' = 0$. Then
$x \cdot y = x \cdot y + 0$ (4a)
$= x \cdot y + x \cdot y'$ $(x \cdot y' = 0)$
$= x \cdot (y + y')$ (3b)
$= x \cdot 1$ (5a)
$= x$ (4b)

b. Let $x \cdot y = x$. Then
$x \cdot y' = (x \cdot y) \cdot y'$ $(x \cdot y = x)$
$= x \cdot (y \cdot y')$ (2b)
$= x \cdot 0$ (5b)
$= 0$ (Practice 3b)

*9. a. Let $x = 0$. Then
$x \cdot y' + x' \cdot y = 0 \cdot y' + x' \cdot y$ $(x = 0)$
$= y' \cdot 0 + x' \cdot y$ (1b)
$= 0 + x' \cdot y$ (Practice 3b)
$= x' \cdot y + 0$ (1a)
$= x' \cdot y$ (4a)
$= 1 \cdot y$ (Practice 4)
$= y \cdot 1$ (1b)
$= y$ (4b)

b. Let $x \cdot y' + x' \cdot y = y$. Then
$x \cdot x' + x' \cdot x = x$ (letting y in the hypothesis have the value x)
$x \cdot x' + x \cdot x' = x$ (1b)
$0 + 0 = x$ (5b)
$0 = x$ (4a)

10.*a. $x \oplus y = x \cdot y' + y \cdot x'$ (definition of $\oplus$)
$= y \cdot x' + x \cdot y'$ (1a)
$= y \oplus x$ (definition of $\oplus$)

b. $x \oplus x = x \cdot x' + x \cdot x'$ (definition of $\oplus$)
$= 0 + 0$ (5b)
$= 0$ (4a)

c. $0 \oplus x = 0 \cdot x' + x \cdot 0'$ (definition of $\oplus$)
$= x' \cdot 0 + x \cdot 0'$ (1b)
$= 0 + x \cdot 0'$ (Practice 3b)
$= 0 + x \cdot 1$ (Practice 4)
$= 0 + x$ (4b)
$= x + 0$ (1a)
$= x$ (4a)

d. $1 \oplus x = 1 \cdot x' + x \cdot 1'$ (definition of $\oplus$)
$= x' \cdot 1 + x \cdot 1'$ (1b)
$= x' + x \cdot 1'$ (4b)
$= x' + x \cdot 0$ (Practice 4)
$= x' + 0$ (Practice 3b)
$= x'$ (4a)

11. a. Let $x + y = 0$. Then
$x = x + 0$ (4a)
$= x + (x + y)$ $(x + y = 0)$
$= (x + x) + y$ (2a)
$= x + y$ (idempotent property)
$= 0$ $(x + y = 0)$
Similarly, if $x + y = 0$ then $y = 0$.

b. Let $x = y$. Then
$x \cdot y' + y \cdot x' = x \cdot x' + x \cdot x'$ $(x = y)$
$= 0 + 0$ (5b)
$= 0$ (4a)

Let $x \cdot y' + y \cdot x' = 0$. Then $x \cdot y' = 0$ and $y \cdot x' = 0$ from part (a). Also,
$x + y' = (x + y') + 0$ (4a)
$= (x + y') + y \cdot x'$ $(x \cdot y' = 0)$
$= ((x + y') + y) \cdot ((x + y') + x')$ (3a)
$= ((x + y') + y) \cdot ((y' + x) + x')$ (1a)
$= (x + (y' + y)) \cdot (y' + (x + x'))$ (2a)
$= (x + (y + y')) \cdot (y' + (x + x'))$ (1a)
$= (x + 1) \cdot (y' + 1)$ (5a)
$= 1 \cdot 1$ (Practice 3a)
$= 1$ (4b)
Because $x \cdot y' = 0$ and $x + y' = 1$, $y' = x'$ by the uniqueness of complements, and therefore $y = x$.

12. a. In any Boolean algebra, take $z = 0$, $x = y \neq 0$. Then
$x + y = x + x = x$ and
$x + z = x + 0 = x$
but $y \neq z$.

b. Let $x + y = x + z$ and $x' + y = x' + z$. Then

$y = y + 0$	(4a)
$= y + x \cdot x'$	(5b)
$= (y + x) \cdot (y + x')$	(3a)
$= (x + y) \cdot (x' + y)$	(1a)
$= (x + z) \cdot (x' + z)$	(hypothesis)
$= (z + x) \cdot (z + x')$	(1a)
$= z + (x \cdot x')$	(3a)
$= z + 0$	(5b)
$= z$	(4a)

13. Suppose $x + 0_1 = x$ for all $x \in B$. Then

$0 + 0_1 = 0$ and $0_1 + 0 = 0_1$ so

$0_1 = 0_1 + 0 = 0 + 0_1 = 0$ and $0_1 = 0$. Then $1 = 0'$, so 1 is unique by the theorem on uniqueness of complements.

*14. a.

b.

c.

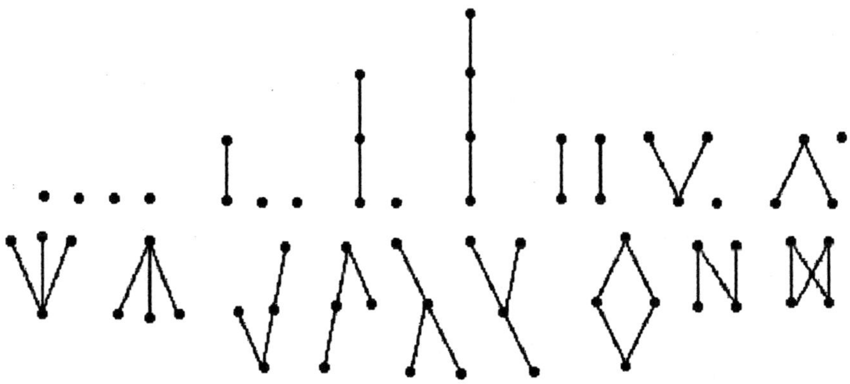

There are 16.

15. For example,

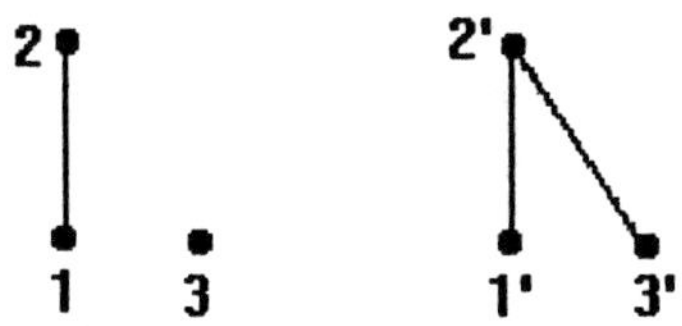

$f(1) = 1'$
$f(2) = 2'$
$f(3) = 3'$

Then $1 \prec 2$ and $f(1) \prec' f(2)$, but $f(3) \prec' f(2)$ and not $3 \prec 2$.

16. a. (i) bijection (ii) for x, y ∈ S, $f(x \cdot y) = f(x) + f(y)$
 b. Let f(0) = 5, f(l) = 7. Then
 $f(0 \cdot 0) = f(1) = 7 = 5 + 5 = f(0) + f(0)$
 $f(0 \cdot 1) = f(0) = 5 = 5 + 7 = f(0) + f(1)$
 $f(1 \cdot 0) = f(0) = 5 = 7 + 5 = f(1) + f(0)$
 $f(1 \cdot 1) = f(1) = 7 = 7 + 7 = f(1) + f(1)$

*17. a. $f: R \to R^+$
 f is onto; given $y \in R^+$, let x = log y; then x ∈ R and $f(x) = 2^x = 2^{\log y} = y$
 f is one-to-one; if f(x) = f(w) then $2^x = 2^w$ and (taking the log of both sides) x = w.
 b. for x, y ∈ R, $g(x + y) = g(x) \cdot g(y)$
 c. f is a bijection from R to R^+ and for x, y ∈ R, $f(x + y) = 2^{x+y} = 2^x \cdot 2^y = f(x) \cdot f(y)$
 d. f^{-1} is the function $f^{-1}(y) = \log y$
 e. f^{-1} is a bijection from R^+ to R and for any x, y $\in R^+$,
 $f^{-1}(x \cdot y) = \log(x \cdot y) = \log x + \log y = f^{-1}(x) + f^{-1}(y)$

18. a. f(1) = {1, 2} and f(a') = {2}. In the Boolean algebra on $\mathcal{P}(\{1, 2\})$,
 $\{1,2\} \cap \{2\} = \{2\}$. Then $f^{-1}(\{2\}) = a'$.
 b. f(a) = {1} and in the Boolean algebra on $\mathcal{P}(\{1, 2\})$, {1}' = {2}.
 Then $f^{-1}(\{2\}) = a'$.
 c. $f^{-1}(\{1\}) = a$ and $f^{-1}(\{2\}) = a'$. In the Boolean algebra on B, a + a' = 1. Then
 f(1) = {1, 2}.

19. For example, $f_1 = f_4 + f_{13}$, therefore $f_1 \to \{1\} \cup \{3\} = \{1, 3\}$.
 Therefore $f_1 \to \{1, 3\}$. Similarly, $f_2 \to \{1, 2\}$, $f_3 \to \{1, 2, 3\}$, $f_5 \to \{1, 2, 4\}$,
 $f_6 \to \{1, 3, 4\}$, $f_7 \to \{2, 3, 4\}$, $f_8 \to \{2, 3\}$, $f_9 \to \{2, 4\}$, $f_{10} \to \{1, 4\}$, $f_{11} \to \{3, 4\}$

*20. a. For any y ∈ b, y = f(x) for some x ∈ B. Then
 y & f(0) = f(x) & f(0) = f(x + 0) = f(x) = y, and f(0) = ϕ because the zero element in any Boolean algebra is unique (see Exercise 13).
 b. f(1) = f(0') = [f(0)]" = ϕ" = 1

21.*a. i. If $x \preccurlyeq y$ then $x \preccurlyeq y$ and $x \preccurlyeq x$ so x is a lower bound of x and y. If $w^* \preccurlyeq x$ and $w^* \preccurlyeq y$, then $w^* \preccurlyeq x$ so x is a greatest lower bound, and $x = x \cdot y$. If $x = x \cdot y$, then x is a greatest lower bound of x and y, so $x \preccurlyeq y$.
 ii. Similar to i.

 b. i. Let x + y = z. Then z is a least upper bound of x and y, which is a least upper bound of y and x, so z = y + x.
 ii. Similar to i.
 iii. Let (x + y) + z = p and x + (y + z) = q. Then $y \preccurlyeq x + y \preccurlyeq p$ and $z \preccurlyeq p$ so p is an upper bound for y and z; because y + z is the least upper bound for y

and z, $y + z \preccurlyeq p$. Also $x \preccurlyeq x + y \preccurlyeq p$. Therefore p is an upper bound for x and $y + z$, and $q \preccurlyeq p$ because q is the least upper bound for x and $y + z$. Similarly $p \preccurlyeq q$, so $p = q$.

iv. Similiar to iii.

c. $x + 0 = x \leftrightarrow 0 \preccurlyeq x$, which is true because 0 is a least element.

$x \cdot 1 = x \leftrightarrow x \preccurlyeq 1$, which is true because 1 is a greatest element.

d. (a) no - no least element
(b) yes
(c) yes
(d) no - not distributive:

$$2 + (3 \cdot 4) = 2 + 1 = 2$$
$$(2 + 3) \cdot (2 + 4) = 5 \cdot 5 = 5$$

Also, both 3 and 4 are complements of 2, so complements are not unique.

22. Suppose B has an odd number of elements. Arrange these distinct elements as follows: $0, 1, x_1, x_1', x_2, x_2', \ldots, x_{n-1}, x_{n-1}', x_n$. Then x_n' exists in B and must be an element of the list. If $x_n' = x_n$, then $x_n + x_n = x_n$ by idempotent property, but $x_n + x_n = 1$ by (5a); thus $x_n = 1$, a contradiction. If $x_n' = x_i$ for $i < n$, then $(x_n')' = x_n = x_i'$, a previous element of the list; contradiction. If $x_n' = x_i'$ for $i < n$, then $(x_n')' = x_n = x_i$, a previous element of the list; again, a contradiction.

EXERCISES 7.2

1. *a.

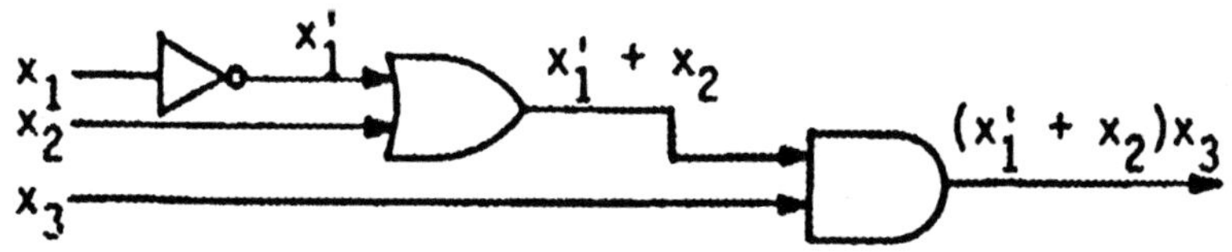

b.

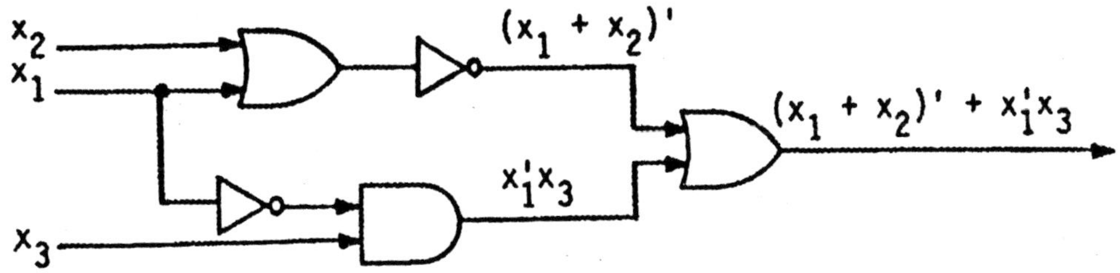

c.

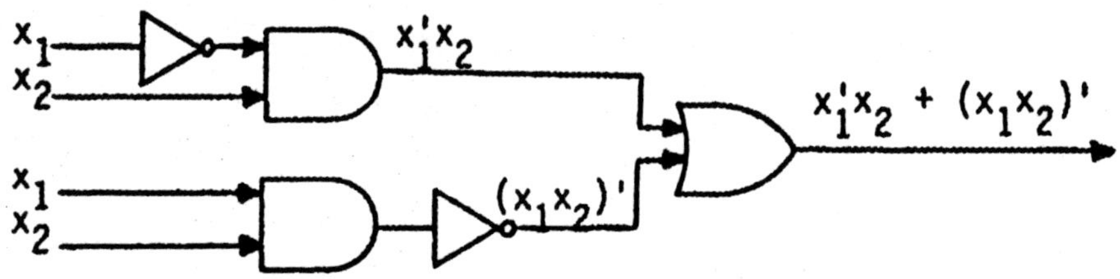

d.

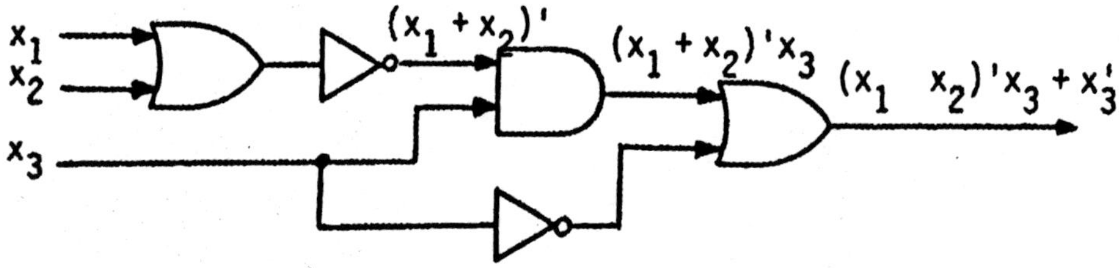

2. $x_1x_2 + x_2'$

x_1	x_2	$f(x_1, x_2)$
1	1	1
1	0	1
0	1	0
0	0	1

*3. $(x_1x_2)'(x_2 + x_3')$

x_1	x_2	x_3	$f(x_1, x_2, x_3$
1	1	1	0
1	1	0	0
1	0	1	0
1	0	0	1
0	1	1	1
0	1	0	1
0	0	1	0
0	0	0	1

4. $[(x_1' + x_2)x_3]'$

x_1	x_2	x_3	$f(x_1, x_2, x_3)$
1	1	1	0
1	1	0	1
1	0	1	1
1	0	0	1
0	1	1	0
0	1	0	1
0	0	1	0
0	0	0	1

5. $x_1(x_1 + x_2')(x_2x_3)'$

x_1	x_2	x_3	$f(x_1, x_2, x_3)$
1	1	1	0
1	1	0	1
1	0	1	1
1	0	0	1
0	1	1	0
0	1	0	0
0	0	1	0
0	0	0	0

6. a.

x	y	f(x,y)
1	1	0
1	0	1
0	1	1
0	0	0

b.

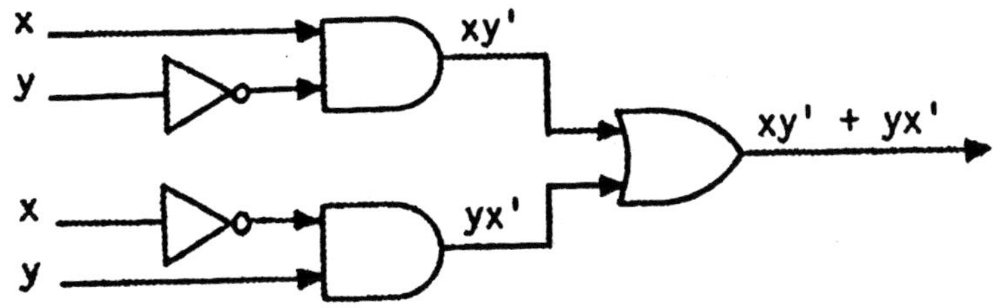

c. The truth function for the network is also

x	y	f(x,y)
1	1	0
1	0	1
0	1	1
0	0	0

The network illustrates "x OR y" and "NOT both x AND y"

*7. $x_1'x_2'$

8. $x_1x_2 + x_1'x_2$

9. $x_1x_2x_3' + x_1x_2'x_3 + x_1'x_2x_3 + x_1'x_2'x_3'$

10. $x_1x_2'x_3 + x_1x_2'x_3' + x_1'x_2x_3'$

11. $x_1x_2x_3x_4 + x_1x_2x_3'x_4 + x_1x_2'x_3x_4 + x_1x_2'x_3'x_4 + x_1'x_2'x_3x_4 + x_1'x_2'x_3x_4'$

12. $x_1x_2x_3x_4 + x_1x_2x_3'x_4 + x_1x_2'x_3x_4 + x_1x_2'x_3x_4' + x_1'x_2x_3x_4 + x_1'x_2x_3'x_4$

*13. $x_1x_2'x_3x_4' + x_1'x_2x_3x_4 + x_1'x_2x_3'x_4 + x_1'x_2'x_3x_4 + x_1'x_2'x_3x_4' + x_1'x_2'x_3'x_4$

14. $x_1x_2x_3x_4 + x_1x_2x_3x_4' + x_1x_2x_3'x_4' + x_1x_2'x_3x_4 + x_1'x_2x_3x_4' + x_1'x_2x_3'x_4'$

*15. a. $x_1x_2x_3' + x_1x_2'x_3'$

b.

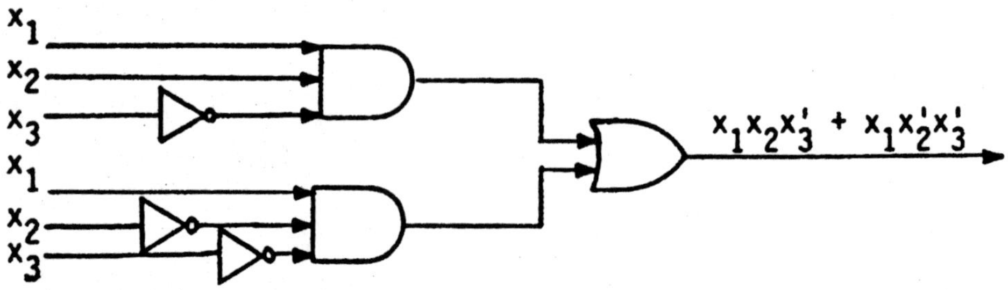

c. $x_1x_2x_3' + x_1x_2'x_3' = x_1x_3'x_2 + x_1x_3'x_2' = x_1x_3'(x_2 + x_2') = x_1x_3'\cdot 1 = x_1x_3'$

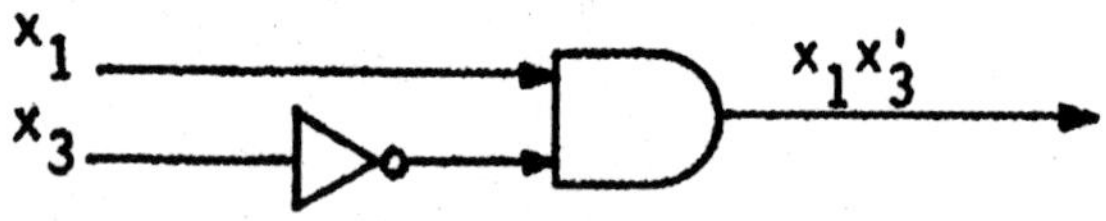

16. a. $x_1x_2x_3 + x_1'x_2x_3 + x_1'x_2x_3'$

b.

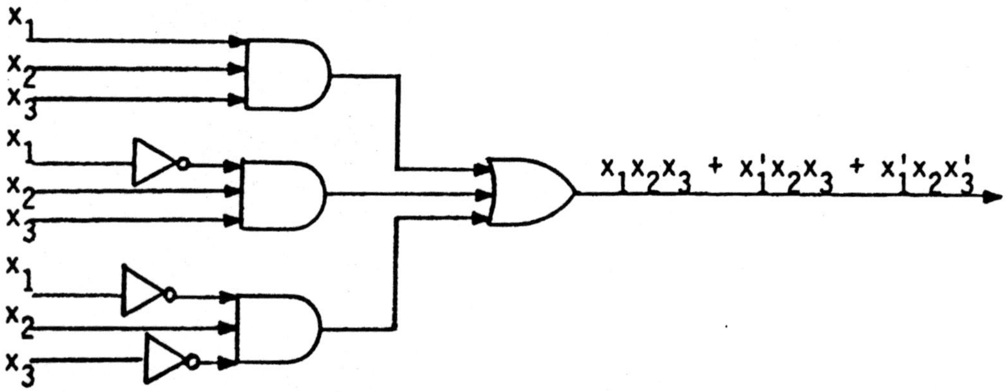

c. $x_1x_2x_3 + x_1'x_2x_3 + x_1'x_2x_3' = x_2x_3x_1 + x_2x_3x_1' + x_1'x_2x_3' + x_1'x_2x_3$
$= x_2x_3(x_1 + x_1') + x_1'x_2(x_3' + x_3)$
$= x_2x_3\cdot 1 + x_1'x_2\cdot 1 = x_2x_3 + x_1'x_2$
$= x_2x_3 + x_2x_1' = x_2(x_3 + x_1')$

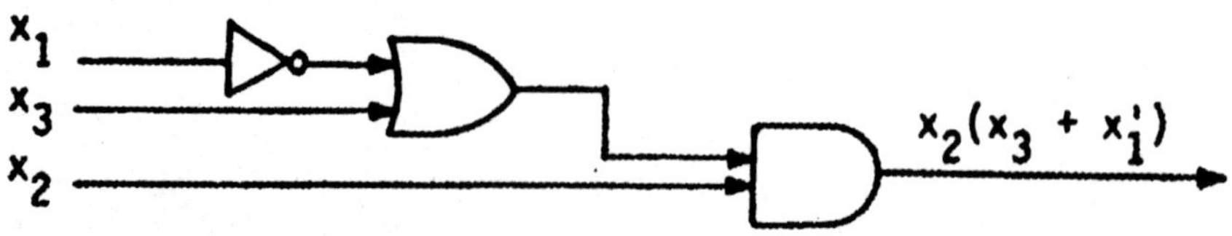

17. a.

x_1	x_2	x_3	$f(x_1,x_2,x_3)$
1	1	1	1
1	1	0	0
1	0	1	1
1	0	0	0
0	1	1	1
0	1	0	1
0	0	1	0
0	0	0	0

b. $x_1x_2x_3 + x_1x_2'x_3 + x_1'x_2x_3 + x_1'x_2x_3'$

c. $x_1x_3 + x_1'x_2 = (x_1x_3 + x_1')(x_1x_3 + x_2)$
$= (x_1' + x_1x_3)(x_2 + x_1x_3) = (x_1' + x_1)(x_1' + x_3)(x_2 + x_1)(x_2 + x_3)$
$= (x_1 + x_1')(x_1' + x_3)(x_1 + x_2)(x_2 + x_3)$
$= (x_1' + x_3)(x_1 + x_2)(x_2 + x_3) = (x_1 + x_2)(x_1' + x_3)(x_2 + x_3)$

*18. a. $(x_1' + x_2')(x_1' + x_2)(x_1 + x_2')$
b. $(x_1' + x_2)(x_1 + x_2)$
c. $(x_1' + x_2' + x_3')(x_1' + x_2 + x_3)(x_1 + x_2' + x_3)(x_1 + x_2 + x_3')$
d. $(x_1' + x_2' + x_3')(x_1' + x_2' + x_3)(x_1 + x_2' + x_3')(x_1 + x_2 + x_3')(x_1 + x_2 + x_3)$

19.

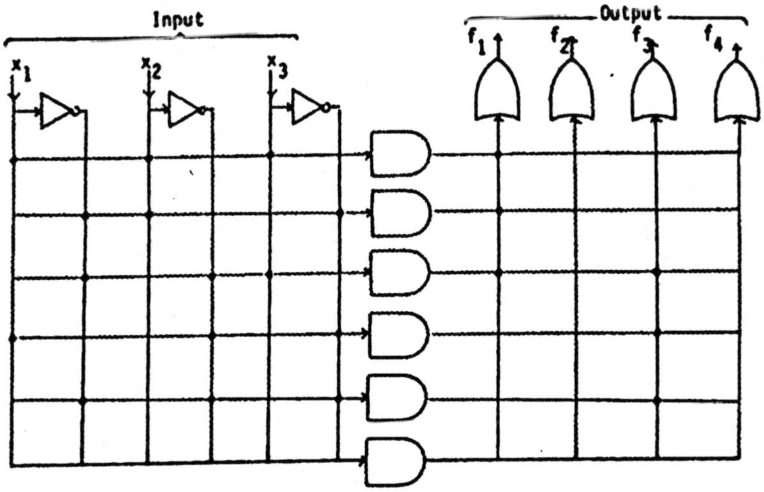

20. a.
```
   1100
   0100
(1)0000
```
b.
```
   1001
   0111
(1)0000
```
c.
```
   001
   111
(1)000
```

21. a.

x_i	c_{i-1}	r_i
0	0	0
0	1	1
1	0	1
1	1	0

If $c_{i-1} = 0$, no 1-digit has yet been seen, and $r_i = x_i$.
If $c_{i-1} = 1$, a 1-digit has been seen and $r_i = x_i'$.

x_i	c_{i-1}	c_i
0	0	0
0	1	1
1	0	1
1	1	1

If $c_{i-1} = 1$, a 1-digit has already been seen, and $c_i = 1$ to convey this information to the next column. If $x_i = c_{i-1} = 0$, no 1 has been seen yet, and does not occur in column i, so $c_i = 0$. If $x_i = 1$ and $c_{i-1} = 0$, this is the first 1, so set $c_i = 1$.

b. $r_i = x_i'c_{i-1} + x_ic_{i-1}' = (x_i + c_{i-1})(x_ic_{i-1})'$

$c_i = x_i'c_{i-1} + x_ic_{i-1}' + x_ic_{i-1}$

$= x_i'c_{i-1} + x_ic_{i-1}' + x_ic_{i-1} + x_ic_{i-1}$

$= c_{i-1}(x_i' + x_i) + x_i(c_{i-1}' + c_{i-1})$

$= c_{i-1} + x_i$

c.

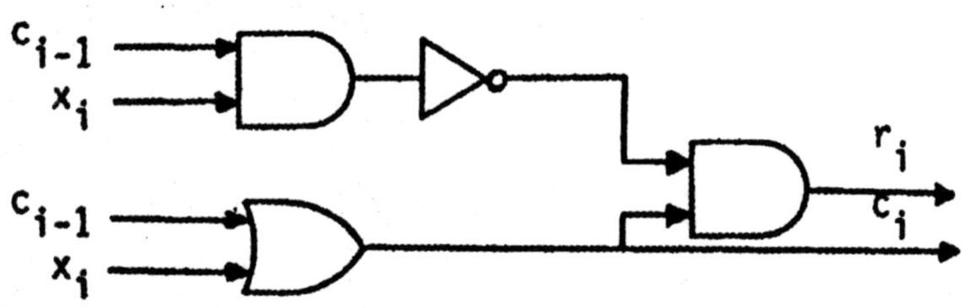

d.

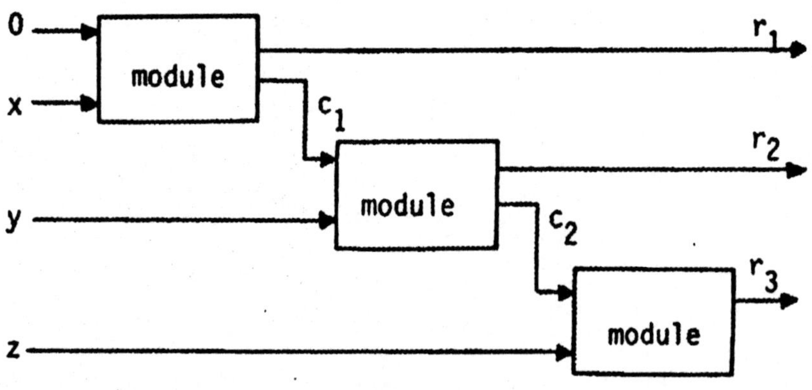

22. a.

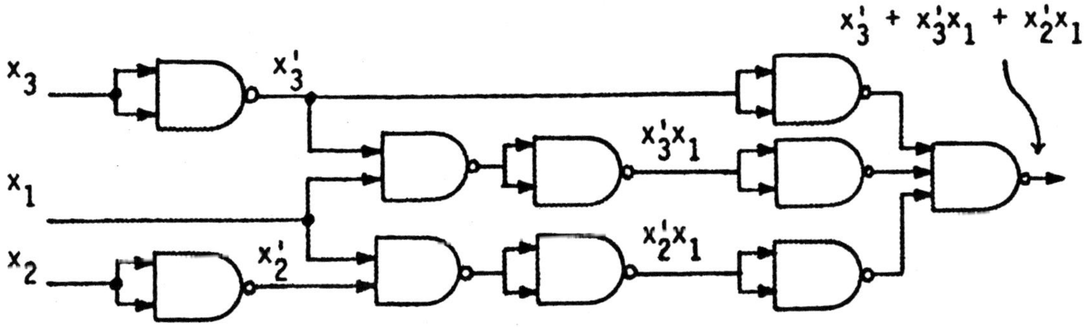

b. $x_3'x_1 + x_2'x_1 + x_3' = x_1x_3' + x_1x_2' + x_3' = x_1(x_3' + x_2') + x_3'$
$= (x_3x_2)'x_1 + x_3' = (((x_3x_2)'x_1)'x_3)'$

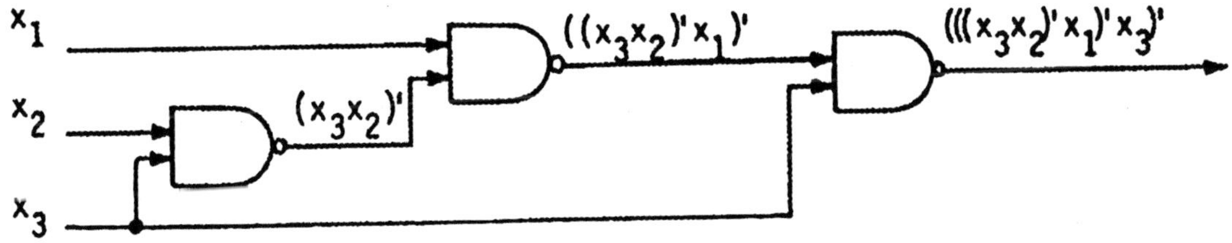

*23. Network is represented by $(x_1'(x_2'x_3)')'$ and $(x_1'(x_2'x_3)')' = x_1 + x_2'x_3$

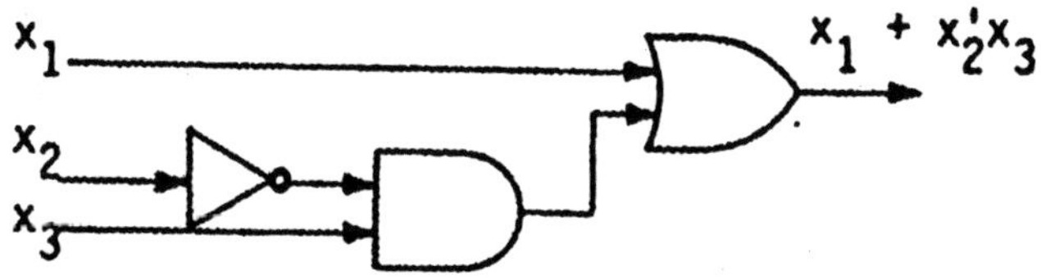

24. a.

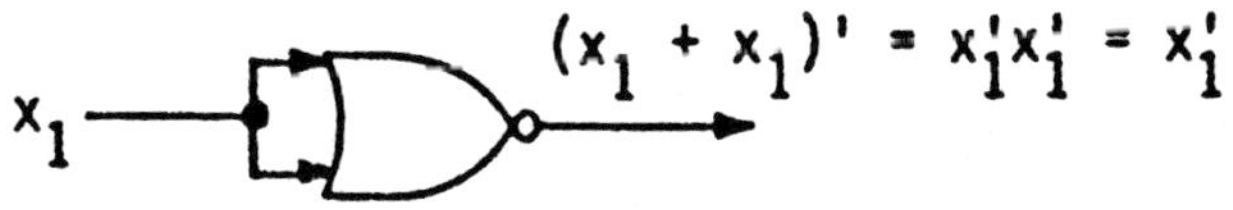

b.

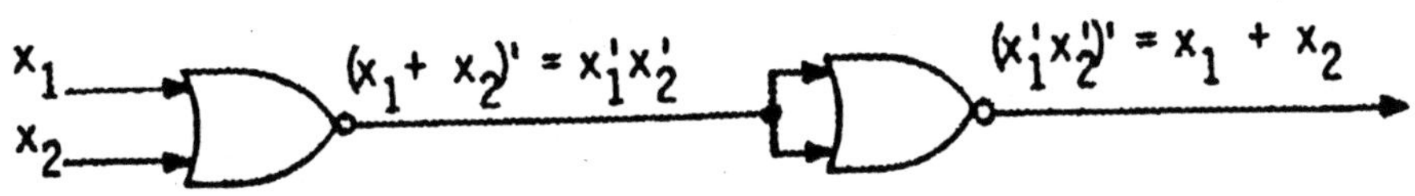

c.

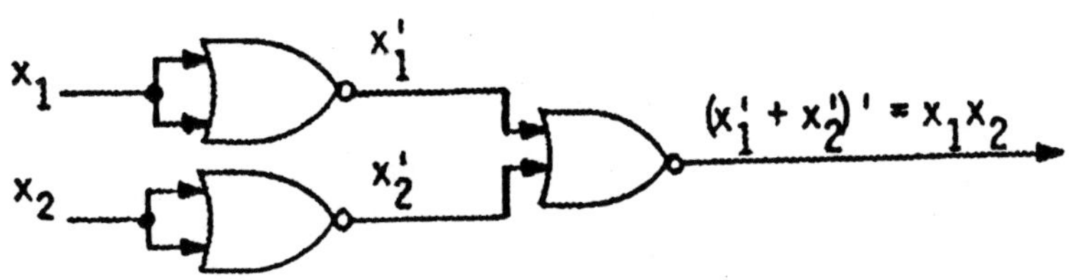

*25. The truth function for | is that of the NAND gate, the truth function for ↓ is that of the NOR gate. In Section 1.1, we learned that every compound statement is equivalent to one using only | or to one using only ↓, therefore any truth function can be realized by using only NAND gates or only NOR gates.

26. $s = (x_1 + x_2)(x_1x_2)' = x_1(x_1x_2)' + x_2(x_1x_2)' = ((x_1(x_1x_2)')' \cdot (x_2(x_1x_2)')')'$
$c = x_1x_2 = ((x_1x_2)')'$

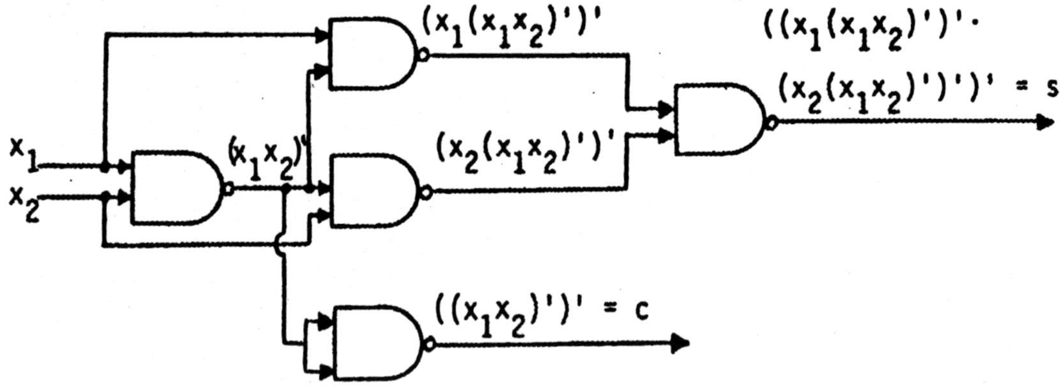

*27. x_1 = neutral
x_2 = park
x_3 = seat belt

x_1	x_2	x_3	$f(x_1,x_2,x_3)$
1	1	1	-
1	1	0	-
1	0	1	1
1	0	0	0
0	1	1	1
0	1	0	0
0	0	1	0
0	0	0	0

$(x_1 + x_2)x_3$

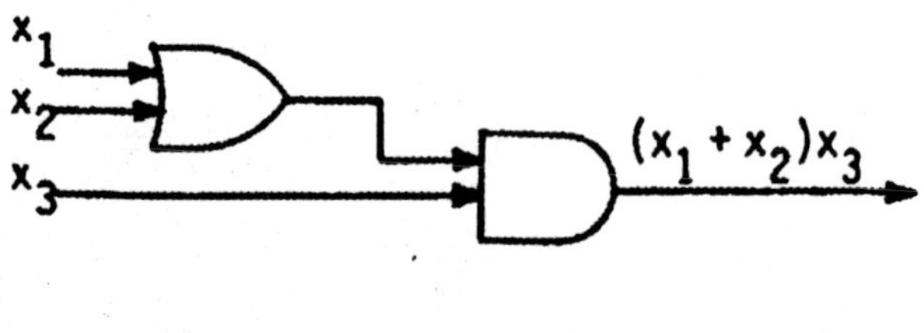

28. a. x_2: 2 + 3 + 6 + 7

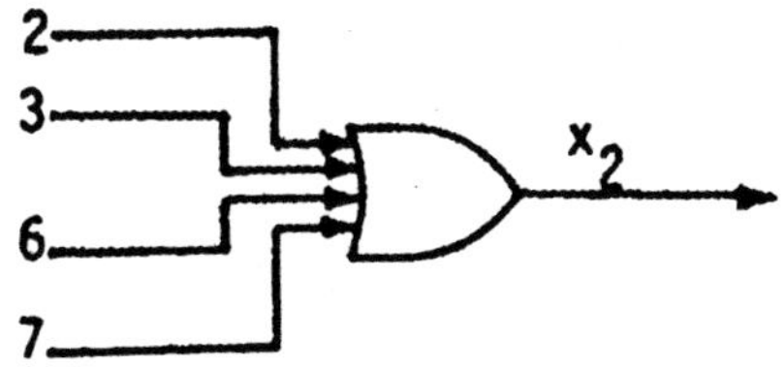

b.

	x_4	x_3	x_2	x_1	y_5	y_6
0	0	0	0	0	1	1
1	0	0	0	1	1	0
2	0	0	1	0	1	1
3	0	0	1	1	1	0
4	0	1	0	0	1	0
5	0	1	0	1	0	0
6	0	1	1	0	0	1
7	0	1	1	1	1	0
8	1	0	0	0	1	1
9	1	0	0	1	1	0

don't care

$$y_5 = (x_4 + x_3' + x_2 + x_1')(x_4 + x_3' + x_2' + x_1)$$
$$= (x_4 + x_3') + (x_2 + x_1')(x_2' + x_1)$$

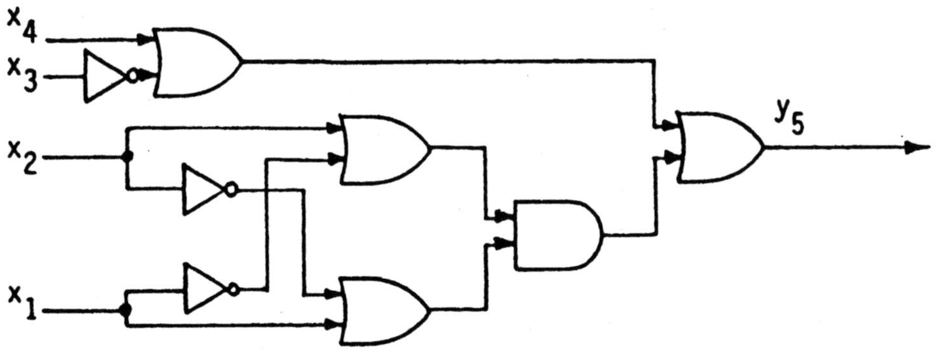

$$y_6 = x_4'x_3'x_2'x_1' + x_4'x_3'x_2x_1' + x_4'x_3x_2x_1' + x_4x_3'x_2'x_1'$$
$$= x_3'x_2'x_1' + x_4'x_2x_1'$$

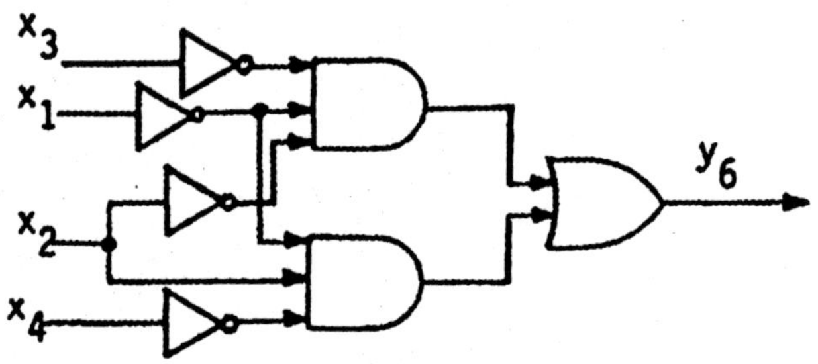

29.

x_1	x_2	x_3	$f(x_1,x_2,x_3)$
1	1	1	1
1	1	0	1
1	0	1	1
1	0	0	0
0	1	1	1
0	1	0	0
0	0	1	0
0	0	0	0

$$x_1x_2x_3 + x_1x_2x_3' + x_1x_2'x_3 + x_1'x_2x_3 = x_1x_2 + (x_1x_2' + x_1'x_2)x_3$$

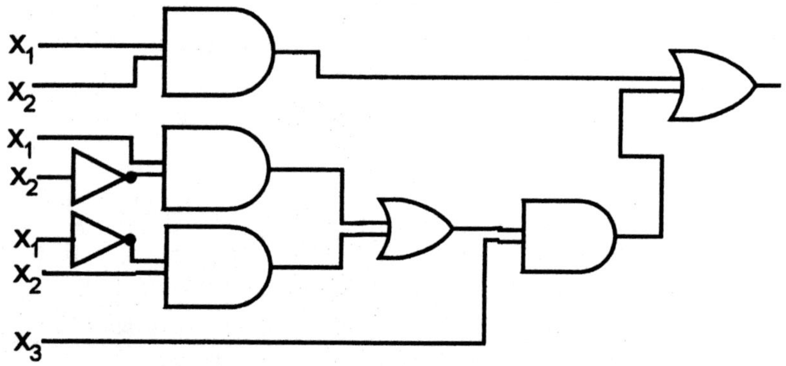

30. The problem can be described by a truth table, where

x_1 = pressure (1 when pressure > 50 psi, otherwise 0)
x_2 = salinity (1 when salinity > 45 g/L, otherwise 0)
x_3 = temperature (1 when temperature > 53° C, otherwise 0)
x_4 = acidity (1 when acidity < 7.0 pH, otherwise 0)

The output for each valve should be 1 when the valve is to open, 0 otherwise. The truth table is

x_1	x_2	x_3	x_4	A	B
1	1	1	1	1	0
1	1	1	0	1	0
1	1	0	1	1	0
1	1	0	0	1	0
1	0	1	1	0	1
1	0	1	0	0	0
1	0	0	1	0	0
1	0	0	0	0	0
0	1	1	1	0	1
0	1	1	0	0	0
0	1	0	1	0	0
0	1	0	0	0	0
0	0	1	1	0	1
0	0	1	0	0	0
0	0	0	1	0	0
0	0	0	0	0	0

$A = x_1x_2x_3x_4 + x_1x_2x_3x_4' + x_1x_2x_3'x_4 + x_1x_2x_3'x_4'$
$B = x_1x_2'x_3x_4 + x_1'x_2x_3x_4 + x_1'x_2'x_3x_4$

Using the canonical sum-of-products form, the circuit for A would require 2 inverters (one for x_3 and one for x_4, assuming we split the output from an inverter into more than one gate), 4 AND gates, and 1 OR gate; B would require 2 inverters, 3 AND gates, and 1 OR gate. It is possible to write simpler equivalent expressions.

EXERCISES 7.3

*1.

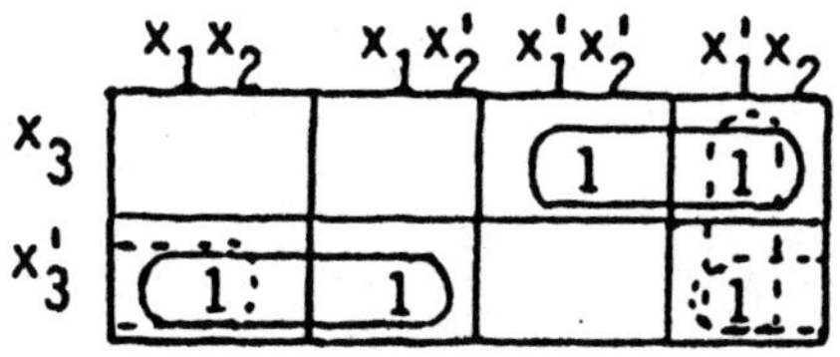

$x_1'x_3 + x_1x_3' + x_1'x_2$

or

$x_1'x_3 + x_1x_3' + x_2x_3'$

2.

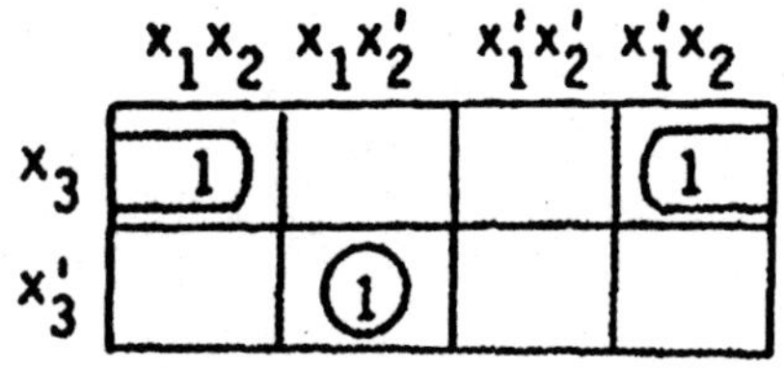

$x_2x_3 + x_1x_2'x_3'$

*3.

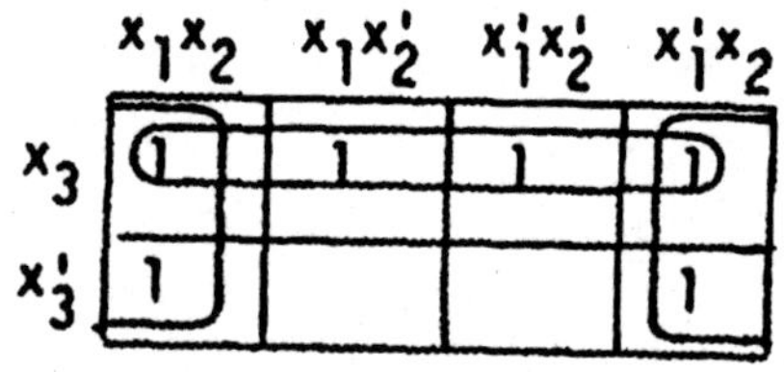

$x_3 + x_2$

4.

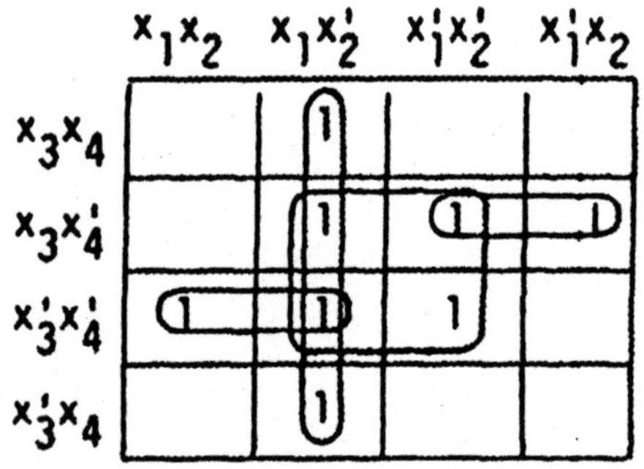

$x_1x_3'x_4' + x_1'x_3x_4' + x_2'x_4' + x_1x_2'$

5.

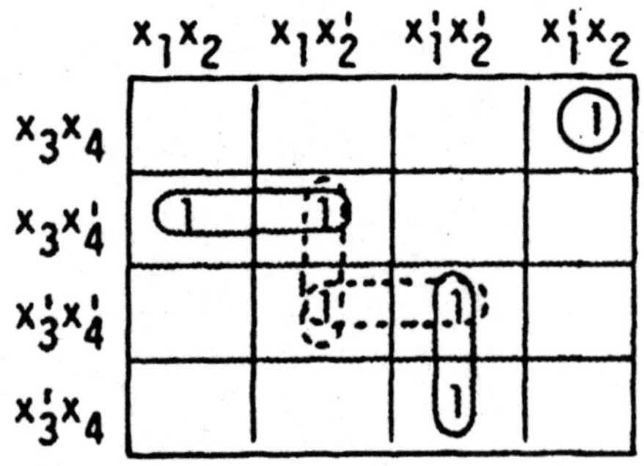

$x_1'x_2x_3x_4 + x_1x_3x_4' + x_1'x_2'x_3' + x_1x_2'x_4'$

or

$x_1'x_2x_3x_4 + x_1x_3x_4' + x_1'x_2'x_3' + x_2'x_3'x_4'$

*6.

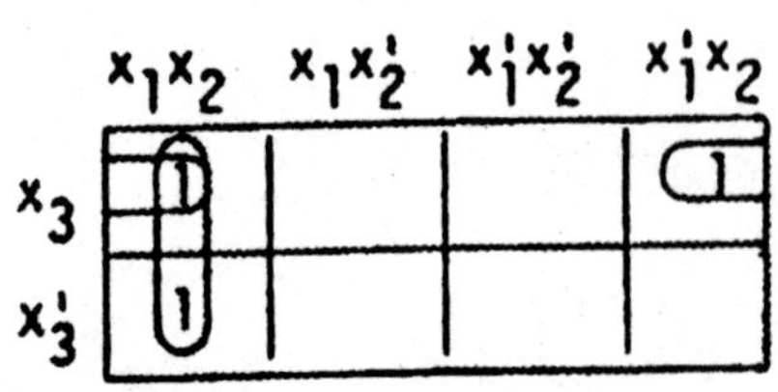

$x_1x_2 + x_2x_3$

7.

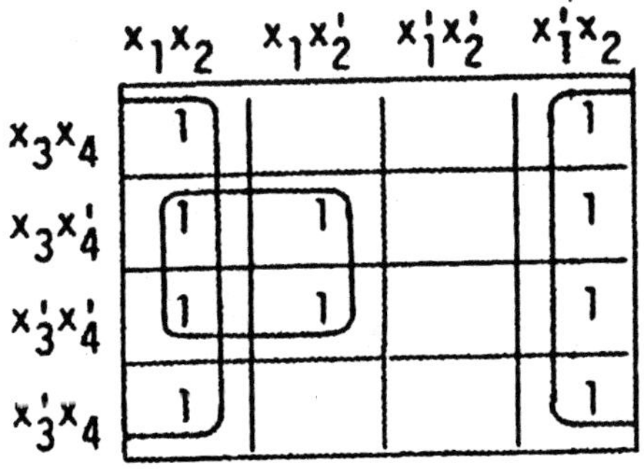

$x_1x_4' + x_2$

8.

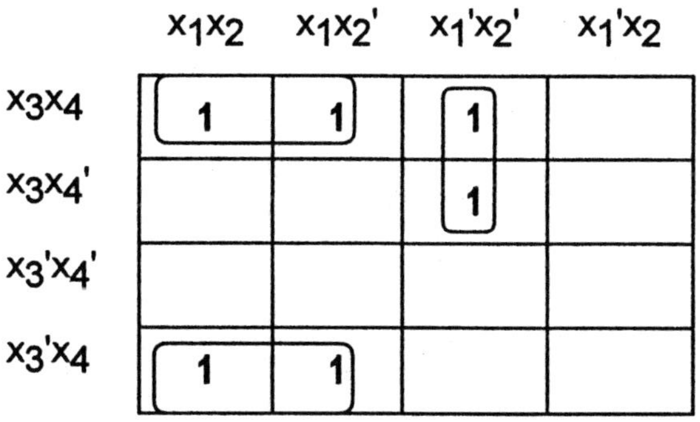

$x_1x_4 + x_1'x_2'x_3$

9.

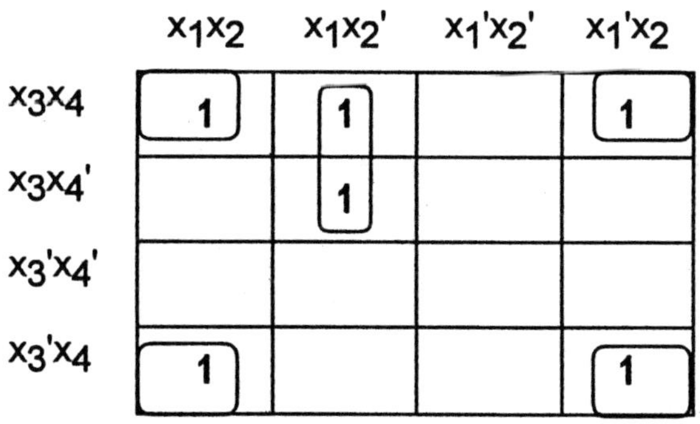

$x_2x_4 + x_1x_2'x_3$

*10. a.

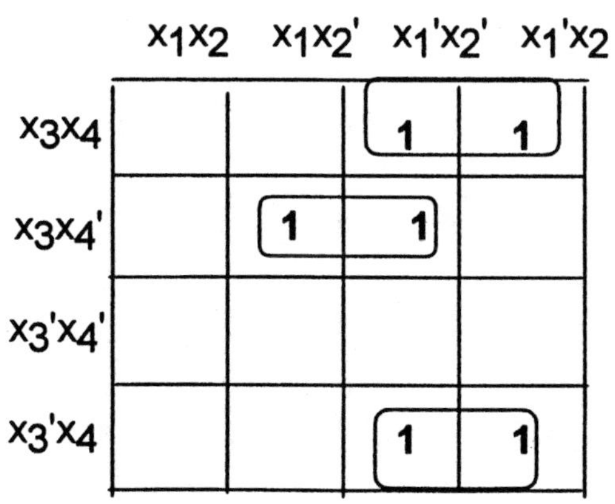

$x_1'x_4 + x_2'x_3x_4'$

b.

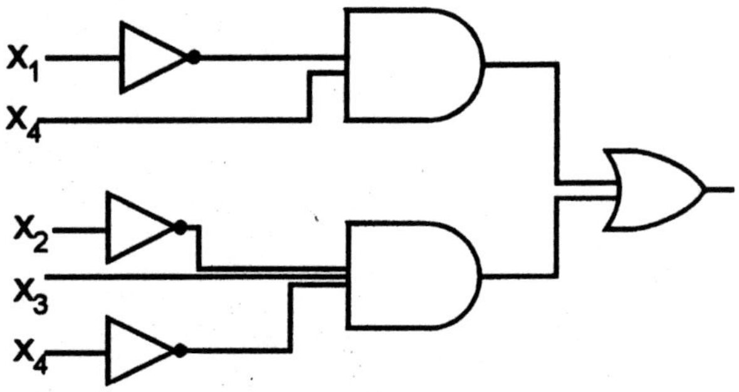

11. a.

	x_1x_2	x_1x_2'	$x_1'x_2'$	$x_1'x_2$
x_3x_4	1	1		
x_3x_4'	1			1
$x_3'x_4'$	1			1
$x_3'x_4$				

$x_1x_3x_4 + x_2x_4'$

b.

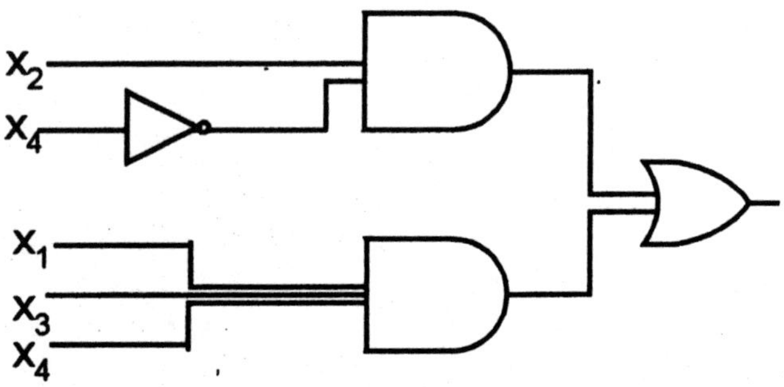

12. a.

	x_1x_2	x_1x_2'	$x_1'x_2'$	$x_1'x_2$
x_3x_4			1	1
x_3x_4'		1	1	
$x_3'x_4'$				
$x_3'x_4$	1		1	1

$x_2x_3'x_4 + x_2'x_3x_4' + x_1'x_4$

b.

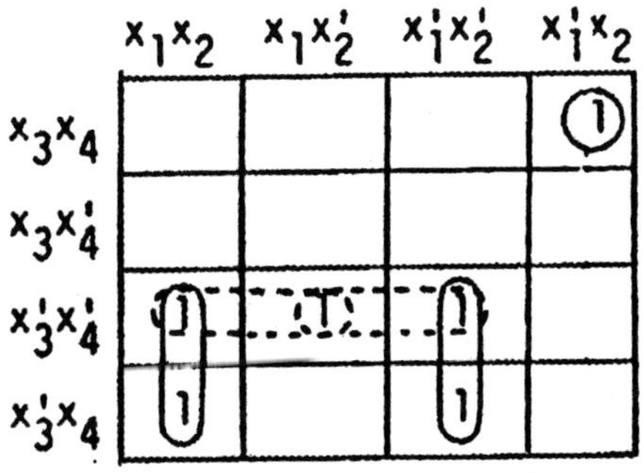

$x_1'x_2x_3x_4 + x_1'x_2'x_3' + x_1x_2x_3' + x_1x_3'x_4'$

or

$x_1'x_2x_3x_4 + x_1'x_2'x_3' + x_1x_2x_3' + x_2'x_3'x_4'$

13. Original expression: $x_1x_2x_3 + x_1x_2'x_3 + x_1x_2'x_3' + x_1'x_2x_3 + x_1'x_2x_3'$

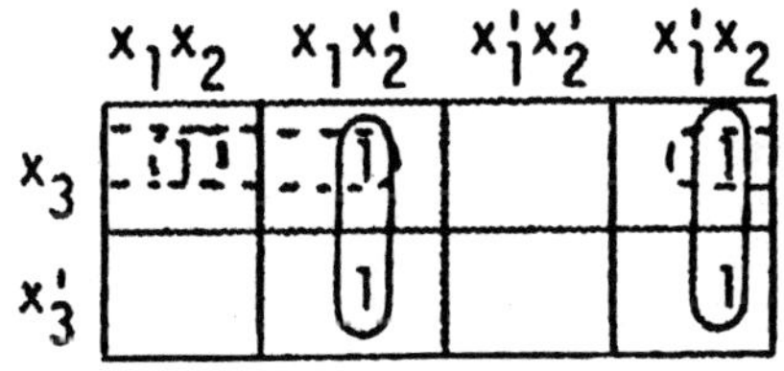

Reduced expression: $x_1x_3 + x_1x_2' + x_1'x_2$ or $x_2x_3 + x_1x_2' + x_1'x_2$

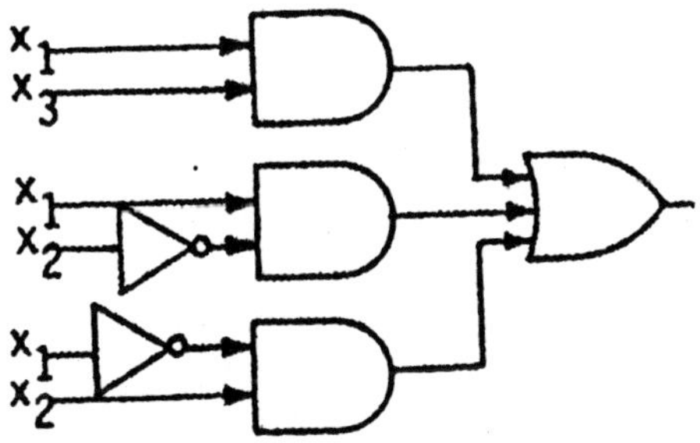

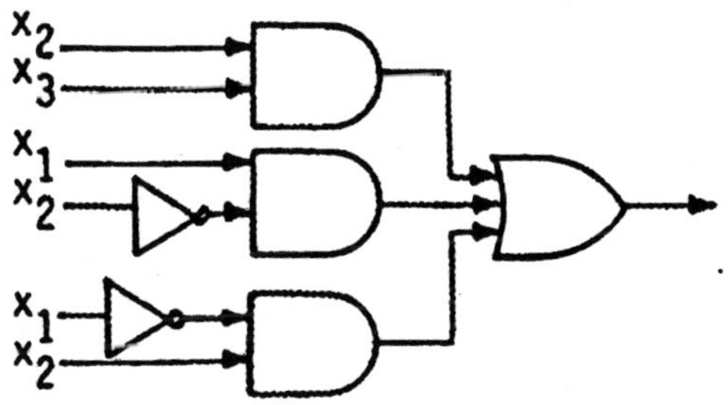

*14.

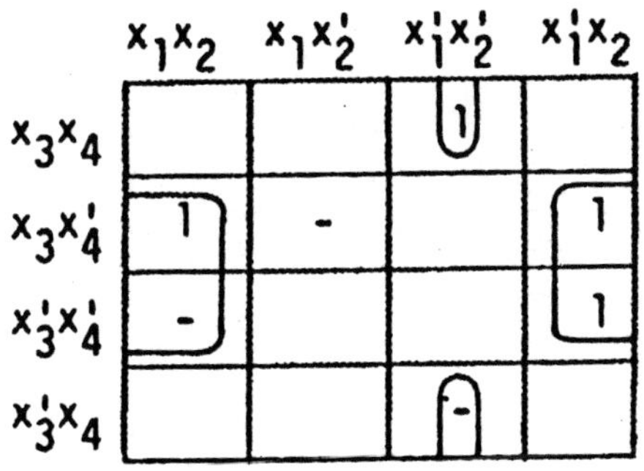

$x_2x_4' + x_1'x_2'x_4$

15. From Exercise 30 of Section 7.2,

$A = x_1x_2x_3x_4 + x_1x_2x_3x_4' + x_1x_2x_3'x_4 + x_1x_2x_3'x_4'$
$B = x_1x_2'x_3x_4 + x_1'x_2x_3x_4 + x_1'x_2'x_3x_4$

Karnaugh maps and minimized expressions are:

	x_1x_2	x_1x_2'	$x_1'x_2'$	$x_1'x_2$
x_3x_4	1			
x_3x_4'	1			
$x_3'x_4'$	1			
$x_3'x_4$	1			

$A = x_1x_2$

	x_1x_2	x_1x_2'	$x_1'x_2'$	$x_1'x_2$
x_3x_4		1	1	1
x_3x_4'				
$x_3'x_4'$				
$x_3'x_4$				

$B = x_2'x_3x_4 + x_1'x_3x_4$

*16.

x_1	x_2	x_3	
1	1	1	1,2,3
1	0	1	1,4
0	1	1	2,5,6
1	1	0	3,7
0	0	1	4,5
0	1	0	6,7

x_1	x_2	x_3	
1	-	1	1
-	1	1	1,2
1	1	-	2
-	0	1	1
0	-	1	1
0	1	-	2
-	1	0	2

x_1	x_2	x_3
-	-	1
-	1	-

	111	101	011	110	001	010
--1	✓	✓	✓		✓	
-1-	✓		✓	✓		✓

- -1 and -1- are essential. The minimal form is $x_3 + x_2$.

17. Canonical sum-of-products form is

$x_1x_2x_3x_4 + x_1x_2'x_3x_4 + x_1x_2'x_3x_4' + x_1'x_2x_3x_4 + x_1x_2x_3'x_4 + x_1'x_2x_3'x_4$

x_1	x_2	x_3	x_4	
1	1	1	1	1,2,3
1	0	1	1	1,4
0	1	1	1	2,5
1	1	0	1	3,6
1	0	1	0	4
0	1	0	1	5,6

x_1	x_2	x_3	x_4	
1	-	1	1	
-	1	1	1	1
1	1	-	1	1
1	0	1	-	
0	1	-	1	1
-	1	0	1	1

x_1	x_2	x_3	x_4
-	1	-	1

	1111	1011	0111	1101	1010	0101
1-11	✓	✓				
101-		✓			✓	
-1-1	✓		✓	✓		✓

101- and -1-1 are essential; 1-11 is redundant. The minimal form is $x_1x_2'x_3 + x_2x_4$.

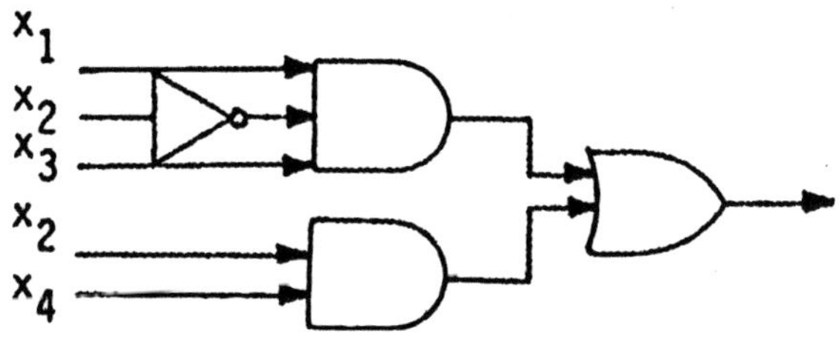

18.

x_1	x_2	x_3	x_4	
1	1	1	0	1
1	0	1	0	1,2,3
1	0	0	1	4
0	0	1	1	5
1	0	0	0	2,4,6
0	1	0	0	7
0	0	1	0	3,5,8
0	0	0	0	6,7,8

x_1	x_2	x_3	x_4	
1	-	1	0	
1	0	-	0	1
-	0	1	0	1
1	0	0	-	
0	0	1	-	
-	0	0	0	1
0	-	0	0	
0	0	-	0	1

x_1	x_2	x_3	x_4
-	0	-	0

	1110	1010	1001	0011	1000	0100	0010	0000
1-10	✓	✓						
100-			✓		✓			
001-				✓			✓	
0-00						✓		✓
-0-0		✓			✓		✓	✓

1-10, 100-, 001-, 0-00 are essential; -0-0 is redundant. The minimal form is $x_1x_3x_4' + x_1x_2'x_3' + x_1'x_2'x_3 + x_1'x_3'x_4'$

19.

x_1	x_2	x_3	x_4	
1	1	1	1	1,2,3
1	1	0	1	1,4,5
1	0	1	1	2,6,7
0	1	1	1	3,8,9
1	0	0	1	4,6,10,11
0	1	0	1	5,8,12,13
0	0	1	1	7,9,14
1	0	0	0	10,15
0	1	0	0	12,16
0	0	0	1	11,13,14,17
0	0	0	0	15,16,17

x_1	x_2	x_3	x_4	
1	1	-	1	1,2
1	-	1	1	1,3
-	1	1	1	2,3
1	-	0	1	1,4
-	1	0	1	2,4
1	0	-	1	1,5
-	0	1	1	3,5
0	1	-	1	2,6
0	-	1	1	3,6
1	0	0	-	7
-	0	0	1	4,5,7
0	1	0	-	8
0	-	0	1	4,6,8
0	0	-	1	5,6
-	0	0	0	7
0	-	0	0	8
0	0	0	-	7,8

x_1	x_2	x_3	x_4	
1	-	-	1	1
-	1	-	1	1
-	-	1	1	1
-	-	0	1	1
-	0	-	1	1
0	-	-	1	1
-	0	0	-	
0	-	0	-	

x_1	x_2	x_3	x_4
-	-	-	1

	1111	1101	1011	0111	1001	0101	0011	1000	0100	0001	0000
-00-					✓			✓		✓	
0-0-						✓			✓	✓	✓
---1	✓	✓	✓	✓	✓	✓	✓			✓	✓

-00-, 0-0-, and - - -1 are essential. The minimal form is $x_2'x_3' + x_1'x_3' + x_4$

20.*a.

x_1	x_2	x_3	x_4	
0	1	1	1	1
1	0	1	0	
0	0	1	1	1,2
0	1	0	0	3
0	0	0	1	2,4
0	0	0	0	3,4

x_1	x_2	x_3	x_4
0	-	1	1
0	0	-	1
0	-	0	0
0	0	0	-

	0111	1010	0011	0100	0001	0000
1010		✓				
0-11	✓		✓			
00-1			✓		✓	
0-00				✓		✓
000-					✓	✓

1010, 0-11, 0-00 are essential. Either 00-1 or 000- can be used as the fourth term. The minimal sum-of-products form is

$$x_1x_2'x_3x_4' + x_1'x_3x_4 + x_1'x_3'x_4' + x_1'x_2'x_4$$

or

$$x_1x_2'x_3x_4' + x_1'x_3x_4 + x_1'x_3'x_4' + x_1'x_2'x_3'$$

b.

x_1	x_2	x_3	x_4	
1	1	1	1	1,2,3
1	0	1	1	1,4
1	1	1	0	2,5,6,7
1	1	0	1	3,8,9
1	0	1	0	4,5
0	1	1	0	6,10
1	1	0	0	7,8,11
0	1	0	1	9,12
0	1	0	0	10,11,12

x_1	x_2	x_3	x_4	
1	-	1	1	1
1	1	1	-	1,2
1	1	-	1	2
1	0	1	-	1
1	-	1	0	1
-	1	1	0	3
1	1	-	0	2,3
1	1	0	-	2,4
-	1	0	1	4
0	1	-	0	3
-	1	0	0	3,4
0	1	0	-	4

x_1	x_2	x_3	x_4
1	-	1	-
1	1	-	-
-	1	-	0
-	1	0	-

	1111	1011	1110	1101	1010	0110	1100	0101	0100
1-1-	✓	✓	✓		✓				
11--	✓		✓	✓			✓		
-1-0			✓			✓	✓		✓
-10-				✓			✓	✓	✓

1-1-, -1-0, -10- are essential; 11- - is redundant. The minimal form is $x_1x_3 + x_2x_4' + x_2x_3'$.

c.

x_1	x_2	x_3	x_4	
1	1	1	1	1,2
1	1	1	0	1,3,4
1	1	0	1	2,5,6
0	1	1	0	3,7
1	1	0	0	4,5,8
1	0	0	1	6,9
0	1	0	0	7,8,10
0	0	0	1	9,11
0	0	0	0	10,11

x_1	x_2	x_3	x_4	
1	1	1	-	1
1	1	-	1	1
-	1	1	0	2
1	1	-	0	1,2
1	1	0	-	1
1	-	0	1	
0	1	-	0	2
-	1	0	0	2
-	0	0	1	
0	-	0	0	
0	0	0	-	

x_1	x_2	x_3	x_4
1	1	-	-
-	1	-	0

	1111	1110	1101	0110	1100	1001	0100	0001	0000
1-01			✓			✓			
-001						✓		✓	
0-00							✓		✓
000-								✓	✓
11--	✓	✓	✓		✓				
-1-0		✓		✓	✓		✓		

11- -, -1-0 are essential. The additional terms can be 1-01 and 000-, or -001 and 0-00, or --001 and 000-. The minimal form is

$$x_1x_2 + x_2x_4' + x_1x_3'x_4 + x_1'x_2'x_3'$$

or

$$x_1x_2 + x_2x_4' + x_2'x_3'x_4 + x_1'x_3'x_4'$$

or

$$x_1x_2 + x_2x_4' + x_2'x_3'x_4 + x_1'x_2'x_3'$$

d.

x_1	x_2	x_3	x_4	x_5	
1	1	1	1	1	1,2
1	0	1	1	1	1,3
1	1	0	1	1	2
0	1	1	0	1	4
1	0	1	0	1	3,5
1	1	1	0	0	
0	1	0	1	0	6
0	0	1	0	1	4,5,7
0	0	0	0	1	7

x_1	x_2	x_3	x_4	x_5
1	-	1	1	1
1	1	-	1	1
1	0	1	-	1
0	-	1	0	1
-	0	1	0	1
0	-	0	1	0
0	0	-	0	1

	11111	10111	11011	01101	10101	11100	01010	00101	00001	00010
11100						✓				
1-111	✓	✓								
11-11	✓		✓							
101-1		✓			✓					
0-101				✓				✓		
-0101					✓			✓		
0-010							✓			✓
00-01								✓	✓	

11100, 11-11, 0-101, 0-010, 00-01 are essential. The only single additional term that works is 101-1. The minimal form is

$$x_1x_2x_3x_4'x_5' + x_1x_2x_4x_5 + x_1'x_3x_4'x_5 + x_1'x_3'x_4x_5' + x_1'x_2'x_4'x_5 + x_1x_2'x_3x_5$$

21.

x_1	x_2	x_3	x_4	
1	1	1	1	1,2
1	0	1	1	1,3
1	1	1	0	2,4
0	1	0	1	5
1	0	1	0	3,4,6
1	0	0	0	6,7
0	1	0	0	5,8
0	0	0	0	7,8

x_1	x_2	x_3	x_4	
1	-	1	1	1
1	1	1	-	1
1	0	1	-	1
1	-	1	0	1
0	1	0	-	
1	0	-	0	
-	0	0	0	
0	-	0	0	

x_1	x_2	x_3	x_4
1	-	1	-

	1111	1011	1110	0101	1010	1000	0100	0000
010-				✓			✓	
10-0					✓	✓		
-000						✓		✓
0-00							✓	✓
1-1-	✓	✓	✓		✓			

1-1- and 010- are essential. Either 10-0 and 0-00, or -000 will cover the remaining columns. Choosing -000, the minimal form is

$$x_1x_3 + x_1'x_2x_3' + x_2'x_3'x_4'$$

CHAPTER 8: Modeling Arithmetic, Computation, and Languages

This final chapter of the book extends the modeling concept introduced with the Boolean algebra structure in Chapter 7. The first section introduces models for ordinary arithmetic, with an emphasis on groups. Numerous examples of semigroups, monoids, and groups are given, many of them familiar to the students. Examples where the elements are things like strings or functions instead of just numbers may be a little harder to grasp. Elementary theorems such as the cancellation laws, and also the idea of group isomorphisms, go fairly smoothly because they are similar to the proofs and isomorphisms done for Boolean algebras. The proof of Cayley's theorem breaks down into many small steps; part of the reason for its inclusion in the text is that the proof utilizes so many of the ideas of this section.

Finite-state machines, with their graphical representations, provide a welcome break from the abstractions of group theory. They represent an initial attempt to model general computation; Kleene's theorem defines the limitations of the model. It is worth the time to point out in class that there is a difference between recognizing, say, a string that *contains* 00 (which, once it's true, remains true as more characters are processed) and a string that *ends* with 00 (which can be true at one time, then false as more characters are processed, then true again at a later time).

Turing machines are discussed as both set recognizers and function computers. Grading Turing machine solutions is like grading programs - students can come up with different "algorithms" that work. Be sure that you require descriptions of groups of quintuples (what I call "comments" for their Turing machine "programs"), otherwise it is too hard to figure out the algorithm the student has in mind. After constructing a number of Turing machines to do various tasks, students are generally willing to accept the Church-Turing thesis. The proof of the unsolvability of the Halting Problem is another matter. It ranks right up there with proofs by induction as something you, the instructor, are trying to put over on them! Computational complexity and the P = NP problem are touched upon briefly.

Formal languages (as models of natural languages or construction mechanisms for programming languages) are discussed in the final section of this chapter. Context-free grammars and parse trees are emphasized, and a connection made between classes of languages and sets recognized by various automata.

This chapter provides an introduction to topics that students will see again in subsequent courses they may take in theory of computation, formal languages and automata theory, compiler theory, or coding theory.

EXERCISES 8.1

1. *a. associative *b. commutative c. neither
   d. commutative, associative e. commutative

*2. a. not commutative: $a \cdot b \neq b \cdot a$
   not associative: $a \cdot (b \cdot d) \neq (a \cdot b) \cdot d$
   b.

·	p	q	r	s
p	p	q	r	s
q	q	r	s	p
r	r	s	p	q
s	s	p	q	r

commutative

3. For example:
   a. $a \cdot b = (a + b)^2$ b. $a \cdot b = a$
   c. $a \cdot b = a + 1$ d. $a \cdot b = a + b$

4. Compute $\left(\begin{bmatrix} a_{11} & a_{12} \\ a_{21} & a_{22} \end{bmatrix} \cdot \begin{bmatrix} b_{11} & b_{12} \\ b_{21} & b_{22} \end{bmatrix}\right) \cdot \begin{bmatrix} c_{11} & c_{12} \\ c_{21} & c_{22} \end{bmatrix}$ and

$$\begin{bmatrix} a_{11} & a_{12} \\ a_{21} & a_{22} \end{bmatrix} \cdot \left(\begin{bmatrix} b_{11} & b_{12} \\ b_{21} & b_{22} \end{bmatrix} \cdot \begin{bmatrix} c_{11} & c_{12} \\ c_{21} & c_{22} \end{bmatrix}\right)$$

The results are equal by the distributive, commutative, and associative properties of Z under addition and multiplication.

5. *a. Semigroup
   *b. Not a semigroup - not associative
   *c. Not a semigroup - S not closed under ·
   *d. Monoid; $i = 1 + 0\sqrt{2}$
   *e. Group; $i = 1 + 0\sqrt{2}$
   *f. Group; $i = 1$
   *g. Monoid; $i = 1$
   h. Monoid; $i = 1$
   i. Semigroup
   j. Monoid; $i = (0,1)$
   k. Group; $i = 0$
   l. Not a semigroup - S not closed under ·
   m. Group; i = zero polynomial
   n. Not a semigroup - S not closed under ·
   o. Group; $i = \begin{bmatrix} 1 & 0 \\ 0 & 1 \end{bmatrix}$
   p. Group; $i = 1$
   q. Group; $i = 0$
   r. Monoid; i = function mapping every x to 0

*6. a. f_0 = identity function $f_1(1) = 2$ $f_1(2) = 1$
$f_2(1) = 1$ $f_2(2) = 1$ $f_3(1) = 2$ $f_3(2) = 2$

∘	f_0	f_1	f_2	f_3
f_0	f_0	f_1	f_2	f_3
f_1	f_1	f_0	f_3	f_2
f_2	f_2	f_2	f_2	f_2
f_3	f_3	f_3	f_3	f_3

b. f_0 and f_1 are the elements

∘	f_0	f_1
f_0	f_0	f_1
f_1	f_1	f_0

*7.

∘	R_1	R_2	R_3	F_1	F_2	F_3
R_1	R_2	R_3	R_1	F_3	F_1	F_2
R_2	R_3	R_1	R_2	F_2	F_3	F_1
R_3	R_1	R_2	R_3	F_1	F_2	F_3
F_1	F_2	F_3	F_1	R_3	R_1	R_2
F_2	F_3	F_1	F_2	R_2	R_3	R_1
F_3	F_1	F_2	F_3	R_1	R_2	R_3

Identity element is R_3; inverse for F_1 is F_1; inverse for R_2 is R_1.

8. $\alpha_1 \to R_3$, $\alpha_2 \to F_3$, $\alpha_3 \to F_2$, $\alpha_4 \to F_1$, $\alpha_5 \to R_1$, $\alpha_6 \to R_2$

9. *a. No - not the same operation
*b. No - zero polynomial (identity) does not belong to P
c. No - not every element of Z has an inverse in Z*
d. Yes
e. No - Z is not a subset of $M_2(Z)$
f. Yes
g. No - {0,3,6} not closed under $+_8$

*10. $[\{0\},+_{12}]$, $[Z_{12},+_{12}]$, $[\{0,2,4,6,8,10\},+_{12}]$, $[\{0,4,8\},+_{12}]$, $[\{0,3,6,9\},+_{12}]$, $[\{0,6\},+_{12}]$

11. In each set closure holds, i is a member, and each element has an inverse.

*12. 4!/2 = 24/2 = 12 elements
$\alpha_1 = i$ $\alpha_2 = (1,2)\circ(3,4)$ $\alpha_3 = (1,3)\circ(2,4)$ $\alpha_4 = (1,4)\circ(2,3)$
$\alpha_5 = (1,3)\circ(1,2)$ $\alpha_6 = (1,2)\circ(1,3)$ $\alpha_7 = (1,3)\circ(1,4)$ $\alpha_8 = (1,4)\circ(1,2)$
$\alpha_9 = (1,4)\circ(1,3)$ $\alpha_{10} = (1,2)\circ(1,4)$ $\alpha_{11} = (2,4)\circ(2,3)$ $\alpha_{12} = (2,3)\circ(2,4)$

13.*a. No - $f(x + y) = 2$, $f(x) + f(y) = 2 + 2 = 4$
*b. No - $f(x + y) = |x + y|$, $f(x) + f(y) = |x| + |y|$
*c. Yes - $f(x \cdot y) = |x \cdot y|$, $f(x) \cdot f(y) = |x| \cdot |y|$
d. Yes - $f(a_nx^n + ... + a_1x + a_0 + b_kx^k + ... + b_1x + b_0)$
$= f(a_nx^n + a_{n-1}x^{n-1} + ... + (a_k + b_k)x^k + ... + (a_1 + b_1)x + (a_0 + b_0))$
$= a_n + a_{n-1} + ... + a_k + b_k + ... + a_1 + b_1 + a_0 + b_0$
$= f(a_nx^n + ... + a_1x + a_0) + f(b_kx^k + ... + b_1x + b_0)$
e. No - take α and β both odd. Then $f(\alpha o \beta) = 1$ but $f(\alpha) +_2 f(\beta) = 0 +_2 0 = 0$
None are isomorphisms.

14. a. Yes; $f: Z \rightarrow 12Z$, $f(x) = 12x$
b. No; Z_5 is finite, $5Z$ is infinite
c. Yes; $f: 5Z \rightarrow 12Z$, $f(x) = \frac{12}{5}x$
d. No; $[S_3, o]$ is noncommutative, $[Z_6, +_6]$ is commutative
e. Yes; $f: \{a_1x + a_0\} \rightarrow C$, $f(a_1x + a_0) = a_0 + a_1i$
f. No; the order of $[Z_6, +_6]$ is 6 but the order of $[S_6, o]$ is 6!
g. Yes; $f: Z_2 \rightarrow S_2$, $f(x) = \begin{cases} i & \text{if } x = 0 \\ (1,2) & \text{if } x = 1 \end{cases}$

15. a. Closure holds, 1 is an identity, each element is self-inverse.
b. Consider four cases:
i. α even, β even: then $f(\alpha) = 1$, $f(\beta) = 1$, and $\alpha o \beta$ is even. Thus $f(\alpha o \beta) = 1 = f(\alpha) \cdot f(\beta)$.
ii. α odd, β even: then $f(\alpha) = -1$, $f(\beta) = 1$, and $\alpha o \beta$ is odd. Thus $f(\alpha o \beta) = -1 = f(\alpha) \cdot f(\beta)$.
iii. α even, β odd: similar to (ii).
iv α odd, β odd: then $f(\alpha) = -1$, $f(\beta) = -1$, and $\alpha o \beta$ is even. Thus $f(\alpha o \beta) = 1 = f(\alpha) \cdot f(\beta)$.

*16. a. Closure: $\begin{bmatrix} 1 & z \\ 0 & 1 \end{bmatrix} \cdot \begin{bmatrix} 1 & w \\ 0 & 1 \end{bmatrix} = \begin{bmatrix} 1 & w+z \\ 0 & 1 \end{bmatrix} \in M_2^0(Z)$
Matrix multiplication is associative.
$\begin{bmatrix} 1 & 0 \\ 0 & 1 \end{bmatrix} \in M_2^0(Z)$
The inverse of $\begin{bmatrix} 1 & z \\ 0 & 1 \end{bmatrix}$ is $\begin{bmatrix} 1 & -z \\ 0 & 1 \end{bmatrix}$ which belongs to $M_2^0(Z)$

b. f is a bijection and

$$f\left(\begin{bmatrix}1 & z\\0 & 1\end{bmatrix}\cdot\begin{bmatrix}1 & w\\0 & 1\end{bmatrix}\right)=f\left(\begin{bmatrix}1 & w+z\\0 & 1\end{bmatrix}\right)=w+z=z+w=f\left(\begin{bmatrix}1 & z\\0 & 1\end{bmatrix}\right)+f\left(\begin{bmatrix}1 & w\\0 & 1\end{bmatrix}\right)$$

c. $f\left(\begin{bmatrix}1 & 7\\0 & 1\end{bmatrix}\right)=7$ and $f\left(\begin{bmatrix}1 & -3\\0 & 1\end{bmatrix}\right)=-3$

$7+(-3)=4$

$f^{-1}(4)=\begin{bmatrix}1 & 4\\0 & 1\end{bmatrix}$

d. $f^{-1}(2)=\begin{bmatrix}1 & 2\\0 & 1\end{bmatrix}$ and $f^{-1}(3)=\begin{bmatrix}1 & 3\\0 & 1\end{bmatrix}$

$\begin{bmatrix}1 & 2\\0 & 1\end{bmatrix}\cdot\begin{bmatrix}1 & 3\\0 & 1\end{bmatrix}=\begin{bmatrix}1 & 5\\0 & 1\end{bmatrix}$ and $f\left(\begin{bmatrix}1 & 5\\0 & 1\end{bmatrix}\right)=5$

17. a. $i \cdot i = i$ so $i = i^{-1}$

b. $x^{-1} \cdot x = x \cdot x^{-1} = i$ so $x = (x^{-1})^{-1}$

18. a.

·	1	a
1	1	a
a	a	1

b.

·	1	a	b
1	1	a	b
a	a	b	1
b	b	1	a

c.

·	1	a	b	c
1	1	a	b	c
a	a	1	c	b
b	b	c	1	a
c	c	b	a	1

·	1	a	b	c
1	1	a	b	c
a	a	b	c	1
b	b	c	1	a
c	c	1	a	b

$1 \leftrightarrow 1$, $a \leftrightarrow b$, $b \leftrightarrow c$, $c \leftrightarrow a$

·	1	a	b	c
1	1	a	b	c
a	a	c	1	b
b	b	1	c	a
c	c	b	a	1

$1 \leftrightarrow 1$, $a \leftrightarrow c$, $b \leftrightarrow b$, $c \leftrightarrow a$

·	1	a	b	c
1	1	a	b	c
a	a	1	c	b
b	b	c	a	1
c	c	b	1	a

*19. a. $i_L = i_L \cdot i_R = i_R$ so $i_L = i_R$ and this element is an identity in [S, ·].

b. For example,

·	a	b
a	a	b
b	a	b

c. For example,

·	a	b
a	a	a
b	b	b

d. For example, $[R^+, +]$

20. a. $x_L^{-1} = x_L^{-1} \cdot i = x_L^{-1} \cdot (x \cdot x_R^{-1}) = (x_L^{-1} \cdot x) \cdot x_R^{-1} = i \cdot x_R^{-1} = x_R^{-1}$
so $x_L^{-1} = x_R^{-1}$ and this element is x^{-1}. Therefore every element has an inverse, and [S, ·] is a group.

b. For $x \in N$, $(g \circ f)(x) = g(f(x)) = g(2x) = 2x/2 = x$.
Therefore $g \circ f = i$, the identity function on N. If f had a right inverse, say h, then $(f \circ h)(x) = x$. But $(f \circ h)(x) = f(h(x)) = 2(h(x))$. For $2(h(x)) = x$ we must have $h(x) = x/2$, but this function is not a member of S.

21. Let $x \in G$. Then $x, x^2, x^3, \ldots$ are all members of G because G is closed under the operation. Because G is finite, these are not all distinct elements of G, and $x^n = x^m$ for some n and m, $m > n$. Then $\underbrace{x \cdot x \cdots x}_{n} \cdot i = \underbrace{x \cdot x \cdots x}_{m}$ and by left cancellation, $i = x^{m-n}$, $m-n > 0$.

*22. a. $x \rho x$ because $i \cdot x \cdot i^{-1} = x \cdot i^{-1} = x \cdot i = x$
If $x \rho y$ then for some $g \in G$, $g \cdot x \cdot g^{-1} = y$ or $g \cdot x = y \cdot g$ or
$x = g^{-1} \cdot y \cdot g = (g^{-1}) \cdot y \cdot (g^{-1})^{-1}$ so $y \rho x$.
If $x \rho y$ and $y \rho z$ then for some $g_1, g_2 \in G$, $g_1 \cdot x \cdot g_1^{-1} = y$ and $g_2 \cdot y \cdot g_2^{-1} = z$
so $g_2 \cdot g_1 \cdot x \cdot g_1^{-1} \cdot g_2^{-1} = z$ or $(g_2 \cdot g_1) \cdot x \cdot (g_2 \cdot g_1)^{-1} = z$ and $x \rho z$.

b. Suppose G is commutative and $y \in [x]$. Then for some $g \in G$,
$y = g \cdot x \cdot g^{-1} = x \cdot g \cdot g^{-1} = x \cdot i = x$. Thus $[x] = \{x\}$. Conversely suppose $[x] = \{x\}$ for each $x \in G$, let $x, y \in G$, and denote the element $y \cdot x \cdot y^{-1}$ by z. Then $x \rho z$, so $z = x$ and $y \cdot x \cdot y^{-1} = x$ or $y \cdot x = x \cdot y$.

23. Let $x \in S$ with left inverse y. Then $y \in S$, so let z be the left inverse of y. Then $x \cdot y = i_L \cdot (x \cdot y) = (z \cdot y) \cdot (x \cdot y) = z \cdot (y \cdot x) \cdot y = z \cdot i_L \cdot y = z \cdot y = i_L$ so y is also a right inverse of x. Also, $x \cdot i_L = x \cdot (y \cdot x) = (x \cdot y) \cdot x = i_L \cdot x = x$ so i_L is also a right identity in S, therefore an identity.

24. For some fixed $a \in S$, let x_1 be the solution to $x \cdot a = a$. Let b be any element of S. Then $a \cdot x = b$ for some $x \in S$ and $x_1 \cdot b = x_1 \cdot (a \cdot x) = (x_1 \cdot a) \cdot x = a \cdot x = b$. Therefore x_1 is a left identity in S. Also for any $b \in S$, there is an x such that $x \cdot b = x_1$, hence every element of S has a left inverse. Result follows from Exercise 23.

25. Let $S = \{x_1, \ldots, x_n\}$. The products $x_1 \cdot x_1, x_1 \cdot x_2, \ldots, x_1 \cdot x_n$ are distinct, for if $x_1 \cdot x_i = x_1 \cdot x_j$, then $x_i = x_j$ by left cancellation. These products are the n elements of S, so $x_1 \cdot x_i = x_1$ for some i. Then for any $x_j \in S$, $x_1 \cdot x_j = x_1 \cdot x_j \rightarrow (x_1 \cdot x_i) \cdot x_j = x_1 \cdot x_j \rightarrow x_1 \cdot (x_i \cdot x_j) = x_1 \cdot x_j \rightarrow x_i \cdot x_j = x_j$ by left cancellation, and x_i is a left identity. Also, for any $x_j \in S$, form the n products $x_1 \cdot x_j, x_2 \cdot x_j, \ldots, x_n \cdot x_j$. These are distinct, by right cancellation, so $x_k \cdot x_j = x_i$ for some k, and x_k is thus a left inverse of x_j. Result follows from Exercise 23.

26. If G is commutative, then $(x \cdot y)^2 = (x \cdot y) \cdot (x \cdot y) = x \cdot (y \cdot x) \cdot y = x \cdot (x \cdot y) \cdot y = (x \cdot x) \cdot (y \cdot y) = x^2 \cdot y^2$. For the converse, let $x, y \in G$; then $x \cdot y \cdot x \cdot y = x \cdot x \cdot y \cdot y$, and by left and right cancellation, $y \cdot x = x \cdot y$, so G is commutative.

27. Let $x, y \in G$. Then $x \cdot y \cdot x \cdot y = i = x \cdot x \cdot y \cdot y$ and by left and right cancellation, $y \cdot x = x \cdot y$.

*28 a. $S \cap T \subseteq G$. Closure: for $x, y \in S \cap T$, $x \cdot y \in S$ because of closure in S, $x \cdot y \in T$ because of closure in T, so $x \cdot y \in S \cap T$.
Identity: $i \in S$ and $i \in T$ so $i \in S \cap T$.
Inverses: for $x \in S \cap T$, $x^{-1} \in S$ and $x^{-1} \in T$ so $x^{-1} \in S \cap T$.

b. No. For example, $[\{0,4,8\},+_{12}]$ and $[\{0,6\},+_{12}]$ are subgroups of $[Z_{12},+_{12}]$ but $[\{0,4,6,8\},+_{12}]$ is not a subgroup of $[Z_{12},+_{12}]$ (not closed).

29. Closure: let $s_1 \cdot t_1, s_2 \cdot t_2 \in ST$. Then $(s_1 \cdot t_1) \cdot (s_2 \cdot t_2) = s_1 \cdot s_2 \cdot t_1 \cdot t_2$ and $s_1 \cdot s_2 \in S$, $t_1 \cdot t_2 \in T$, so $s_1 \cdot s_2 \cdot t_1 \cdot t_2 \in ST$.
Identity: $i \in S$ and $i \in T$ so $i = i \cdot i \in ST$.
Inverses: let $s \cdot t \in ST$. Then $s^{-1} \in S$, $t^{-1} \in T$, so $s^{-1} \cdot t^{-1} \in ST$ and $s \cdot t \cdot s^{-1} \cdot t^{-1} = s \cdot s^{-1} \cdot t \cdot t^{-1} = i$, also $s^{-1} \cdot t^{-1} \cdot s \cdot t = i$.

*30. Closure: let $x, y \in B_k$. Then $(x \cdot y)^k = x^k \cdot y^k = i \cdot i = i$, so $x \cdot y \in B_k$.
Identity: $i^k = i$ so $i \in B_k$.
Inverses: for $x \in B_k$, $(x^{-1})^k = (x^k)^{-1} = i^{-1} = i$, so $x^{-1} \in B_k$.

31. a. Closure: let $x, y \in A$. Then for any $g \in G$, $(x \cdot y) \cdot g = x \cdot (y \cdot g) = x \cdot (g \cdot y) = (x \cdot g) \cdot y = (g \cdot x) \cdot y = g \cdot (x \cdot y)$.
Identity: $i \cdot g = g \cdot i = g$ for all $g \in G$.
Inverses: let $x \in A$. Then for any $g \in G$, $x \cdot g = g \cdot x$. Multiplying twice by x^{-1}, $g = x^{-1} \cdot g \cdot x$ and $g \cdot x^{-1} = x^{-1} \cdot g$ so $x^{-1} \in A$.

b. $\{R_3\}$

c. Let G = A; clearly G is then commutative. Now let G be commutative. We have $A \subseteq G$. Let $x \in G$. Then for any $g \in G$, $x \cdot g = g \cdot x$ so $x \in A$ and $G \subseteq A$.

d. $y \in G$ so $x \cdot y^{-1} \cdot y = y \cdot x \cdot y^{-1}$ or $x = y \cdot x \cdot y^{-1}$ or $x \cdot y = y \cdot x$.

32. a. Closure: let $f, g \in H_a$. Then $(f \circ g)(a) = f(g(a)) = f(a) = a$, so $f \circ g \in H_a$.
Identity: the identity mapping on A leaves a fixed.
Inverses: let $f \in H_a$. Then $f(a) = a$ so $f^{-1}(a) = a$, and $f^{-1} \in H_a$.

b. $(n - 1)!$

33. a. Identity: Let $x \in A$ ($A \neq \varnothing$). Then $x \cdot x^{-1} = i \in A$.
Inverses: let $x \in A$. Then $i \cdot x^{-1} = x^{-1} \in A$.
Closure: let $x, y \in A$. Then $y^{-1} \in A$, and $x \cdot (y^{-1})^{-1} = x \cdot y \in A$.

b. $B_k \neq \varnothing$ since $i \in B_k$. Let $x, y \in B_k$. Then $(x \cdot y^{-1})^k = x^k \cdot (y^{-1})^k = i \cdot (y^k)^{-1} = i \cdot i^{-1} = i$, so $x \cdot y^{-1} \in B_k$.

34. a. Let $x = a^{z_1}, y = a^{z_2} \in A$. Then $x \cdot y^{-1} = a^{z_1} \cdot (a^{z_2})^{-1} = a^{z_1} \cdot (a^{-1})^{z_2} = a^{z_1 - z_2} \in A$. By Exercise 33, A is a subgroup.

b. $2^0 = 0$, $2^1 = 2$, $2^2 = 2 +_7 2 = 4$, $2^3 = 6$, $2^4 = 1$, $2^5 = 3$, $2^6 = 5$

c. $5^0 = 0$, $5^1 = 5$, $5^2 = 5 +_7 5 = 3$, $5^3 = 1$, $5^4 = 6$, $5^5 = 4$, $5^6 = 2$.

d. $3^0 = 0$, $3^1 = 3$, $3^2 = 3 +_4 3 = 2$, $3^3 = 1$

*35. Let $x = a^{z_1}, y = a^{z_2} \in G$. Then $x \cdot y = a^{z_1} \cdot a^{z_2} = a^{z_1+z_2} = a^{z_2+z_1} = a^{z_2} \cdot a^{z_1} = y \cdot x$.

36 $f(x \cdot y) = (x \cdot y)^{-1} = y^{-1} \cdot x^{-1} = x^{-1} \cdot y^{-1} = f(x) \cdot f(y)$
f is onto: let $g \in G$. Then $g^{-1} \in G$ and $f(g^{-1}) = (g^{-1})^{-1} = g$.
f is one-to-one: if $f(x) = f(y)$ then $x^{-1} = y^{-1}$; because inverses are unique, $(x^{-1})^{-1} = (y^{-1})^{-1}$ or $x = y$.

*37. a. [Aut(S),o] is closed because composition of isomorphisms is an isomorphism (Practice 30). Associativity always holds for function composition. The identity function i_S is an automorphism on S. Finally, if f is an automorphism on S, so is f^{-1}.

b.

i: $0 \to 0$, $1 \to 1$, $2 \to 2$, $3 \to 3$

f: $0 \to 0$, $1 \to 3$, $2 \to 2$, $3 \to 1$

o	i	f
i	i	f
f	f	i

38. Let i_G and i_H denote the identity elements of G and H, respectively. Let f be an isomorphism, $f: G \to H$. Then $f(i_G) = i_H$ and since f is one-to-one, i_G is the only such element. Now let f be a homomorphism from G onto H; then $f(i_G) = i_H$. Suppose i_G is the only such element, and let $f(g_1) = f(g_2)$ for $g_1, g_2 \in G$. Then $f(g_1 \cdot g_2^{-1}) = f(g_1) \cdot f(g_2^{-1}) = f(g_1) \cdot (f(g_2))^{-1} = f(g_1) \cdot (f(g_1))^{-1} = i_H$. Therefore $g_1 \cdot g_2^{-1} = i_G$ and $g_1 = i_G \cdot g_2 = g_2$. Thus f is one-to-one; f is already an onto homomorphism, so it is an isomorphism.

39. To show that f is onto, let $y \in G$. Then $g^{-1} \cdot y \cdot g \in G$ and $f(g^{-1} \cdot y \cdot g) =$ $g \cdot (g^{-1} \cdot y \cdot g) \cdot g^{-1} = (g \cdot g^{-1}) \cdot y \cdot (g \cdot g^{-1}) = y$. To show that f is one-to-one, $f(x) = f(y) \rightarrow g \cdot x \cdot g^{-1} = g \cdot y \cdot g^{-1} \rightarrow x = y$ by cancellation. To show that f is a homomorphism, $f(x \cdot y) = g \cdot (x \cdot y) \cdot g^{-1} = g \cdot x \cdot (g^{-1} \cdot g) \cdot y \cdot g^{-1}$ $= (g \cdot x \cdot g^{-1}) \cdot (g \cdot y \cdot g^{-1}) = f(x) \cdot f(y)$.

EXERCISES 8.2

*1. a. 0001111110 b. aaacaaaa c. 00100110

2. a. 110100, 111010 b. none c. $0a_1a_2a_3a_4a_5$

3.

	Next state		
	Present input		
Present state	0	1	Output
s_0	s_1	s_1	0
s_1	s_2	s_1	1
s_2	s_2	s_0	0

Output is 010010

4.

	Next state		
	Present input		
Present state	0	1	Output
s_0	s_3	s_0	1
s_1	s_2	s_0	0
s_2	s_2	s_1	1
s_3	s_1	s_2	0

Output is 11101011

*5.

	Next state		
	Present input		
Present state	0	1	Output
s_0	s_1	s_2	a
s_1	s_2	s_3	b
s_2	s_2	s_1	c
s_3	s_2	s_3	b

Output is abbcbb

6.

	Next state			
	Present input			
Present state	a	b	c	Output
s_0	s_0	s_1	s_2	0
s_1	s_0	s_2	s_2	1
s_2	s_0	s_2	s_1	0

Output is 0000010

7.

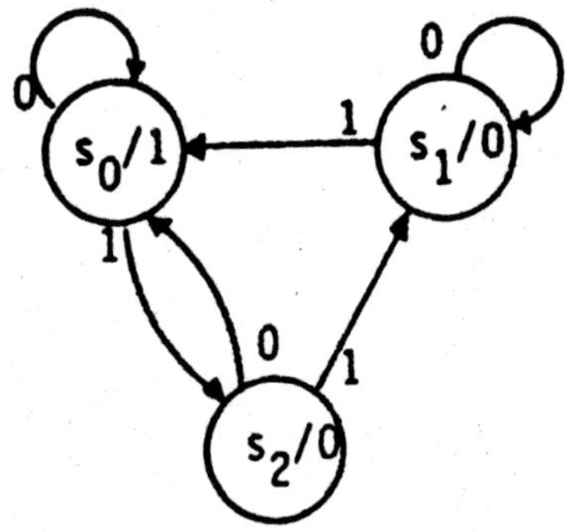

Output is 101110

8.

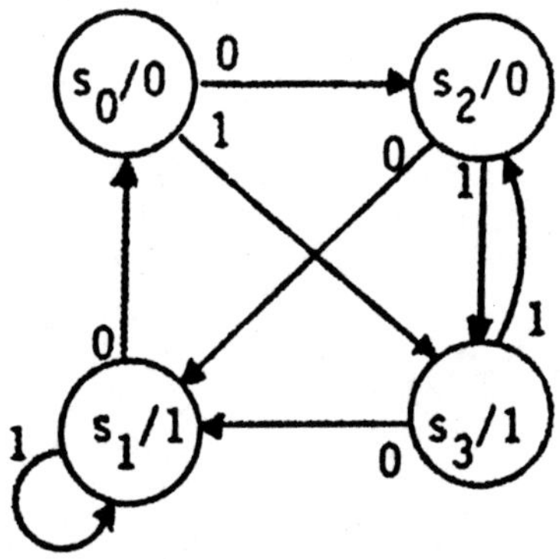

Output is 00111

*9.

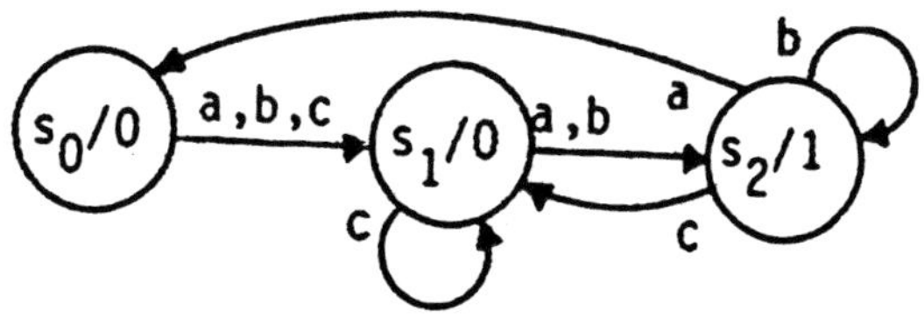

Output is 0001101

10.

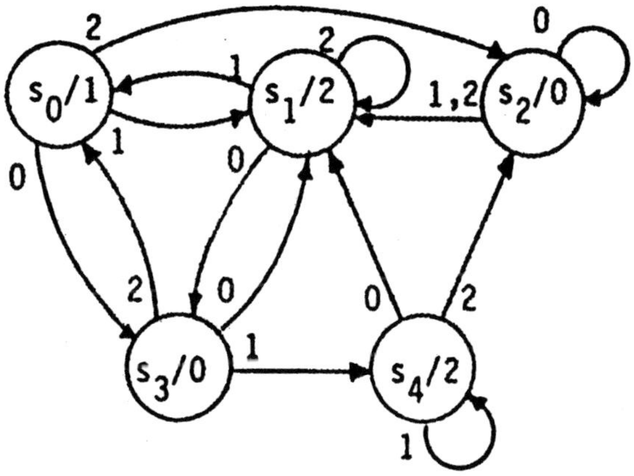

Output is 102012

*11. a.

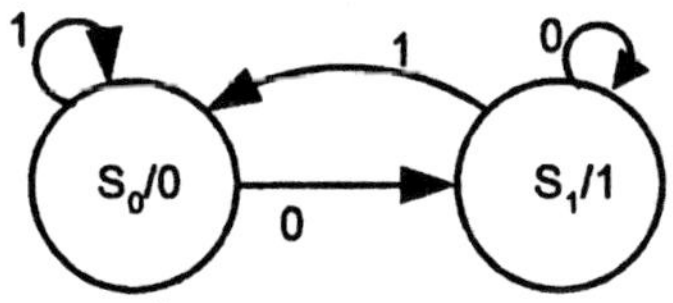

b. Output is 010100

12.

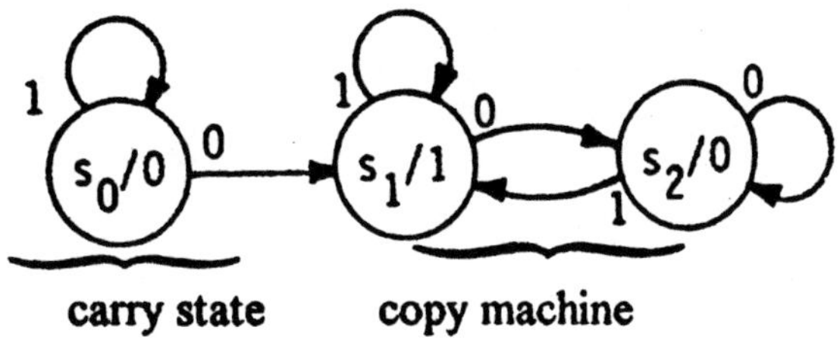

13. a.

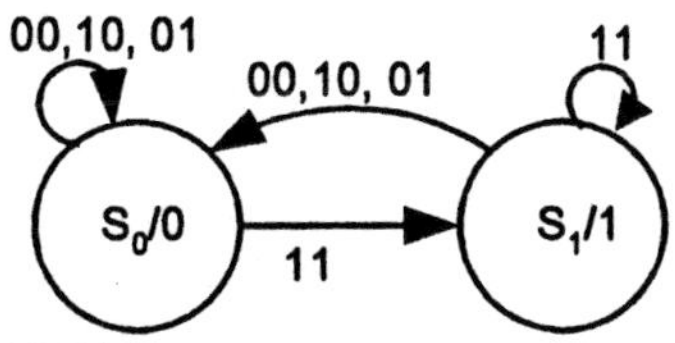

b. 010010

14. a.

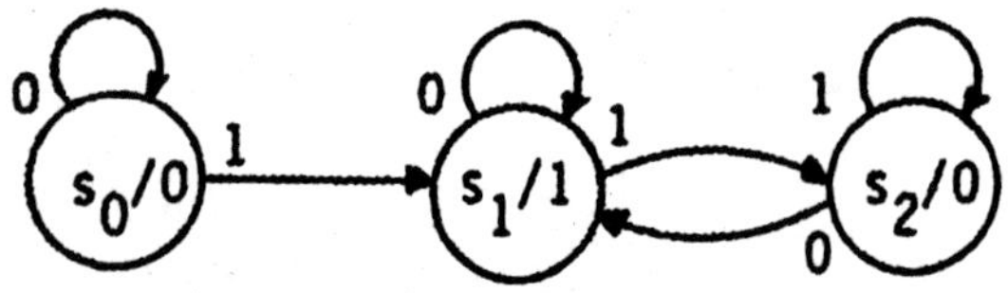

b. 0100(0)
0101(0)

*15. a.

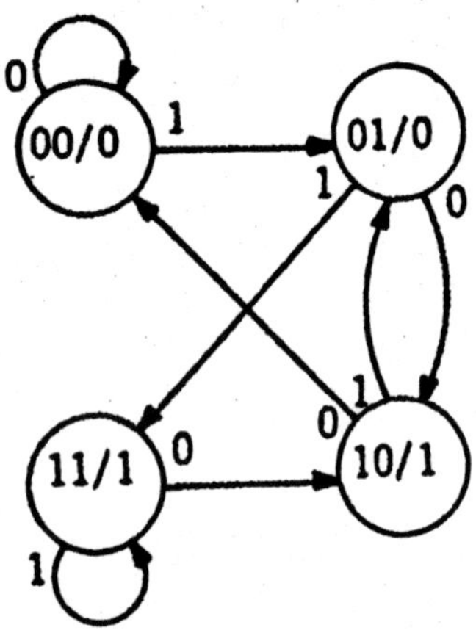

b. The length of time required to remember a given input grows without bound and eventually would exceed the number of states.

16.

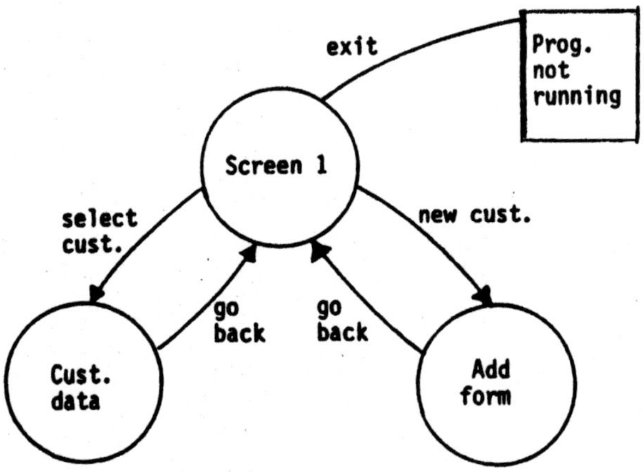

17.

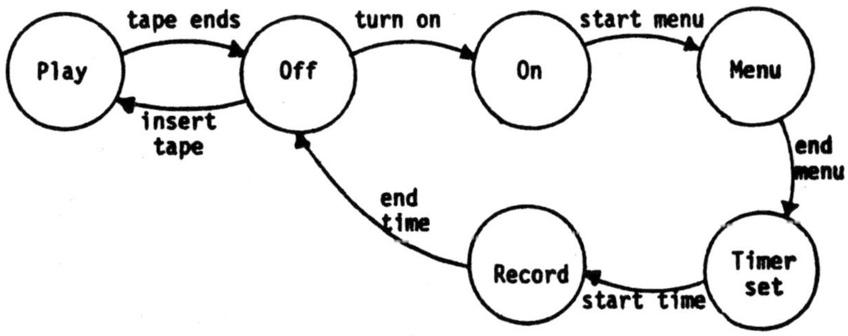
Play
tape ends
Off
insert tape
turn on
On
start menu
Menu
end menu
Timer set
start time
Record
end time

*18.

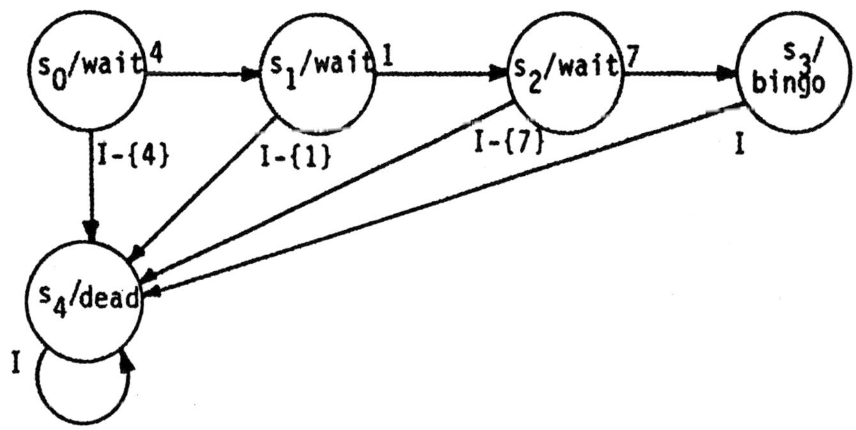
s0/wait
4
s1/wait
1
s2/wait
7
s3/ bingo
I-{4}
I-{1}
I-{7}
I
s4/dead
I

19.*a.

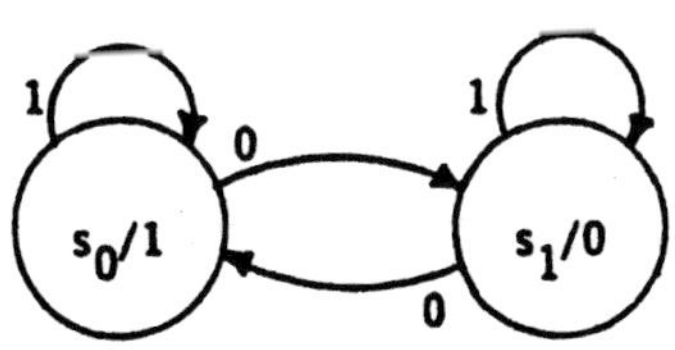
1
0
1
s0/1
s1/0
0

*b.

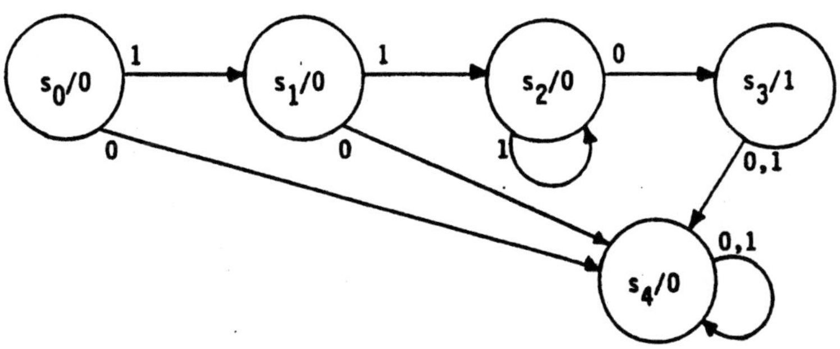
s0/0
1
s1/0
1
s2/0
0
s3/1
0
0
1
0,1
s4/0
0,1

*c.

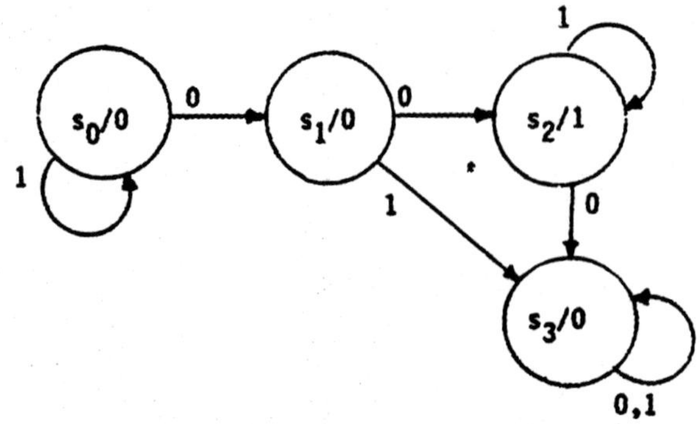
s0/0
s1/0
s2/1
s3/0
0
0
1
1
1
0
0,1

d.

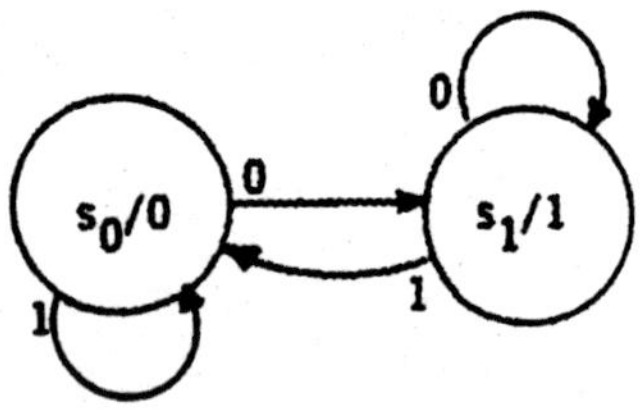
s0/0
s1/1
0
0
1
1

e.

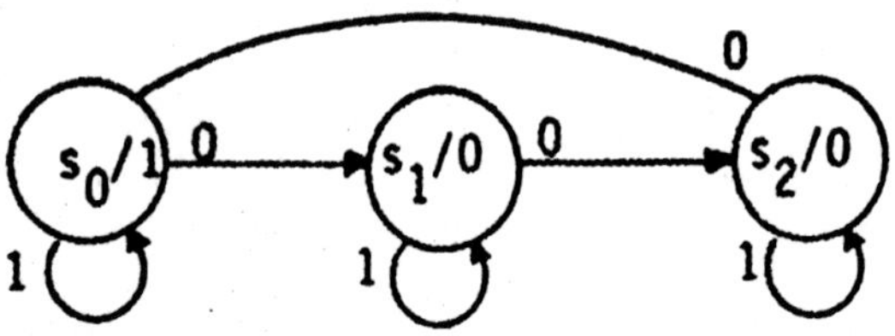
0
s0/1
s1/0
s2/0
0
0
1
1
1

f.

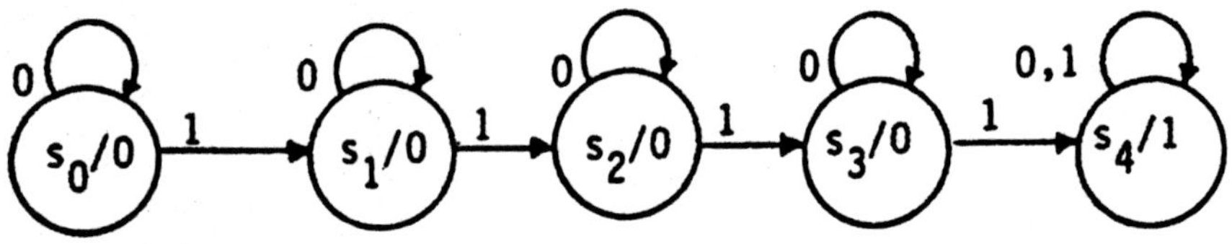
0
0
0
0
0,1
s0/0
s1/0
s2/0
s3/0
s4/1
1
1
1
1

20. a.

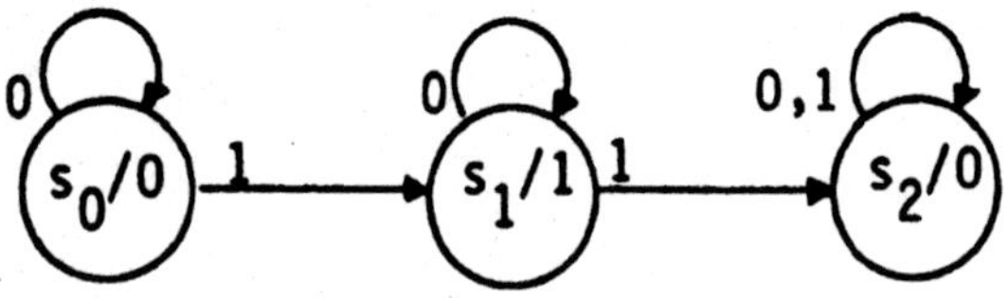
0
0
0,1
s0/0
s1/1
s2/0
1
1

b.

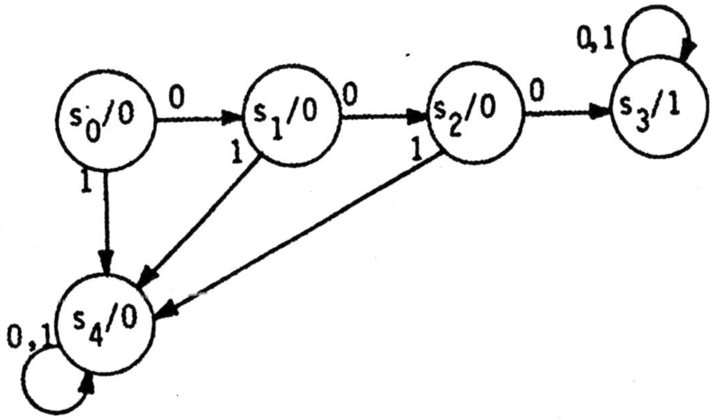
s0/0
s1/0
s2/0
s3/1
s4/0
0
0
0
1
1
1
0,1
0,1

c.

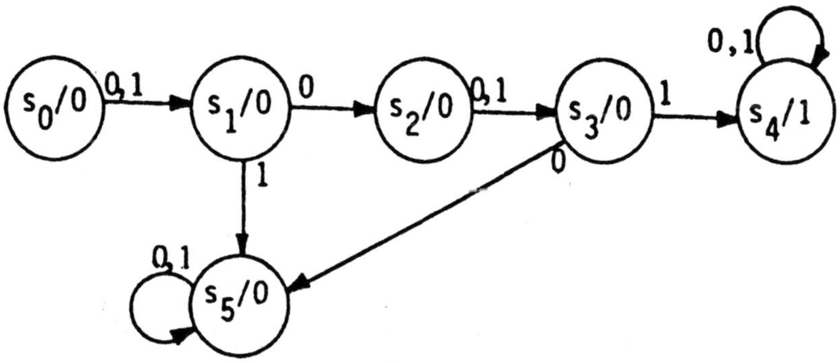
s0/0
s1/0
s2/0
s3/0
s4/1
s5/0
0,1
0
0,1
1
1
0
0,1
0,1

d.

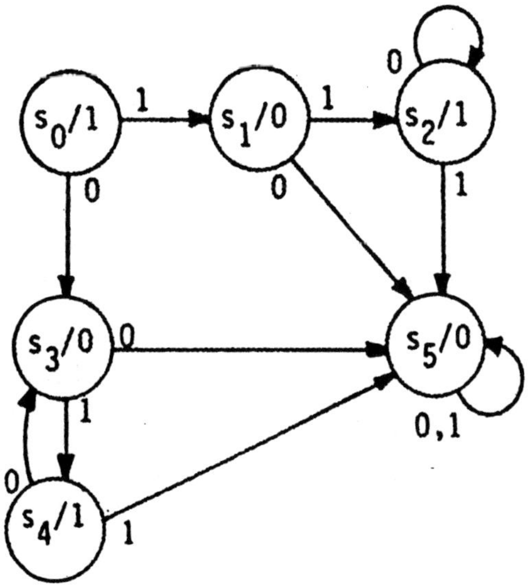
s0/1
s1/0
s2/1
s3/0
s4/1
s5/0
1
1
0
0
0
1
0
1
0
1
0,1

e.

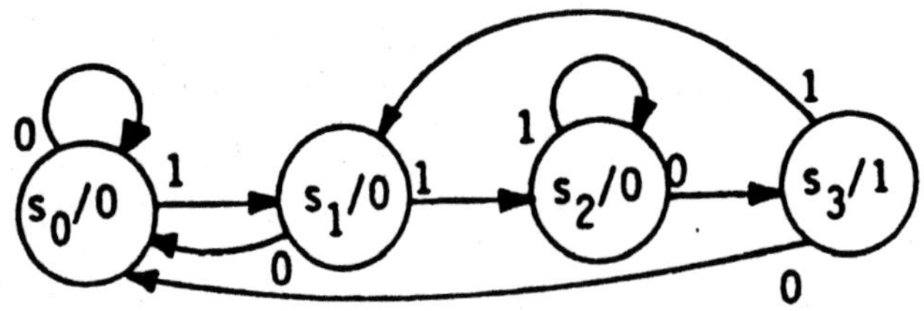

f.

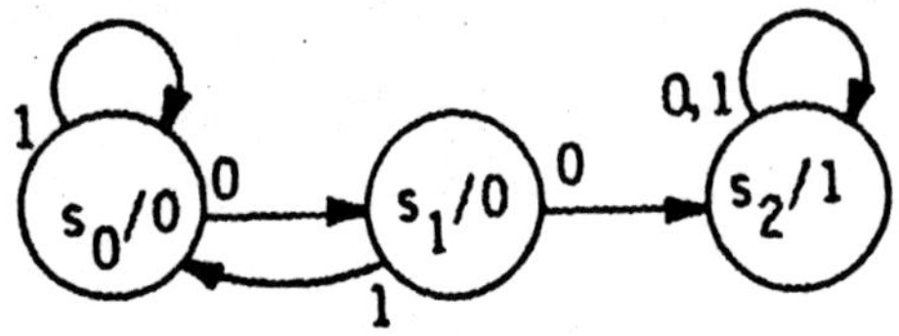

21. The object is to recognize the substring βcon.

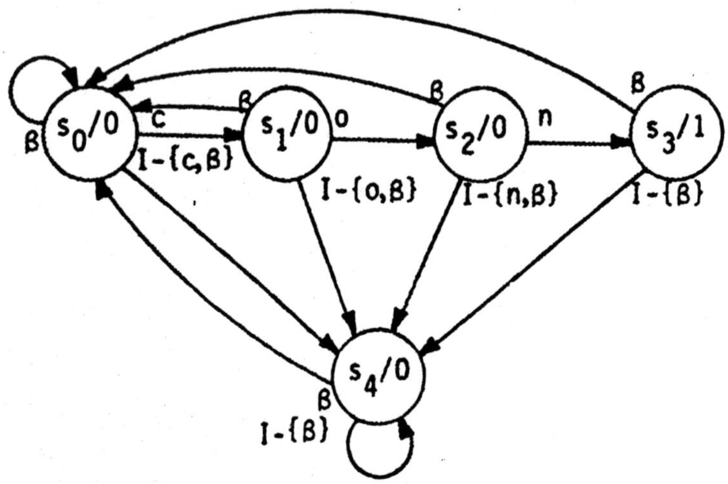

22. a.

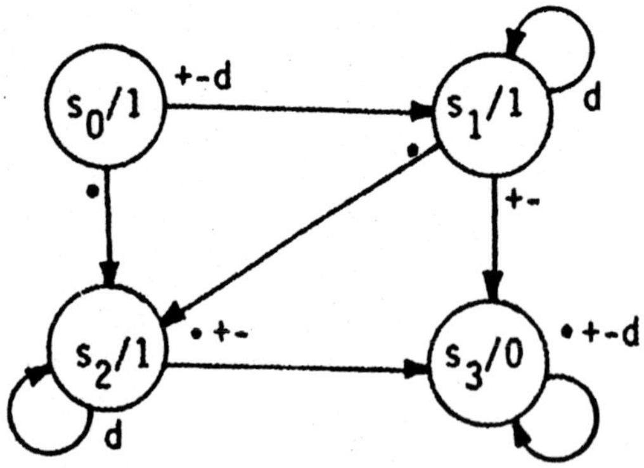

b.

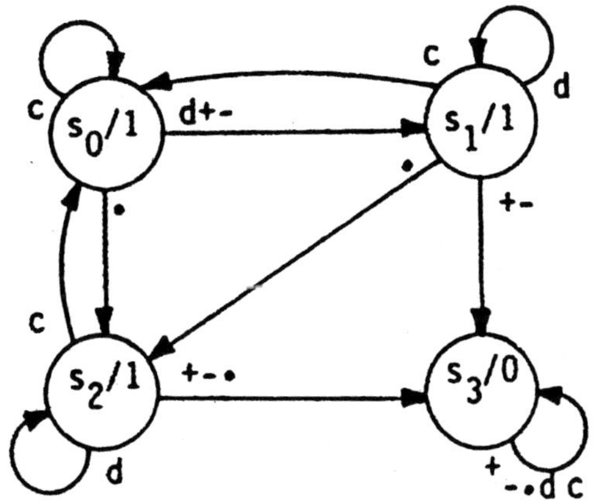

c. sd* or sdd*.dd* or d* or dd*.dd*
There should be extra states around "." to guarantee at least one d.

*23. Once a state is revisited, behavior will be periodic since there is no choice of paths from a state. The maximum number of inputs which can occur before this happens is n - 1 (visiting all n states before repeating). The maximum length of a period is n (output from all n states, with the last state returning to s_0).

24.

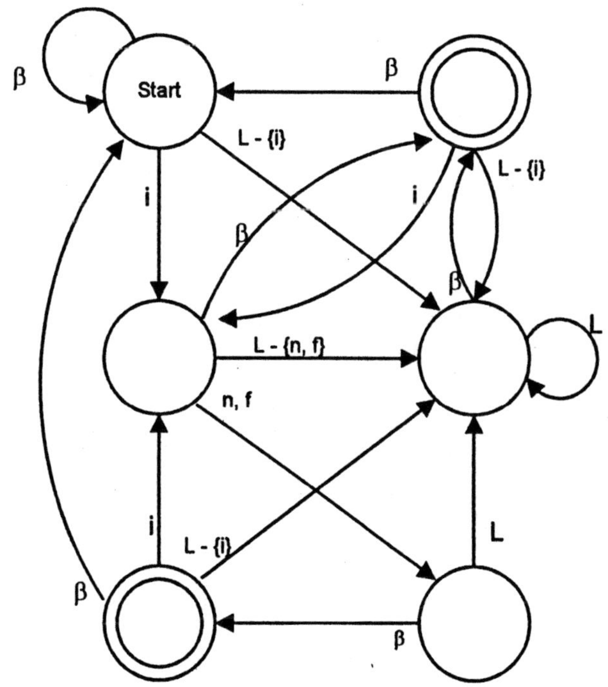

25. 0*

26. 1*000*

27. 01 ∨ (110)*

28. 11*(01)* 29. $(1 \vee 01)(01)^*$ 30. $1(1 \vee 00^*11)^*$

31. a. 10*1 b. $(00)^* \vee (00)^*10^*$ c. $1^* \vee (010)^*$

32. a. $0(0 \vee 1)^*1$ b. 1*01*(01*0)*1* *c. 100*1
d. $(0 \vee 1)^*0(0 \vee 1)^*$ e. b*(abbb*)* f. 1*01*01*

*33. a. Yes b. No c. No d. Yes

34. $L(L \vee d)^*$ where L stands for any letter, d stands for any digit

*35. $dd^*(+ \vee -)dd^*$

36. a. (00)*
b. 111*0
c. 1*001*
d. $(1 \vee 0)^*00^*$
e. 1*(01*01*0)*1*
f. $0^*10^*10^*10^*1(1 \vee 0)^*$

37. a. 0*10*
b. $000(1 \vee 0)^*$
c. $(1 \vee 0)0(1 \vee 0)1(1 \vee 0)^*$
d. $(01)^* \vee 110^*$
e. $(1 \vee 0)^*110$
f. $(1 \vee 0)^*00(1 \vee 0)^*$

38. a. Proof is by induction on the length of the regular expression. For the base step, if $A = \phi$, λ, or i, then $A^R = \phi$, λ, or i. Assume that for all expressions of length $\leq k$, A regular $\to$ A^R regular. Let A be a regular expression of length k + 1. If A = BC, where B and C are regular, then B^R and C^R are regular by inductive hypothesis, and $A^R = C^R B^R$ so A^R is regular. Similarly, if $A = B \vee C$, then $A^R = B^R \vee C^R$ (regular), and if $A = B^*$, then $A^R = (B^R)^*$ (regular).
b. No, cannot write a regular expression for this set.

39. Let M be a finite-state machine recognizing A (M exists by Kleene's Theorem). Interchange final and nonfinal states of M to make a new machine M'. M' recognizes I* - A, so I* - A is regular by the other half of Kleene's Theorem.

*40. a. s_1 b. s_2

*41. A = {0}, B = {1,2,5}, C = {3,4}, D = {6}

Present state	Next state Present input 0	1	Output
A	C	D	1
B	C	B	0
C	B	A	1
D	C	B	1

42. A = {0,3,4}, B = {1}, C = {2,5}

Present state	Next state Present input 0	1	Output
A	C	A	1
B	C	C	0
C	B	A	0

43. A = {0}, B = {5}, C = {2}, D = {7,8}, E = {1,3}, F = {4,6}

Present state	Next state Present input 0	1	Output
A	E	C	0
B	F	D	0
C	E	F	0
D	D	E	0
E	C	E	1
F	B	F	1

44. A = {0,3}, B = {1,6}, C = {2,4}, D = {5,7}

	Next state		
	Present input		
Present state	0	1	Output
A	D	B	1
B	A	A	1
C	D	B	0
D	C	A	0

45. A = {0}, B = {2}, C = {1,4}, D = {3}, E = {5}

	Next state		
	Present input		
Present state	0	1	Output
A	C	D	0
B	E	C	0
C	B	C	1
D	C	B	2
E	C	A	2

*46. A = {0,2}, B = {1,3}, C = {4}

	Next state			
	Present input			
Present state	a	b	c	Output
A	B	C	A	1
B	C	A	B	0
C	B	A	A	0

47. A = {0,1,2,3}

	Next state		
	Present input		
Present state	0	1	Output
A	A	A	0

48. M is already minimized.

49. A = {0}, B = {2,4}, C = {1,5}, D = {3}

	Next state		Output
	Present input		
Present state	0	1	
A	D	A	0
B	C	B	0
C	B	D	1
D	A	B	1

50. A = {0}, B = {1,3}, C = {6}, D = {2}, E = {4,5}

	Next state		Output
	Present input		
Present state	0	1	
A	B	E	1
B	B	C	1
C	D	E	1
D	A	E	0
E	E	B	0

*51. Possible answer:

	d_1	d_2
s_0	0	0
s_1	0	1
s_2	1	0
s_3	1	1

$x(t)$	$d_1(t)$	$d_2(t)$	$y(t)$	$d_1(t+1)$	$d_2(t+1)$
0	0	0	0	1	0
1	0	0	0	1	1
0	0	1	1	0	0
1	0	1	1	0	1
0	1	0	0	0	1
1	1	0	0	1	1
0	1	1	1	0	1
1	1	1	1	1	0

$y(t) = d_1'd_2 + d_1d_2 = d_2$
$d_1(t + 1) = x'd_1'd_2' + xd_1'd_2' + xd_1d_2' + xd_1d_2 = d_1'd_2' + xd_1$
$d_2(t + 1) = xd_1'd_2' + xd_1'd_2 + x'd_1d_2' + xd_1d_2' + x'd_1d_2 = x(d_1' + d_2') + x'd_1$

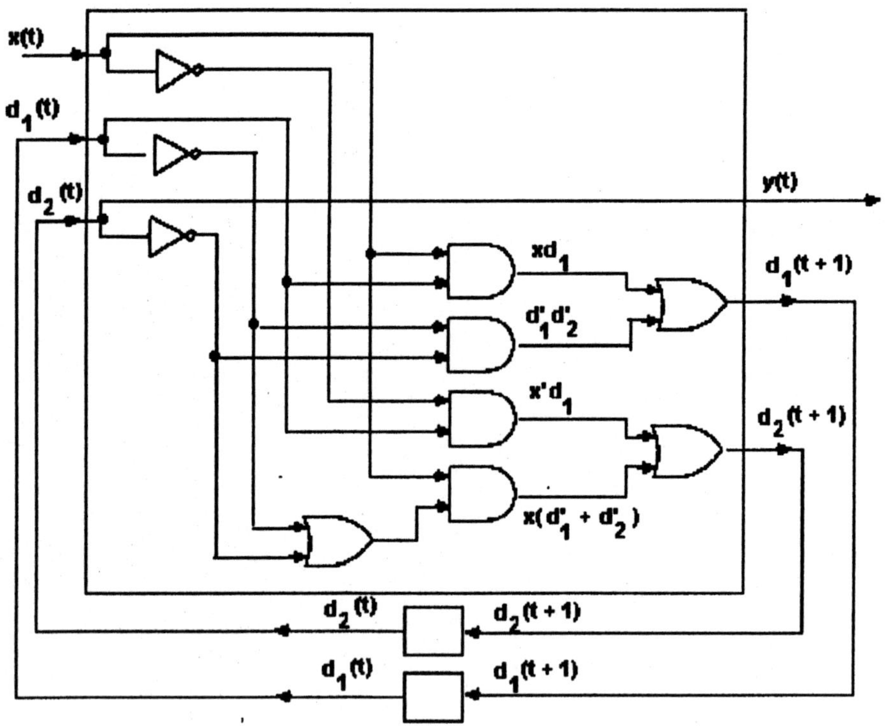

52. Possible answer:

	d_1	d_2	d_3
s_0	0	0	0
s_1	0	0	1
s_2	0	1	0
s_3	0	1	1
s_4	1	0	0

x(t)	$d_1(t)$	$d_2(t)$	$d_3(t)$	y(t)	$d_1(t + 1)$	$d_2(t + 1)$	$d_3(t + 1)$
0	0	0	0	0	0	0	0
1	0	0	0	0	0	1	1
0	0	0	1	1	0	0	0
1	0	0	1	1	1	0	0
0	0	1	0	1	0	1	0
1	0	1	0	1	0	0	1
0	0	1	1	0	0	0	0
1	0	1	1	0	1	0	0
0	1	0	0	1	0	0	1
1	1	0	0	1	0	1	0
0	1	0	1	1	0	0	1
1	1	0	1	1	1	0	0
0	1	1	0	1	0	1	1
1	1	1	0	1	0	0	0
0	1	1	1	1	0	0	1
1	1	1	1	1	1	0	0

$y(t) = d_1 + d_2d_3' + d_2'd_3$
$d_1(t + 1) = xd_3$
$d_2(t + 1) = d_3'(xd_2' + x'd_2)$
$d_3(t + 1) = xd_1'd_3' + x'd_1$

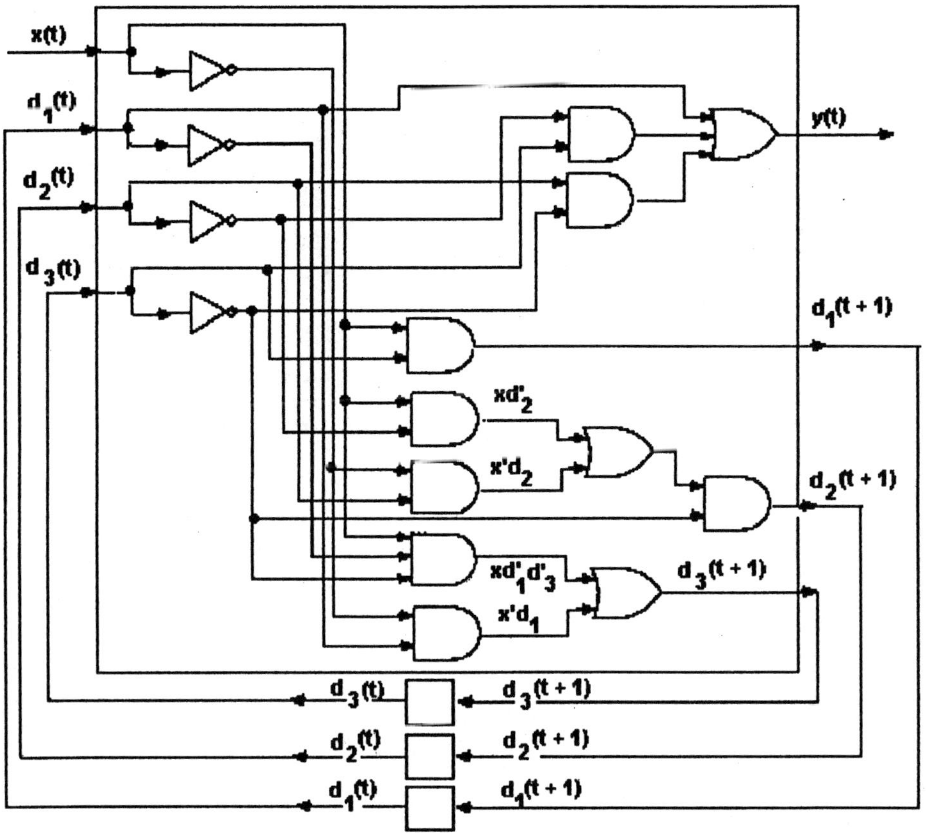

EXERCISES 8.3

*1. a. halts with final tape

⋯	*b*	0	0	0	0	0	*b*	⋯

b. does not change the tape and moves forever to the left

2. a. halts with final tape

⋯	*b*	1	0	1	0	*b*	⋯

b. moves forever back and forth adding a 1 at each end of the nonblank portion of the tape.

3. (0,0,1,0,R)
   (0,1,0,0,R)

4. One answer: State 1 is a final state

(0,0,0,0,R)	passes over 0's
(0,1,1,1,R)	first 1, go to final state, halt and accept

*5. One answer: State 2 is a final state

(0,b,b,2,R)	blank tape or no more 1's
(0,1,1,1,R)	has read odd number of 1's
(1,1,1,0,R)	has read even number of 1's

6. One answer: State 3 is a final state

(0,0,0,0,R) (0,1,1,1,R)	pass over 0's to first 1
(1,0,0,1,R) (1,1,1,2,R)	pass over 0's to second 1
(2,b,b,3,R)	end of string, halt and accept

7. One answer: State 3 is a final state

(0,(,(,0,R) (0,),X,1,L) (0,X,X,0,R)	looks for leftmost), replaces it with X
(0,b,b,2,L)	no more)
(1,(,X,0,R) (1,X,X,1,L)	looks for matching (
(2,b,b,3,R) (2,X,X,2,R)	no more (, halts and accepts

*8. One answer: State 9 is a final state

(0,b,b,9,R)	accepts blank tape
(0,0,0,0,R) (0,1,X,1,R)	finds first 1, marks with X
(1,1,1,1,R) (1,Y,Y,1,R)	searches right for 2's
(1,2,Y,3,R) (3,2,Y,4,L)	pair of 2's, marks with Y's

(4,Y,Y,4,L) (4,X,X,4,L) (4,1,1,4,L) (4,Z,Z,4,L)	searches left for 0's
(4,0,Z,5,L) (5,0,Z,6,R)	pair of 0's, marks with Z's
(6,Z,Z,6 R) (6,X,X,6,R) (6,1,X,1,R)	passes right to next 1
(6,Y,Y,7,R)	no more 1's
(7,Y,Y,7,R) (7,b,b,8,L)	no more 2's
(8,Y,Y,8,L) (8,X,X,8,L) (8,Z,Z,8,L) (8,b,b,9,L)	no more 0's, halts and accepts

9. One answer: State 7 is a final state

(0,0,b,1,R)	0 is leftmost symbol
(1,0,0,1,R) (1,1,1,1,R) (1,*,*,1,R) (1,b,b,2,L)	finds right end
(2,0,b,3,L)	match, erases right symbol
(3,0,0,3,L) (3,1,1,3,L) (3,*,*,3,L) (3,b,b,0,R)	finds left end and begins again
(0,1,b,4,R)	1 is leftmost symbol
(4,0,0,4,R) (4,1,1,4,R) (4,*,*,4,R) (4,b,b,5,R)	finds right end
(5,1,b,3,L)	match, erases right symbol
(0,*,*,6,R)	word left of * is empty
(6,b,b,7,R)	word right of * is empty, halts and accepts

10. One answer: State 8 is a final state

(0,0,b,1,R)	0 read on left of w_1
(1,0,0,1,R)	
(1,1,1,1,R)	moves right to *
(1,*,*,2,R)	
(2,X,X,2,R)	passes over X's
(2,1,1,8,R)	nonzero on left of w_2, halts and accepts
(2,b,b,8,R)	
(2,0,X,3,L)	left symbols match
(3,X,X,3,L)	moves left to *
(3,*,*,4,L)	
(4,1,1,4,L)	
(4,0,0,4,L)	finds leftmost symbol
(4,b,b,0,R)	
(0,1,b,5,R)	1 read on left of w_1
(5,0,0,5,R)	
(5,1,1,5,R)	moves right to *
(5,*,*,6,R)	
(6,X,X,6,R)	passes over X's
(6,0,0,8,R)	non-one on left of w_2, halts and accepts
(6,b,b,8,R)	
(6,1,X,3,L)	left symbols match
(0,*,*,7,R)	word left of * is empty
(7,X,X,7,R)	
(7,0,0,8,R)	word right of * nonempty, halts and accepts
(7,1,1,8,R)	
(0,b,b,0,R)	w_1 initially empty

11. A modification of the machine for Exercise 9 above will do the job.

*12. One approach uses the following general plan: Put a marker at the right end of the original string X_1 and build a new string X_2 to the right of the marker. Working from the lower order end of X_1, for each new symbol in X_1, put a block of that symbol on the end of X_2 twice as long as the previous block of symbols added to X_2. When all symbols in X_1 have been processed, erase X_1 and the marker. Working from left to right in X_2, replace any 0's with 1's from the end of the string until there are no 0's left in X_2. (My implementation of this approach required 23 states and 85 quintuples; can you improve upon this solution?)

13. One answer:

(0,1,X,1,R)	X's leftmost 1 of original string
(1,1,1,1,R) (1,b,*,2,R) (1,*,*,2,R) (2,1,1,2,R) (2,b,1,3,L)	carries 1 to right end of new string and tacks it on
(3,1,1,3,L) (3,*,*,4,L) (4,1,1,4,L) (4,X,X,0,R)	locates leftmost 1 of original string
(0,*,*,5,L) (5,X,1,5,L) (5,b,b,6,R)	original string all copied, changes its X's back to l's and halts

*14.

$$f(n_1, n_2, n_3) = \begin{cases} n_2 + 1 & \text{if } n_2 > 0 \\ \text{undefined} & \text{if } n_2 = 0 \end{cases}$$

15.

(0,1,1,1,R)	this is an odd 1
(1,1,1,0,R)	this is an even 1
(0,b,1,2,R)	even number of 1's on tape, n odd, add 1 and halt
(1,b,b,2,R)	odd number of 1's on tape, n even, halt

*16.

(0,1,1,1,R)	
(1,b,1,4,R)	n = 0, add 1 and halt
(1,1,1,2 R) (2,b,1,4,R)	n = 1, add additional 1 and halt
(2,1,1,3,R) (3,1,b,3,R) (3,b,b,4,R)	n > 2, erase extra 1's and halt

17. One answer:

Quintuples	
(0,1,1,1,R) (1,b,b,8,R)	$n = 0, 2 \cdot 0 = 0$
(1,1,1,2 R) (2,1,1,2,R) (2,b,b,3,L)	$n > 0$, finds end of $\overline{n}$
(3,1,X,4,R) (4,X,X,4,R) (4,1,1,4,R) (4,b,1,5,L)	changes 1 to X, adds 1 at right end of string
(5,1,1,5,L) (5,X,X,6,L) (6,X,X,6,L) (6,1,X,4,R)	goes left to next 1 of $\overline{n}$
(6,b,b,7,R) (7,X,1,7,R)	$\overline{n}$ is doubled, changes X's to 1's
(7,1,1,7,R) (7,b,b,8,L) (8,1,b,9,L)	finds right end, erases extra 1, halts

18. One answer:

Quintuples	
(0,1,1,1,R)	ignores leading 1
(1,1,X 2 R) (2,1,X,3,R) (3,1,1,1,R)	counts XX1
(2,b,b,2,R) (3,b,b,3,R)	$3 \nmid n$ so moves forever right
(1,b,b,4,L)	$3 \mid n$
(4,b,b,8,L)	$n = 0$
(4,1,1,4,L) (4,X,X,5,L) (5,X,X,5,L) (5,1,X,6,R) (6,X,X,6,R) (6,1,1,6,R) (6,b,1,4,L)	collects 1's at right end
(5,b,b,7,R) (7,X,b,7,R) (7,1,1,8,L)	1's all collected, erases X's and halts

*19. One answer:

(0,1,b,1,R)	erases one extra 1
(1,*,b,3,R)	$n_1 = 0$
(1,1,b,2,R) (2,1,1,2,R) (2,*,1,3,R)	$n_1 > 0$, replaces * with leftmost 1 of $\overline{n_1}$, halts

20. One answer:

(0,1,1,0,R) (0,*,*,0,R) (0,b,b,1,L) (0,X,X,1,L)	move to right end of 1's for n_2
(1,1,X,2,L)	X's rightmost 1 of n_2
(2,1,1,2,L) (2 *,*,2 L) (2,b,b,3,R) (2,X,X,3,R) (3,1,X,0,R)	move to left end of 1's for n_1, X leftmost 1
(3,*,X,4,L)	$n_1 < n_2$
(4,X,X,4,L) (4,b,1,5,R) (5,X,b,5,R) (5,1,b,5,R) (5,b,b,9,R)	write 0 on tape and halt
(1,*,*,6,R)	all of n_2 used, now write $n_1 - n_2$ on tape
(6,X,X,6,R) (6,b,b,7,L) (7,X,b,7,L)	erase n_2
(7,*,1,8,L) (8,1,1,8,L) (8,X,b,8,L) (8,b,b,9,R)	clean up $n_1 - n_2$ and halt

21. One answer:

(0,1,X,1,R)	changes first 1 of $\overline{n_1}$ to X
(1,1,1,1,R) (1,*,*,2,R) (2,X,X,2,R) (2,1,X,3,L)	finds leftmost 1 of $\overline{n_2}$, changes to X
(3,X,X,3,L) (3,*,*,4,L) (4,1,1,4,L) (4,X,X,5,R) (5,1,X,1,R)	finds leftmost 1 of $\overline{n_1}$, changes to X
(5,*,*,6,R)	$\overline{n_1}$ exhausted, $n_2 \geq n_1$
(6,X,1,6,R) (6,1,1,7,L) (6,b,b,7,L)	changes X's in $\overline{n_2}$ to 1's
(7,1,1,7,L) (7,*,b,7,L) (7,X,b,7,L) (7,b,b,10,L)	erases $\overline{n_1}$ and halts
(2,b,b,8,L)	$\overline{n_2}$ exhausted, $n_1 > n_2$
(8,X,b,8,L) (8,*,6,9,L)	erases $\overline{n_2}$
(9,X,1,9,L) (9,b,b,10,L)	changes X's in $\overline{n_1}$ to 1's, halts

22. invoke T_1, invoke T_2

23. T erases the leading 1 of $\overline{n_1}$ and moves right; if * is immediately encountered, $n_1 = 0$, so move right erasing $*\overline{n_2}$, finally resulting in a single 1, then halt ($0 \cdot n_2 = 0$). If * is not immediately encountered, move right to $\overline{n_2}$ and count the leading 1's; if there is only one 1 in $\overline{n_2}$, then $n_2 = 0$, go left erasing $\overline{n_1 - 1}*$ and halt ($n_1 \cdot 0 = 0$). If $n_2 \neq 0$, go left to check if $n_1 = 1$. If so, erase $\overline{n_1 - 1}*$ and halt ($1 \cdot n_2 = n_2$). Otherwise, sweep back and forth by Xing out $\overline{n_1 - 1}$ and moving right to the left end of the rightmost block, then COPYING. This creates $\overline{n_1 - 1}$ copies of $\overline{n_2}$. Erase $\overline{n_1 - 1}$ and move right to the left end of the second block from the right. Do a succession of ADDS, moving the resulting string left to close up the gap in the tape. When only a single block of 1's remains on the tape, halt; the result is $\overline{n_1 \cdot n_2}$.

EXERCISES 8.4

*1. a. L(G) = {a}
b. L(G) = {010101, 010111, 011101, 011111, 110101, 110111, 111101, 111111}
c. L(G) = 0(10)*
d. L(G) = 0*1111*

2. a. (c) is regular, (b) and (d) are context-free
b. For example:
for (a): G = (V, V_T, S, P) where V = {a, S}, V_T = {a}, and P = {$S \to a$}.
for (b): G = (V, V_T, S, P) where V = {0, 1, A, B, C, D, E, , G, S}, V_T = {0, 1}, and P consists of the productions

$S \to 0A$	$C \to 1D$
$S \to 1A$	$D \to 1E$
$A \to 1B$	$D \to 1F$
$A \to 1C$	$E \to 0G$
$B \to 0D$	$F \to 1G$
	$G \to 1$

for (d): G = (V, V_T, S, P) where V = {0, 1, A, B, S}, V_T = {0, 1}, and P consists of the productions

$S \to 0S$	$B \to 1B$
$S \to 1A$	$B \to 1$
$A \to 1B$	

3. L(G) = (ab)*

4. L(G) = {1^n01^n | $n \geq 0$}

*5. L(G) = aa*bb*. G is context-sensitive. An example of a regular grammar that generates L(G) is G' = (V, V_T, S, P) where V = {a, b, A, B, S}, V_T = {a, b}, and P consists of the productions

$S \to aA$	$A \to aA$	$B \to bB$
$S \to aB$	$A \to aB$	$B \to b$

6. a. <S> : : = λ | 0<A>
<A>: : = 0<B>
<B>: : = 0 | 0<C>
<C>: : = 0<B>
b. <S> : : = 0<A> | 1<A>
<A>: : = 1<B><B>
<B>: : = 01| 11

c. <S> : : = 0 | 0<A>
<A> : : = 1<B>
<B> : : = 0<A> | 0

d. <S> : : = 0<S> | 11<A>
<A> : : = 1<A> | 1

7. a. 0*011*
b. $S \to 0S$
$S \to 0A$
$A \to 1$
$A \to 1A$

*8. a. 1*1(00)*
b. $S \to 1$
$S \to 1S$
$S \to 0A$
$A \to 0$
$A \to 0B$
$B \to 0A$

9. For example, $G = (V, V_T, S, P)$ where $V = \{(,), S\}$, $V_T = \{(,)\}$, and P consists of the productions
$S \to \lambda$
$S \to (S)S$

10. a. For example, $G = (V, V_T, S, P)$ where $V = \{a, b, A, B, S\}$, $V_T = \{a, b\}$ and P consists of the productions
$S \to \lambda$
$S \to aSa$
$S \to bSb$
$S \to a$
$S \to b$

b. Let $w \in L$. Then $w = w^R$, or $w^R = (w^R)^R$. Therefore all the members of L^R are palindromes.

c. Let $x \in w^*$. Then $x = \lambda$ (a palindrome), or x is the concatenation of k copies of w, $x = w^1 \cdots w^k$, $k \geq 1$. If $x = w^1 \cdots w^k$, then $x^R = (w^1 \cdots w^k)^R = (w^k)^R \cdots (w^1)^R = w \cdots w = x$. Therefore x is a palindrome.

*11. For example, $G = (V, V_T, S, P)$ where $V = \{0, 1, A, S\}$, $V_T = \{0, 1\}$ and P consists of the productions

$S \to 01$	$A \to A0$	$A \to 0$
$S \to A01$	$A \to A1$	$A \to 1$

12. a. For example, $G = (V, V_T, S, P)$ where $V = \{0, 1, A, S\}$, $V_T = \{0, 1\}$, and P consists of the productions

$S \rightarrow 11A \qquad A \rightarrow 1A$
$A \rightarrow \lambda \qquad A \rightarrow 0A$

b. For example, $G = (V, V_T, S, P)$ where $V = \{0, 1, A, B, S\}$, $V_T = \{0, 1\}$, and P consists of the productions

$S \rightarrow 1A \qquad B \rightarrow 0$
$A \rightarrow 1 \qquad B \rightarrow 1$
$A \rightarrow 1B \qquad B \rightarrow 0B$
$\qquad B \rightarrow 1B$

13. a. For example, $G = (V, V_T, S, P)$ where $V = \{1, S\}$, $V_T = \{1\}$, and P consists of the productions

$S \rightarrow \lambda$
$S \rightarrow 11S$

b. $L = (11)^*$
For example, $G = (V, V_T, S, P)$ where $V = \{1, A, S\}$, $V_T = \{1\}$, and P consists of the productions

$S \rightarrow \lambda$
$S \rightarrow 1A$
$A \rightarrow 1S$

*14. For example, $G = (V, V_T, S, P)$ where $V = \{0, 1, S\}$, $V_T = \{0, 1\}$, and P consists of the productions

$S \rightarrow \lambda \qquad S \rightarrow A$
$A \rightarrow 01$
$A \rightarrow 0S1$

15. For example, $G = (V, V_T, S, P)$ where $V = \{0, S, A, B, X\}$, $V_T = \{0\}$, and P consists of the productions

$S \rightarrow A0B \qquad XB \rightarrow B$
$A0 \rightarrow A00X \qquad A \rightarrow \lambda$
$X0 \rightarrow 00X \qquad B \rightarrow \lambda$

16. For example, $G = (V, V_T, S, P)$ where $V = \{0, 1, S\}$, $V_T = \{0, 1\}$, and P consists of the productions

$S \rightarrow SS$
$S \rightarrow 01$
$S \rightarrow 10$
$S \rightarrow 0S1$
$S \rightarrow 1S0$

17. For example, $G = (V, V_T, S, P)$ where $V = \{0,1,S\}$, $V_T = \{0,1\}$, and P consists of the productions

$S \to 001$	$S \to 1S00$
$S \to 100$	$S \to 10S0$
$S \to 010$	$S \to 0S10$
$S \to 0S01$	$S \to 01S0$
$S \to 00S1$	$S \to SS$

*18. For example, $G = (V, V_T, S, P)$ where $V = \{0, 1, S, S_1\}$, $V_T = \{0, 1\}$, and P consists of the productions

$S \to \lambda$	$S_1 \to 1S_11$
$S \to S_1$	$S_1 \to 00$
$S_1 \to 0S_10$	$S_1 \to 11$

19. For example, $G = (V, V_T, S, P)$ where $V = \{0, 1, S, S_1, A, B, M\}$, $V_T = \{0, 1\}$, and P consists of the productions

$S \to \lambda$	$S_1 \to 0S_1A$	$MA \to M0$	$0A \to A0$
$S \to 0S_1A$	$S_1 \to 1S_1B$	$MB \to M1$	$0B \to B0$
$S \to 1S_1B$	$S_1 \to M$	$M \to \lambda$	$1A \to A1$
			$1B \to B1$

20. For example, $G = (V, V_T, S, P)$ where $V = \{a, b, S, G, E, D, F, L, B, M, N, Q, R, Z, X, P, U\}$, $V_T = \{a, b\}$, and P consists of the productions

$S \to GFDE$	
$D \to a$	D generates any number of a's
$D \to Da$	
$Fa \to aLB$	construct the same number of b's as a's
$Ba \to aB$	
$Bb \to bB$	
$BE \to bMNE$	
$BN \to B$	
$bM \to Mb$	
$aM \to Ma$	
$LM \to F$	
$Fb \to Qb$	
$aQ \to RZ$	erase one a
$Zb \to bLX$	copy same number of a's as b's on other side of b's
$Xb \to bX$	
$XN \to PNa$	
$bP \to Pb$	
$LP \to Z$	
$ZN \to QN$	marks finish of a generation of a's
$bQ \to Qb$	
$RQ \to Q$	back at original a's, loop
$GQ \to U$	all original a's have been erased
$Ub \to U$	erase all b's
$UN \to U$	erase end of B's
$Ua \to aU$	leave a's
$UE \to \lambda$	erase end marker

21. *a.

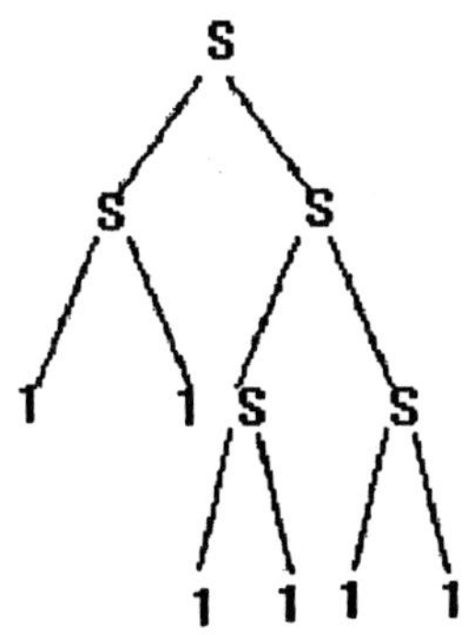

b.

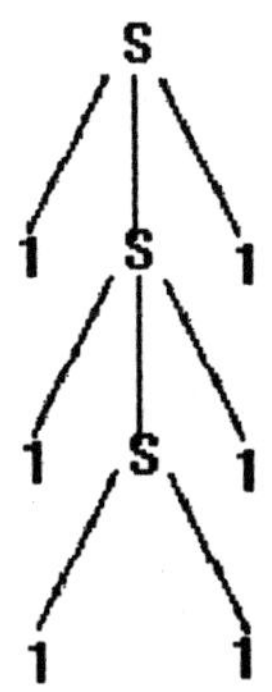

c.

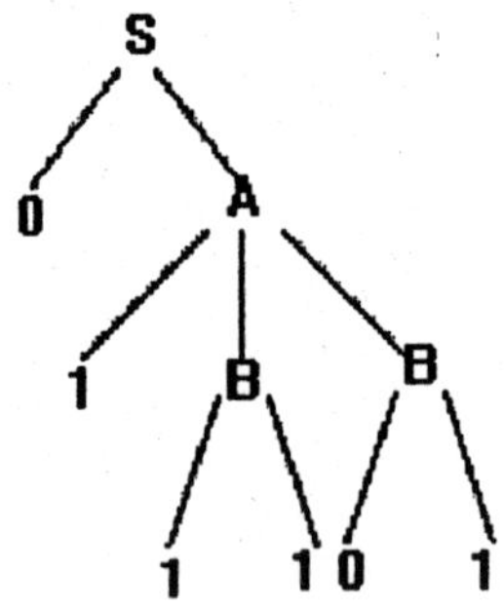

d.

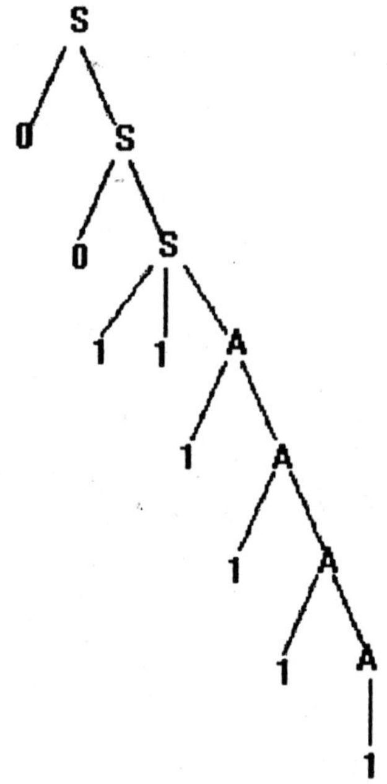

*22.

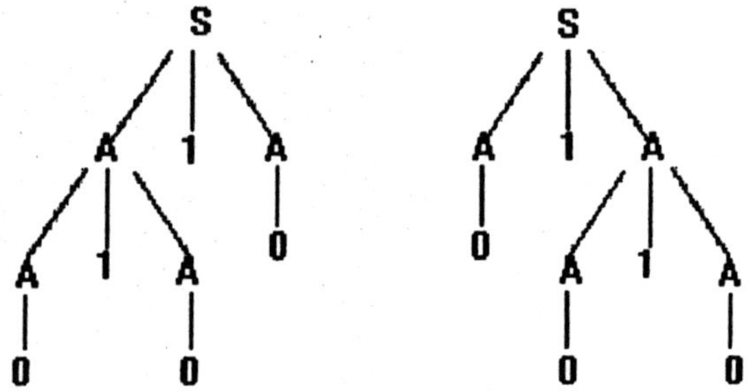

23. One interpretation is that any rider of the escalator must wear shoes and, if traveling with a dog, must carry the dog. Another interpretation is that anyone riding the escalator must do so with worn (as in heavily used) shoes, and must in all cases be carrying a dog.

24. The set of productions for G' is formed from the set of productions for G as follows:

For A and B nonterminals, whenever $A \stackrel{*}{\Rightarrow} B$ in G and $B \rightarrow \alpha$ is a production in G with $|\alpha| \geq 2$, add the production $A \rightarrow \alpha$ to the set, then eliminate all productions of the form $A \rightarrow B$. For any productions of the form $A \rightarrow a$, $a \in V_T$, add to the set of productions those obtained by replacing any A on the right of an existing production by a, then eliminate all productions of the form $A \rightarrow a$. Eliminate $S \rightarrow \lambda$. The remaining productions all have right side with length ≥ 2 and $L(G') \subseteq L(G)$. Only λ and a finite number of one-length words may have been eliminated, so $L(G) - L(G')$ is a finite set.

25. a. Suppose that L is context-free, and let k be the constant of the pumping lemma. Let n be such that $|a^nb^nc^n| \geq k$. Then $a^nb^nc^n = w_1w_2w_3w_4w_5$. Because $|w_2w_4| \geq 1$, not both w_2 and w_4 are empty. Neither w_2 nor w_4 can contain more than one symbol; if, for example, w2 = aab then $w_1w_2^2w_3w_4^2w_5$, which is in L, would contain the string aabaab, a contradiction. If w_2 contains only a's, then the word $w_1w_2^2w_3w_4^2w_5$ contains more than n a's; if w_4 is empty, contains only a's or contains only b's, then $w_1w_2^2w_3w_4^2w_5$ does not contain more than n c's, so does not belong to L, a contradiction, while if w_4 contains only c's, then $w_1w_2^2w_3w_4^2w_5$ does not contain more than n b's, a contradiction. The other cases are similar.

b. Suppose that L is context-free, and let k be the constant of the pumping lemma. Let n be such that $|a^{n^2}| \geq k$. Then $a^{n^2} = w_1w_2w_3w_4w_5$ or $w_1 = a^u$, $w_2 = a^v$, $w_3 = a^w$, $w_4 = a^x$, $w_5 = a^y$ with v and x not both zero. Then $w_1w_2w_3w_4w_5 = a^ua^va^wa^xa^y = a^{u+v+w+x+y} \in L$, so $u + v + w + x + y$ is a perfect square, call it s. Using $i = 2$ in the pumping lemma, $u + 2v + w + 2x + y$ is a perfect square; using $i = 3$, $u + 3v + w + 3x + y$ is a perfect square, etc. Thus there exist a perfect square s and a fixed number $d = v + x > 0$ such that $s + jd$ is a perfect square for any $j \geq 0$. This is a contradiction, since perfect squares grow progressively farther apart.